ADVANCED TOPICS
IN SCIENCE AND TECHNOLOGY IN CHINA

ADVANCED TOPICS IN SCIENCE AND TECHNOLOGY IN CHINA

Zhejiang University is one of the leading universities in China. In *Advanced Topics in Science and Technology in China*, Zhejiang University Press and Springer jointly publish monographs by Chinese scholars and professors, as well as invited authors and editors from abroad who are outstanding experts and scholars in their fields. This series will be of interest to researchers, lecturers, and graduate students alike.

Advanced Topics in Science and Technology in China aims to present the latest and most cutting-edge theories, techniques, and methodologies in various research areas in China. It covers all disciplines in the fields of natural science and technology, including but not limited to, computer science, materials science, life sciences, engineering, environmental sciences, mathematics, and physics.

国家科学技术学术著作出版基金资助出版

Weicheng Cui
Xiaoping Huang
Fang Wang

Towards a Unified Fatigue Life Prediction Method for Marine Structures

With 138 figures

ZHEJIANG UNIVERSITY PRESS
浙江大学出版社

Springer

图书在版编目 (CIP) 数据

海洋结构物疲劳寿命预报的统一方法 = Towards a Unified Fatigue Life Prediction Method for Marine Structures / 崔维成，黄小平，王芳著. —杭州：浙江大学出版社，2013.11
ISBN 978-7-308-10450-0

I. ①海… II. ①崔… ②黄… ③王… III. ①海洋建筑物-建筑结构-疲劳寿命估算 IV. ①P754

中国版本图书馆 CIP 数据核字(2012)第 197646 号

Not for sale outside Mainland China
此书仅限中国大陆地区销售

海洋结构物疲劳寿命预报的统一方法
崔维成 黄小平 王 芳 著

责任编辑 张 琛
文字编辑 陈丽勋
封面设计 俞亚彤
出版发行 浙江大学出版社
网址：http://www.zjupress.com
Springer-Verlag GmbH
网址：http://www.springer.com
排 版 杭州中大图文设计有限公司
印 刷 浙江印刷集团有限公司
开 本 710 mm×1000 mm 1/16
印 张 18.5
字 数 559 千
版 印 次 2013 年 11 月第 1 版 2013 年 11 月第 1 次印刷
书 号 ISBN 978-7-308-10450-0 (浙江大学出版社)
ISBN 978-3-642-41830-3 (Springer-Verlag GmbH)
定 价 150.00 元

浙江大学出版社发行部联系方式：0571-88925591；http://zjdxcbs.tmall.com

Preface

Traditional strength assessment procedure implemented in marine structural design rules is highly experience-based due to the complexity of the structure and its operational environment. With the fast development of computer technology, software and hardware, the possibility to accurately assess the marine structural strength based on the strict principles of mechanics increases. The criteria which allow the direct calculations of loading and strength are often termed as first-principle-based design criteria.

However, it is widely recognized that even the latest classification rules for marine structures are still far from the "real" first-principle-based criteria. Important evidence for this statement is that the strength is assessed at different global (hull girder) and local (stiffened panel and welded joints) levels and in different failure modes (yielding, buckling and fatigue). The relationship among them is not taken into consideration and the relative success of this strength assessment procedure is largely based on past experiences. Furthermore, in most of the fatigue strength assessment methods which are *S-N* curve based, the effects of initial defects and load sequence have been ignored and the damage state has not been specified. These, together with some other factors, lead to large scatter of the predicted fatigue lives. Significant improvements with regard to the fatigue strength assessment methodology for marine structures are required.

The effect of fatigue damage on ultimate strength is not considered too. Thus, for existing marine structures operated for certain period, the strength analyzed may not represent the actual strength a marine structure possessed. Risk analysis based on the current strength analysis procedure is then rather uncertain. Inspection and maintenance decisions based on the assessment may not reflect the actual "optimum".

In order to apply the damage tolerance design philosophy to design marine structures such as ships and offshore platforms in place of the safe life philosophy currently adopted, accurate prediction of fatigue crack growth under service conditions is required. Fatigue Crack Propagation (FCP) theory could fulfil that purpose. Now, more and more people have realized that only a Fatigue Life Prediction (FLP) method based on FCP theory has the potential to explain various fatigue phenomena observed. In this book, the issues leading towards the development of a Unified Fatigue Life Prediction (UFLP) method based on FCP theory are addressed. Based on the philosophy of the unified approach, the current inconsistency between fatigue design and inspection of marine structures could be resolved.

The book is divided into eight chapters. Chapter 1 is an introduction, in which the brief history of fatigue study is summarized and the general layout of the book is described. Since most marine structures are made of metals, in Chapter 2 the latest understanding of fatigue mechanisms of metals is summarized. Chapter 3 introduces the current state-of-the-art of UFLP and the meaning of the unified approach used in this book is explained. In Chapter 4, the basic elements of fracture mechanics are introduced which is the important foundation for understanding the later chapters. Chapter 5 presents the general procedure for a UFLP method for marine structures. Chapter 6 discusses various issues related to fatigue loading, which is the main cause of fatigue failure of marine structures. In Chapter 7, some applications and demonstrations of UFLP are given, which would be helpful in the understanding of the philosophy presented in the book and further recognize the problems to be solved for engineering applications. Chapter 8 discusses the relevant issues related to the code development based on UFLP for marine structures.

This book presents the state-of-the-art and recent advances, including those by the authors, in fatigue studies. It is designed to lead the future direction and to provide a useful tool in many practical applications.

The methodologies and information in this book are addressed to engineers, naval architects, research staff, professionals and graduates engaged in fatigue prevention design and survey of marine structures, in fatigue studies of materials and structures, in experimental laboratory research, in planning the repair and maintenance of affected structures, and in rule development including in-service reliability systems, etc. The book is also an effective educational aid in naval architecture, marine, civil and mechanical engineering.

Weicheng Cui
Wuxi, China

Contents

1

Introduction

1.1 Fatigue Problems in Marine Structures

Marine structures such as ships and offshore platforms are subjected to complex loading histories and one of the most significant failure modes is fatigue (ISSC, 2009). Fatigue is defined as a process of cycle by cycle accumulation of damage in a material undergoing fluctuating stresses and strains (Almar-Naess, 1985). A significant feature of fatigue is that the load is not large enough to cause immediate failure. Instead, failure occurs after a certain number of load fluctuations have been experienced, *i.e.* after the accumulated damage has reached a critical level. The propagation of fatigue cracks may eventually compromise the structural integrity and water-tightness of marine structures, so fatigue prevention has been paid great attention by all the stakeholders of marine structures.

Historically, ship structures have been designed to meet minimum scantling requirements that have included allowances for general corrosion and uncertainties in design methods (Glen, *et al.*, 2000). Until the 1970s, fatigue cracking was not explicitly considered by designers because fatigue cracking was rarely detected in ships less than 10 years old and because the frequency and costs of repairing fatigue cracks in older ships was acceptable to owners. Since the late 1970s, however, fatigue cracking has occurred more frequently in relatively new ships.

This change has been attributed to the design and construction of more structurally optimized ships with thinner scantlings (Glen, *et al.*, 2000). This optimization, which has been motivated by commercial demands to reduce the fabrication costs and weight of hull structures, has been achieved through the greater use of high strength steels, the use of more sophisticated design tools, and the greater exploitation of classification society rules which have permitted design stresses to increase with tensile strength up to a fraction of the tensile strength defined by the so-called material factor. Unfortunately, stress concentrations of structural details have not been adequately reduced to compensate for the higher design stresses and higher local bending stresses associated with thinner scantlings.

Furthermore, the fatigue strength of as-welded steel joints is essentially independent of tensile strength. Therefore, local cyclic stresses in structural details have been permitted to increase without a matching increase in fatigue strength of these details. In addition, corrosive environments have exacerbated this mismatch since the flexibility of a thin structure promotes the flaking of rust which accelerates the wastage process and further increases the flexibility of the thin structure (Liu and Thayamballi, 1995; Skaar, *et al.*, 1987; LR, 1992).

In response to safety concerns and escalating maintenance costs for owners, classification society rules have introduced explicit fatigue design criteria for welded structural details in steel ships (Cramer, *et al.*, 1994; LR, 1996; ABS, 1992). These criteria, which are largely based on well-established fatigue design procedures for welded joints in bridges and offshore structures (UKDE, 1984; BSI, 1980; 1993), are intended to ensure that there is a low probability of fatigue failure occurring during the design life of a ship, where failure is generally considered to be the initiation of a through-thickness crack several inches long. However, premature fatigue cracking as a result of fabrication or design errors can still occur. Furthermore, some fatigue cracking can still be expected in properly designed ships. Therefore, quantitative techniques for predicting the residual life and residual strength of cracked structural welded details are needed, to develop safe but cost-effective inspection schedules at the design stage. These techniques can also be used to optimize the scheduling of repairs for cracks found in service and to assess whether the operation of existing ships can be extended beyond their original design lives (Glen, *et al.*, 2000).

1.2 Current Practices of Fatigue Strength Assessments and Their Deficiencies

Current Fatigue Strength Assessment (FSA) methods for marine structures such as ships are all stress-based using nominal, hot-spot or notch stresses (*e.g.* CSR of IACS, 2006). A linear damage accumulation rule is used which ignores the load sequence effect and the long term distribution of the random loading is often assumed to follow the Weibull distribution (1951). The damage index at failure is always assumed to be a constant no matter what the initial state of the structure and the final state of the fatigue failure are. Due to other differences in the FSA process, such as definition of loading, choice of *S-N* curves and calculation of Stress Concentration Factors (SCFs), different FSA procedures result in large scatter of fatigue lives for the same detail. A recent comparative study on fatigue strength assessment procedures adopted by different ship classification societies indicated a large difference (from 1.8 years to 20.7 years) in the predicted fatigue life for a very simple detail (Fricke, *et al.*, 2002).

The current approach adopted by the International Association of Classification Societies (IACS) for reducing the scatter is unification (*e.g.* CSR of

IACS, 2006). It is the authors' belief that unification will not completely solve the actual problem of scatter. The Cumulative Fatigue Damage (CFD) theory itself is also subjected to some theoretical flaws and is responsible for the main scatter. This is based on the following arguments.

A fatigue life scatter of 10 times for practical structures is certainly not acceptable because a fatigue life of 3 years has a completely different interpretation to a fatigue life of 30 years in a ship. However, if we realized that this scatter of 10 times also exists in well-controlled "identical" small specimens (*e.g.* Hanaki, *et al.*, 2010), the scatter can only be attributed to the sizes of initial defects or the competing failure modes induced separately by surface- and internally-initiated cracks (Ravi Chandran, *et al.*, 2010; Cashman, 2010).

Much experimental evidence (Skorupa, 1998; 1999; Qian and Cui, 2010) has also shown that for variable amplitude loading, the load sequence effect is also very significant. It is well known that for high-low loading, the damage index is less than 1 due to the underload acceleration while for low-high loading, the damage index is greater than 1 due to the overload retardation.

The total error between prediction and experiment can be written as Eq. (1.1) where each term represents an individual factor. From the above arguments, we can say that the scatter induced by initial defects and the load sequence is of the same order as that induced by differences in definition of loading, calculation of SCFs and choice of *S-N* curves. Thus, attention only paid to the calculation accuracy of SCFs and choice of *S-N* curves will not be able to reduce the total scatter significantly. Therefore, currently-used stress-based approaches for FLP of ship structures are subjected to theoretical flaws.

$$\varepsilon_{\text{tot}} = \sqrt{\varepsilon_{\text{SCF}}^2 + \varepsilon_{S\text{-}N}^2 + \underbrace{\varepsilon_{\text{sequence}}^2 + \varepsilon_{\text{initial crack}}^2 + \varepsilon_{\text{final crack}}^2 + \cdots}_{\text{ignored by CFD theory}}} \tag{1.1}$$

Now, more and more people have realized that only an FLP method based on FCP theory has the potential to satisfy the accuracy requirement and to explain various fatigue phenomena observed. This is the fundamental philosophy for the development of a UFLP method for marine structures.

1.3 Historical Overview of Metal Fatigue

Marine structures are mostly made of metals, so a historical overview of metal fatigue would be helpful in understanding the UFLP method. According to Schütz (1996), the history of fatigue begins with Albert. In 1837, he published the first known fatigue-test results. In 1842, Rankine discussed the fatigue strength of railway axles. The term "fatigue" was mentioned for the first time by the Englishman Braithwaite in 1854. In his paper, Braithwaite described many service fatigue failures of brewery equipment, water pumps, propeller shafts, crankshafts, railway axles, levers, cranes, *etc*. Beginning in 1860, Wöhler published the results of fatigue tests with railway axles. In 1870, he presented a final report containing the following conclusions, often called "Wöhler's laws": "Material can be induced to fail by many repetitions of stresses, all of which are lower than the static strength. The stress amplitudes are decisive for the destruction of the cohesion of the material. The maximum stress is of influence in so far as the higher it is, the lower are the stress amplitudes which lead to failure. Wöhler therefore stated the stress amplitude to be the most important parameter for fatigue life, but a tensile mean stress also has a detrimental influence".

Wöhler incidentally represented his test results in the form of tables. Not until 1910 did Basquin (1910) represent the finite life region in the form "$\log \sigma_a$ on the ordinate, $\log N$ on the abscissa" and these curves were called "Wöhler's curves" since 1936 and now they are called *S-N* curves.

The next progress was the discovery of the Bauschinger effect in the period 1870–1905, "the change of the elastic limit by often repeated stress cycles", which is the basis for the hypotheses of Coffin and Manson which originated in the 1950s and which are still being utilized today in the Low Cycle Fatigue (LCF) field and for FLP according to the local concept.

In the period 1920–1945, the foundations were laid for almost all the fatigue knowledge we enjoy today. These include the fatigue strength under variable amplitudes, the measurement of fatigue loads and load spectra, the realization that higher-strength materials do not result in higher fatigue strengths of components, the mechanical methods to improve fatigue strength by inducing compressive residual stresses, the damage accumulation hypotheses of Palmgren-Miner for FLP under variable amplitudes, the statistical scatter of the static and fatigue strengths of materials, the realization that it is difficult, if not impossible, to transfer the fatigue behavior of specimens to that of an actual component, which is true even today.

The period 1945–1960 when the knowledge gained in the years 1920–1945 was utilized. In all industrial countries fatigue was investigated. Due to the well known crashes of the Comet airliner, many full-scale fatigue tests were carried out on fuselages. In this period, the "safe life" concept was gradually replaced by the "fail safe" concept. The scatter of the number of cycles to failure and of the fatigue limit was treated with the help of mathematical statistics.

From about 1950 onwards, doubts about the validity of Miner's rule began to appear in the literature. Damage sums between 0.1 and 10 were found in fatigue tests which, however, were entirely unsuitable for this purpose. A genuine check of Miner's rule was only possible at the end of the 1950s, as only then had suitable test machines been developed.

Toward the end of the period 1945–1960, a large number of crack propagation tests were carried out and crack propagation hypotheses were developed, still without employing Fracture Mechanics (FM). However, in 1958, Irwin had realized that the stress-intensity factor $K = \sigma\sqrt{\pi a}$ was the determining factor for static strength in the cracked state. With this concept, Linear Elastic Fracture

Mechanics (LEFM) was born.

From 1960 onwards, the most significant progress in fatigue was made by applying LEFM. Various FCP theories were proposed although CFD theories were further developed and widely applied in most industries except the aircraft industry (Fatemi and Yang, 1998; Yang and Fatemi, 1998; Cui, 2002; 2008).

Paris, *et al.* (1961; 1963) established that FCP could be described by the following equation, soon erroneously called a "law",

$$\frac{\mathrm{d}a}{\mathrm{d}N} = C\left(\Delta K\right)^n \tag{1.2}$$

This equation was widely accepted and is used even today. However, it contains neither the influence of mean stress on crack propagation, nor the final fracture on reaching the unstable fracture condition, $K_{max}=K_{Cf}$, nor the fatigue threshold ΔK_{th}, which is the SIF range below which no FCP occurs. The complex process of crack propagation is undoubtedly described much too simply by this equation.

In order to explain the stress ratio effect $R(=K_{min}/K_{max})$, Elber (1970) introduced the "crack closure" concept based on his experimental observation that after a high tensile load the crack closes before the load is reduced to zero.

This concept had been highly appraised in the 1980s and 1990s and it was used to explain various other effects including the load sequence effect (*e.g.* McEvily, *et al.*, 1999; McEvily and Ishihara, 2001; Newman Jr., *et al.*, 2006) but now it is subjected to some challenges (*e.g.* Hertzberg, *et al.*, 1988). In particular, many people have agreed that the physical effects of crack closure have been greatly over-estimated in the past (*e.g.* Vasudevan, *et al.*, 1994). The partial crack closure model (Donald and Paris, 1999; Kujawski, 2001a; 2001b) was proposed to overcome the difficulty the crack closure model met. Later, Kujawski (2001c; 2001d) further found that without using the crack closure concept, it is possible to explain the stress ratio effect, even better than by using the concept. Vasudevan, Sadananda and their co-workers (*e.g.* Vasudevan, *et al.*, 2001) had proposed the two-parameter unified approach. According to the unified approach it is both K_{max} (σ_{max}) and ΔK ($\Delta\sigma$) which are responsible for the FCP. Through their efforts the unified approach has demonstrated that all the recognized special phenomena can be explained. Now, more and more people tend to believe that this unified approach might be a correct way forward (Noroozi, *et al.*, 2005; Stoychev and Kujawski, 2005; Kujawski, 2005; Zhang, *et al.*, 2005; Maymon, 2005a; 2005b).

The load sequence effect might be the most challenging problem for the FLP of engineering structures. Although the problem has been studied since the 1960s, the mechanism which causes the load sequence effect has not been understood yet (Skorupa, 1998; 1999). Generally speaking, three classes of mechanisms acting before, at or after the crack tip have been proposed in the literature. De Castro, *et al.* (2005) have discussed extensively the three classes of mechanisms and they found that "plasticity induced crack closure (behind the crack tip) cannot explain load sequence effects in various important problems and a damage accumulation model ahead of the crack tip based on ε-N cyclic properties can explain those effects in the absence of closure". This point of view is similar to the two-parameter unified approach.

1.4 FLP Methods

Since cracks can conceivably lead to catastrophic failure of a structure, it is essential to properly consider and establish relevant crack tolerant design procedures for structures, in addition to the implementation of close-up surveys and maintenance strategies (Paik, *et al.*, 2006). For reliability assessment of aging structures under extreme loads, it is often necessary to take into account a known (existing or assumed or anticipated) crack for the ultimate limit state analysis as a parameter of influence (Paik and Thayamballi, 2003). To make this possible, it is necessary to develop a time-dependent fatigue crack growth model which can predict crack damage in location and size as the ship ages.

Fatigue cracks propagate with time progressively in a ductile material. The time-variant cracking damage model is composed of two separate models, namely: 1) crack initiation assessment or detection; 2) crack growth assessment. Crack initiation at a critical structural detail is typically predicted using CFD theories. Crack growth is often assessed by fracture mechanics based approaches (hereafter called FCP approaches).

The main purpose of this section is to introduce various FLP methods belonging to two theories: CFD theory and FCP theory.

1.4.1 CFD Theories

Even today, CFD analysis still plays a key role in life prediction of components and structures subjected to field load histories. Since the introduction of the damage accumulation concept by Palmgren (1924) and "linear damage rule" by Miner (1945), the treatment of CFD has received increasingly more attention. A recent comprehensive review of the development of CFD theories has been conducted by Fatemi and Yang (1998) and Yang and Fatemi (1998).

Fundamentally speaking, fatigue damage is a result of material structural changes at the microscopic scale level such as dislocations of the atomic structures. While bridging microscopic quantities and macroscopic experimental observations are still long-term endeavors, it is reasonable to believe that microscopic parameters governing fatigue damage have an inherent relationship with the macroscopic stress and strain quantities based upon the continuum mechanics concepts. These macroscopic quantities can be used to account for crack nucleation and early growth. By choosing different macroscopic quantities such as stress, strain, energy density or combination, different CFD formulas are derived.

1.4.1.1 Stress-based approach (*S-N* curve approach)

A stress-based approach was the earliest but even now the most frequently used approach for FLP. In this approach, the fatigue life (number of cycles N) is related to the applied stress range ($\Delta\sigma$ or S) or stress amplitude (σ_a). A plot of the fatigue life vs. true stress amplitude for a metal gives, in general, a curve of Basquin form (Basquin, 1910),

$$\sigma_a = \frac{E \cdot \Delta\varepsilon_e}{2} = \sigma'_f \cdot (2N)^b \tag{1.3}$$

where N is the number of cycles to failure; $2N$ is the number of load reversals to failure; σ'_f is the fatigue strength coefficient; b is the fatigue strength exponent; the sign of b is negative.

In a component or structure, there are two types of stress concentrations. One is due to the structural geometrical change or discontinuity and the other is due to welding. Depending on how to account for the stress concentration effect, stress-based approaches can be further divided into the nominal stress approach, hot-spot stress approach and notch stress approach. Currently, the hot spot stress approach seems to be the most favorable one used by ship classification societies (Wästberg, *et al*., 2006).

Traditional fatigue tests indicated a limit at about N=10^7 cycles. If the applied stress range is lower than the limit, there will be no fatigue failure. However, recent giga-cycle fatigue tests performed on a cold rolled steel sheet (the Young's modulus E=203 GPa, the yield strength σ_y=225 MPa, the ultimate tensile strength (UTS) σ_u=340 MPa, density ρ=7.83 g/cm^3 and the Vickers hardness HV=95) (Wang, *et al*., 1999), a high carbon chromium steel (σ_u=2316 MPa and HV=750–795) (Sakai, *et al.,* 1999) and an aluminium wrought alloy EN AW 6082 (σ_y=341 MPa, σ_u=356 MPa, elongation δ=11.2%, reduction of area ϕ=47.5%) (Berger, *et al*., 2006) showed that the fatigue limit does not appear until 10^9 cycles in the *S-N* curve or the *S-N* curve tends to come down again in the long life region of N>10^7 cycles. This poses questions about the existence of infinite fatigue life (Bathias, 1999). Marines, *et al*. (2003) have summarized the state of the art in this field and their conclusion is that the *S-N* curve in the giga-cycle regime must be determined in order to guarantee the real fatigue strength and safe life of mechanical components. This problem has also been extensively studied by many researchers (Wang, *et al*., 2010; Qian, *et al*., 2010; Pyttel, *et al*., 2010; 2011; Nakajima, *et al*., 2010; Li, *et al*., 2010; Liu, *et al*., 2010; Furuya, 2010).

If a fatigue limit is assumed to exist, the relation expressed by Eq. (1.3) is only valid for the middle part. The stress amplitude is larger than the limit but the maximum stress should not exceed the UTS. Recently, a new function for the description of fatigue curves for both low and high fatigue regions, *i.e.* for the whole cycle region from the UTS to fatigue limit was proposed by Kohout and Vechet (1999). The function takes the following form,

$$\sigma(N)=\sigma_\infty\left[\frac{N+10^7\alpha\beta}{N+10^7\alpha}\right]^{-b}$$
$$\alpha=\frac{\sigma_c^{-1/b}-\sigma_\infty^{-1/b}}{\sigma_u^{-1/b}-\sigma_c^{-1/b}}\quad \beta=\frac{\sigma_u^{-1/b}}{\sigma_\infty^{-1/b}} \tag{1.4}$$

where σ_u is the UTS, σ_∞ is the fatigue limit. Both of them can be measured accurately. σ_c is the fatigue strength at 10^7 cycles and $-b$ is the slope in the middle cycle region. σ_c and b can be determined by the least squares method.

Four parameters have to be determined for a complete *S-N* curve according to Eq. (1.4). In some practical situations, not so much information may be available. Wang, *et al*. (1999) proposed a simple and practical prediction method to estimate both the *S-N* curve and the crack growth rate curve using only the tensile strength.

In the high-cycle fatigue regime, Castillo and Fernandez-Canteli (2006) presented a similar general parametric lifetime model as Eq. (1.4) for predicting fatigue behavior at any stress level and amplitude. The problem of the design of the experimental laboratory test required for such a prediction is dealt with. The surprising and relevant result is that running two groups of tests for two different constant stress levels of σ_{max} or σ_{min} is sufficient to predict the whole collection of Wöhler fields for any possible stress level. However, some combinations of tests, such as one for a fixed value of σ_{max} and one for a fixed value of σ_{min}, are shown to be insufficient.

Generally, the basic fatigue behavior of materials and structures is determined by adopting constant amplitude loading resulting in various characteristic quantities and relationships such as *S-N* curves, cyclic stress-strain curves, fatigue crack growth rates, threshold values, *etc*. While constant amplitude loading is fully described by quite a few parameters (maximum/minimum or range/mean and number of load cycles), most structures experience in-service loading environments with variable amplitudes and mean loads. It is well established that fatigue response may be very sensitive to the specifics of the loading encountered. Therefore, tests with realistic load sequences are often required in order to assess any susceptibility to that kind of phenomenon and to demonstrate the in-service integrity for given materials and structures. For this purpose, Standardized Load-time Histories (SLHs) have been studied for 30 years, since it has been recognized that the use of SLH provides a series of advantages—both for studies of a more generic nature and practical applications. Heuler and Klätschke (2005) have presented an overview of the SLH available. It details the principles applied for collection and analysis of appropriate load data, assessment of operating profiles and generation of the respective load spectra and sequences. The principles may also serve as guidelines for those cases where new load spectra and load-time histories have to be created, for example for a given design problem. This topic will be discussed in detail in Chapter 6.

1.4.1.2 Strain-based approach

In most practical cases of fatigue design, the critical location will be a notch in which plastic strains are imposed by surrounding elastic material. Thus, the situation will be strain controlled with a total strain range composed of an elastic and a plastic part.

The plastic strain resistance is best described by the Coffin-Manson relationship (Coffin and Tavernelli, 1959; Manson and Hirschberg, 1964),

$$\frac{\Delta\varepsilon_p}{2} = \varepsilon_f' \cdot (2N)^c \tag{1.5}$$

where ε_f' is the fatigue ductility coefficient; c is the fatigue ductility exponent—the sign of c is negative.

Manson and Hirschberg (1964) proposed that a metal's resistance to total strain cycling can be considered as a superposition of its elastic and plastic strain resistance. By combining Eqs. (1.3) and (1.5),

$$\frac{\Delta\varepsilon_T}{2} = \varepsilon_a = \frac{\Delta\varepsilon_e}{2} + \frac{\Delta\varepsilon_p}{2} = \frac{\sigma_f'}{\mathrm{E}}(2N)^b + \varepsilon_f' \cdot (2N)^c \tag{1.6}$$

The total strain life curve approaches the plastic strain life curve in the low cycle region and the stress life curve in the high cycle region as shown in Fig. 1.1.

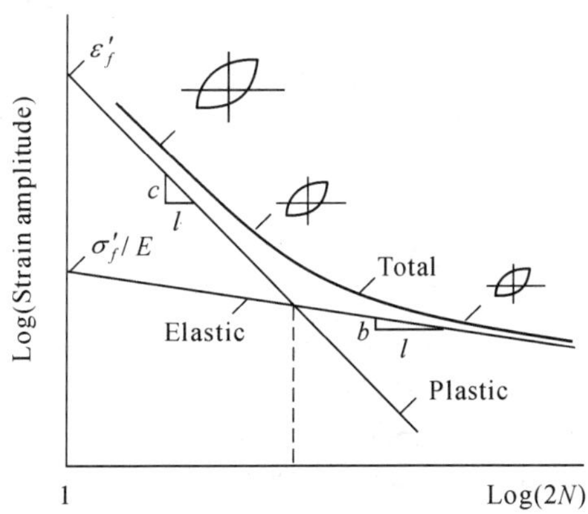

Fig. 1.1 Representation of elastic, plastic and total strain resistance to fatigue loading

For general low cycle and high cycle fatigue, the Coffin-Manson relationship, Eq. (1.6), has a strong curve-fitting ability but it needs to determine 5 material properties. Manson (1965) has simplified the equation even further with his method of universal slopes,

$$\Delta\varepsilon = 3.5\frac{\sigma_u}{E}(N)^{-0.1} + \varepsilon_f^{0.6}(N)^{-0.6} \tag{1.7}$$

where σ_u, E and ε_u are all obtained from a monotonic tensile test. He assumed the two exponents are fixed for all materials and only σ_u, E and ε_f control the fatigue behavior.

Later on, the above equation was further modified to the following expression (Muralidharan and Manson, 1988),

$$\Delta\varepsilon = 0.0266D^{0.115}\left[\sigma_u / E\right]^{-0.53} N_f^{-0.56} + 1.17\left[\sigma_u / E\right]^{0.832} N_f^{-0.09} \tag{1.8}$$

where σ_u is the ultimate strength of metal, D is the ductility of the metal, E is the modulus of elasticity, N_f is the fatigue life. Good correlation between the fatigue life predicted by this equation and the fatigue test data has been found (Muralidharan and Manson, 1988).

Based on a detailed correlation study between monotonic tensile data and constant amplitude strain-controlled fatigue properties, the following simple strain-life formula is proposed by Roessle and Fatemi (2000),

$$\frac{\Delta\varepsilon}{2} = \frac{4.25(HB)+225}{E}(2N_f)^{-0.09} + \frac{0.32(HB)^2 - 487(HB) + 191000}{E}(2N_f)^{-0.56} \tag{1.9}$$

This approximation uses only hardness and modulus of elasticity as inputs for strain-life approximation, both of which are either commonly available, or easily measurable.

The inclusion of mean stress or mean strain effects in FLP methods involving strain-life data is very complex. One method is to replace σ_f' with $\sigma_f' - \sigma_m$ in Eq. (1.6), where σ_m is the mean stress such that

$$\frac{\Delta\varepsilon}{2} = \frac{(\sigma_f' - \sigma_m)(2N)^b}{E} + \varepsilon_f'(2N)^c \tag{1.10}$$

Here σ_m is taken as positive for tensile values and negative for compressive values. Another equation suggested by Smith, *et al.* (1970), based on strain-life test data at fracture obtained with various mean stresses is

$$\sigma_{\max}\varepsilon_a E = \left(\sigma_f'\right)^2 (2N)^{2b} + \sigma_f'\varepsilon_f' E (2N)^{b+c} \quad (1.11)$$

where $\sigma_{\max}=\sigma_{m}+\sigma_a$ and ε_a is the alternating strain. If $\sigma_{\max}$ is zero, Eq. (1.11) predicts infinite life, which implies that tension must be present for fatigue fractures to occur. Both Eqs. (1.10) and (1.11) have been used to handle mean stress effect.

In Ong (1993a), fatigue lives were calculated for 49 steels using published values of σ_f', ε_f', b, c. These lives were compared with lives calculated using some of the approximation methods described above. These include the original and modified versions of the Four-Point Correlation method, the Original Universal Slopes method, and the Mitchell method. The lives covered a range from 10 to 10^7 reversals. The steels covered values of UTS from 345 MPa to 2585 MPa, and Brinell hardness values from 80 to 660. They included the steel SAE1005, SAE1015 and SAE1045.

In all cases correlation between the "experimental" and the "estimated" lives was poor. The modified Four-Point Correlation method was found to be slightly better than the original Universal Slopes method and to be the best within the methods studied.

In another study conducted by Park and Song (1995), six such methods were evaluated and compared. These consisted of the Universal Slopes and Four-Point Correlation methods by Manson (1965), the Modified Universal Slopes method by Muralidharan and Manson (1988), the Uniform Material method by Baumel and Seeger (1990), the Modified Four-Point Correlation method by Ong (1993b), and the method proposed by Mitchell, *et al.* (1977). A total of 138 materials were used in the study including unalloyed steels, low-alloy steels, high-alloy steels, aluminum alloys and titanium alloys, with low-alloy steels providing the most data. Amongst the correlations compared, those proposed by Muralidharan and Manson (1988), Baumel and Seeger (1990) and Ong (1993b) yielded good predictions according to Park and Song (1995). The modified Universal Slopes method was concluded to provide the best correlation.

In the study carried out by Roessle and Fatemi (2000), they also compared their simple formula which uses only hardness and modulus of elasticity for estimation of the strain-life curve with the Modified Universal Slopes method and found their simple formula result is somewhat better and more conservative for predictions over the entire fatigue life regime. A similar study was carried out by Lee and Song (2006) and they found that the (direct) hardness method proposed by Roessle and Fatemi (2000) provides excellent estimation results for steels but the indirect hardness method utilizing the UTS predicted from hardness was proposed by them and successfully applied to estimate fatigue properties for aluminum alloys and titanium alloys. The medians method proposed by Meggiolaro and Castro (2004) is found to provide the best estimation results for aluminum alloys. Based on the results obtained, some guidelines are provided for estimating fatigue properties from simple tensile data or hardness. In addition, a new relationship of UTS versus hardness is proposed for titanium alloys.

Most of the existing methods for estimating ε-N parameters are based on a relatively limited amount of experimental data. In addition, sound statistical evaluation of the popular rules of thumb used in practice to estimate fatigue properties is scarce, if available. Meggiolaro and Castro (2004) presented an extensive statistical evaluation of the existing Coffin-Manson parameter estimates based on monotonic tensile and uniaxial fatigue properties of 845 different metals, including 724 steels, 81 aluminum alloys and 15 titanium alloys. The studied Coffin-Manson estimates include the methods proposed by Muralidharan and Manson, Baumel and Seeger, Roessle and Fatemi, Mitchell, Ong, Morrow, Raske, as well as Manson's universal slope and four-point correlation methods. From the collected data, it is shown that all correlations between the fatigue ductility coefficient ε_f' and the monotonic tensile properties are very poor, and that it is statistically sounder to estimate ε_f' based on constant values for each alloy family. Based on this result, a new estimation method which uses the medians of the individual parameters of the 845 materials is proposed.

1.4.1.3 Energy-based approach

A historical description of the energy-based approaches is given by Fatemi and Yang (1998). For using this type of failure criteria, it was realized that an energy-based damage parameter can unify the damage caused by different types of loading such as thermal cycling, creep and fatigue. In conjunction with Glinka's rule (Glinka, 1988), it is possible to analyze the damage accumulation of notched specimen or components with the energy approach. Energy based damage models can also include mean stress and multiaxial loads since multiaxial fatigue parameters based on strain energy have been developed.

Recently, Pan, *et al.* (1999) proposed the following fatigue strain energy density parameter for the critical plane to predict the fatigue life of various materials under multiaxial loading,

$$W^* = \frac{\Delta\sigma_{12}}{2}\frac{\Delta\gamma_{12}}{2} + k_1 k_2 \frac{\Delta\varepsilon_{22}}{2}\frac{\Delta\sigma_{22}}{2} \tag{1.12}$$

where $\Delta\sigma_{12}$ and $\Delta\sigma_{22}$ are the shear and normal stress ranges in the critical plane, $\Delta\gamma_{12}$ and $\Delta\varepsilon_{22}$ are the shear and normal strain ranges in the critical plane, k_1 and k_2 are two weight constants for strain and stress amplitudes which are defined as follows,

$$k_1 = \frac{\gamma_f'}{\varepsilon_f'}, \qquad k_2 = \frac{\sigma_f'}{\tau_f'} \tag{1.13}$$

where σ_f' is the uniaxial fatigue strength coefficient; ε_f' is the uniaxial fatigue ductility coefficient, τ'_f is the torsional fatigue strength coefficient; γ'_f is the torsional fatigue ductility coefficient.

By using the concept of volume influencing fatigue crack initiation proposed by Palin-Luc and Lasserre (1998) with an energy based approach, a new criterion is presented by Banvillet, *et al.* (2003). Based on the strain-work density given to the material, this proposal is usable whatever the constant amplitude loading is, whether in and out-of-phase combined loadings, with or without mean stress. Its predictions are compared both with a total of 38 experiments on four materials (a mild steel, two high strength steels and a spheroidal graphite cast iron) and with the predictions of local criteria. The comparison shows that the predictions of the volumetric proposal are very good and less scattered than those of the local approaches, especially for loadings with mean stresses or under non-proportional loadings.

Gasiak and Pawliczek (2003) applied an energy model for FLP of construction steels under bending, torsion and synchronous bending and torsion and a good agreement between the calculated and experimental results has been obtained.

An energy parameter accounting for normal and shear deformation is used by Lee, *et al.* (2003) to predict fatigue lives under variable amplitude multiaxial loading. Fatigue damage is computed as a function of orientation of the plane, and the maximum damage plane is considered as the critical plane. The linear damage rule is employed with the rainflow cycle counting either on the shear or on the normal strain history depending on which gives more damage when evaluated individually. The method yields fatigue life correlations comparable to those of existing methods for materials failing in shear mode (S45C, SNCM439, SNCM630). However, a noticeable improvement is found for stainless steel 304 that exhibits a mixed normal and shear fracture behavior.

On the basis of the analysis of the energy of the elastic-plastic body subjected to monotonic and cyclic loading, a physical relationship between Neuber's rule and the Equivalent Strain Energy Density (ESED) method was found by Ye, *et al.* (2004). It is shown that Neuber's rule is actually a particular case of the ESED method, namely when the dissipation of the plastic strain energy at the notch root is neglected in the ESED method. The reason for the overestimation of the local strains using Neuber's rule is thus explained essentially and the physical meaning of the ESED method in both monotonic and cyclic form is further defined. In terms of the real physical behavior occurring at the notch root during cyclic plastic deformation, a modified version of the ESED method, in which only the heat energy is considered as dissipation and the stored energy is regarded as a contribution to local stress and strain ranges, has been developed by Ye, *et al.* (2004). It is shown that, for the case of cyclic loading, the modified ESED method further improves the accuracy of the original ESED method in prediction of the nonlinear stress/strain behavior of notches. It is also shown that the relation can easily be used for a simulation of the local strain-stress history near a notch root.

Jahed and Varvani-Farahani (2006) addressed the problem using an energy-based parameter to evaluate the fatigue life of metallic components subjected to multiaxial loading conditions. Cyclic strain-life data and corresponding Coffin-Manson coefficients for both normal and shear strain-lives were first defined. Energy-fatigue life curves were then generated from strain-fatigue life properties. The upper and lower limits of life are estimated using the proposed life equations. The upper life limit is obtained by assuming that the dominant cracking mechanism is Case A and the lower life limit is obtained by assuming that the dominant cracking mechanism is Case B. The proposed method was developed based on physical evidence of crack initiation and growth as well as the amount of dissipated energy over life cycles. The fatigue life data fall between the upper and the lower limits resulting in a promising life prediction. The proposed method has been used to evaluate the fatigue life of various metallic materials of SAE 1045, AISI 304, Inc 718 and Haynes 188 reported in the literature. Results of FLP were found to be in good agreement with experimental life data.

1.4.1.4 Continuum Damage Mechanics (CDM) approaches

CDM is a relatively new subject in engineering mechanics and deals with the mechanical behavior of a deteriorating medium on the continuum scale. The general concepts and fundamental aspects of this subject can be found in Kachanov (1986). Chaboche and Lesne (1988) were the first to apply CDM to FLP. For the one dimensional case, he postulated that fatigue damage evolution per cycle can be generalized by a function of the load condition and damage state. By measuring the changes in tensile load- carrying capacity and using the effective stress concept, he formulated a nonlinear damage evolution equation as

$$D = 1 - [1 - r^{1/(1-\alpha)}]^{1/(1+\beta)} \tag{1.14}$$

where β is a material constant, α is a function of the stress state and r is the damage state. This damage model is highly nonlinear in damage evolution and is able to account for the mean stress effect. It is, therefore, called a Nonlinear Continuous Damage (NLCD) model. The main features, advantages and some deficiencies of the NLCD model, are summarized by Chaboche and Lesne (1988). Based on the CDM concept, many other forms of fatigue damage equations have been developed. More information can be found in Fatemi and Yang (1998). Basically, all these CDM-based approaches are very similar to the Chaboche NLCD model in both form and nature. The main differences lie in the number and the characteristics of the parameters used in the model, in the requirements for additional experiments, and in their applicability.

Dattoma, *et al.* (2006) presented a non-linear continuum damage mechanics model based on a general thermodynamic framework developed by Lemaitre and Chaboche (1990). The proposed model has been formulated to take into account the material damage evolution at different load levels and it allows the effect of the loading sequence to be included. By means of a recurrence formula derived for multilevel loading, complex load sequences can be considered. The fatigue test data for hardened and tempered steel have been used to verify the model proposed and the results show a good agreement in predicting fatigue life under complex load sequences.

The CDM models were mainly developed for uniaxial fatigue loading. Some difficulties arise when these models are extended to multiaxial loading. Due to the complexity of nonproportional multiaxial fatigue problems, a 3D anisotropic CDM model does not yet exist. Great efforts are still needed to obtain an appropriate generalized prediction model for CFD under multiaxial loading.

Memon, *et al.* (2002) discussed the prediction of fatigue lives for 3D problems in the elasto-plastic range with variable load amplitude. The effect of the loading sequence on the fatigue life of a structure member and the validity of Miner's rule are studied. The computational method is derived according to damage mechanics, theory of plasticity and Finite Element Analysis (FEA). A two-block cyclic loading is considered with high-low and low-high load sequence. To consider the effect of the high stress level beyond the yielding point of material in one load block, deformation theory and the iteration method are applied to the stress analysis of the first load cycle. The damage evolution under given stress and the damage field for each block loading are determined by a damage evolution equation. Furthermore, an additional loading method is introduced to perform the stress analysis with a given damage field to avoid the reassembling of the stiffness matrix of the structure member. In this research, the damage increment at the critical element is considered as a step length instead of load cycle increment.

1.4.1.5 A brief summary of CFD methods

For many years, design engineers used the Palmgren-Miner law, Linear Damage Rule (LDR) and its modifications to predict the fatigue life of components in the case of variable loading (Dattoma, *et al.*, 2006). This method is based on the assumption of constant work absorption per cycle independent of the loading levels. The main deficiencies with LDR are the load-level independence character, the lack of damage contribution for stress below the fatigue limit and lack of load interaction accountability; this can produce a discrepancy of up to an order of magnitude between the predicted and the experimented life and this calculation may be non-conservative. Many researchers tried to modify Miner's rule but, due to its intrinsic deficiencies, no matter which version is used, life prediction based on this rule is often unsatisfactory (Fatemi and Yang, 1998).

In summary, no matter what quantities are used, fatigue criteria based on CFD theory suffer from another significant deficiency in addition to the above. That is, they have not consistently defined that failure may be life to a small detectable crack, life to a certain percentage decrease in load amplitude, or life to a fracture. Differences in fatigue life depending on these three criteria may be small or appreciable.

In general, the following factors will affect the fatigue life:

- quality of material processing (size and distribution of inclusions, voids, *etc*.);
- procedure of material processing (annealed, quenched, tempered…);
- procedure of specimen manufacture (specimen shape, machining method);
- quality of specimen manufacture (scratched, surface condition);
- material properties (yield strength, ultimate strength, strain at failure, σ-ε curve);
- geometry (length, width, thickness, diameter, transition radius, constraint effect);
- stress state (uniaxial, multiaxial, stress ratio, mean stress);
- effect of environment (temperature, corrosive environment).

Accurate prediction of fatigue life must take all these factors into account. For example, the effects of different surface conditions on the fatigue properties of cyclically loaded bending specimens of case-hardened steel SAE 5115 were investigated by Laue, *et al*. (2006). On the basis of a well-defined adjustment of different surface conditions, the effect of internal oxidation, surface roughness and residual stresses on fatigue crack initiation and growth were assessed. The effect of different parameters on the endurance limit has been quantified by the application of a fracture mechanics and a weakest-link approach. The result of fatigue testing was that internal oxidation has a detrimental influence on the endurance limit, which becomes stronger when the surface roughness is raised. Microhardness measurements within the oxidized layer showed a strong decrease in hardness due to the oxidation of alloying elements. The removal of the damaged zone results in an increase in the endurance limit although the newly generated surface is more sensitive to surface roughness due to its greater hardness. In the case of the electrolytically polished variant, it became apparent that it is very difficult to examine the influence of a single parameter on the endurance limit. The removal of the oxidized layer resulted in a modification of the surface hardness, surface roughness and residual stresses all at the same time. Shot peening of carburized specimens resulted in an increase in the endurance limit due to the generation of a pronounced maximum of compressive residual stresses. An explanation for this was found by the observation of the arrest of small fatigue cracks in front of the maximum compressive residual stresses. The experimental results suggested the use of a fracture mechanics approach to fatigue of case-hardened steels. If the internal oxidation is taken as the initial crack for fatigue crack growth and the residual stress state at the crack tip is taken into consideration as well, one can calculate the endurance limit of case-hardened steels. The calculations contributed to a deeper insight into the complex interaction between the parameters governing the fatigue process.

Many critical mechanical components experience multiaxial cyclic loading during their service life. In recent decades, numerous attempts to develop multiaxial fatigue damage modeling have been reported. The multiaxial fatigue problem of engineering materials mainly occurs for two reasons. In isotropic materials, the multiaxial stress within the material is due to the complex applied loading history. In anisotropic materials, a multiaxial stress state is obtained even if the applied loading is uniaxial. Different from the uniaxial fatigue problem, the multiaxial fatigue problem is more complex due to the complex stress states, loading histories and possible anisotropy of the material. A unified multiaxial fatigue damage model based on a characteristic plane approach is proposed by Liu and Mahadevan (2007), integrating both isotropic and anisotropic materials into one framework. Compared with most available critical plane-based models for a multiaxial fatigue problem, the physical basis of the characteristic plane does not rely on the observations of the fatigue crack in the proposed model. The cracking information is not required for multiaxial fatigue analysis and the proposed model can automatically adapt to very different materials experiencing different failure modes. The effect of the mean normal stress is also included in the proposed model. The results of the proposed FLP model are validated using experimental results of metals as well as unidirectional and multidirectional composite laminates.

Up to now, no CFD models are available which can consider all these influencing factors. As a result, the fatigue data are subjected to large scatter. However, the fuzzy definition of fatigue failure in CFD theory can be overcome through a fracture mechanics approach. FCP theory introduced in the next section could eliminate this deficiency.

1.4.2 FCP Theories

The basic assumption of the FCP theory used for predicting fatigue life or failure analysis is that initial defects always exist in engineering structures and fatigue is purely a crack propagation problem. Actually, in addition to cracks themselves, other types of flaws that are crack like in form may easily develop into cracks, and these need to be treated as if they were cracks (Dowling, 2007). This assumption has been confirmed due to progress in two areas: 1) FM provides the possibility for specific analysis of cracks in engineering components but not only based on the methods for uncracked material and FM has been successfully applied at the microstructural level in recent years, and 2) microstructural level cracks can now be accurately measured and the interest and advances in structural integrity monitoring technology are increasing. For the theory of FM, many books have given an historical review (*e.g.* Perez, 2004).

According to the manner of the cracked members' failure or the magnitude of the plastic deformation induced by the load applied to the cracked members, FM theory can be branched to LEFM and Plastic Fracture Mechanics (PFM). In the effective practical use of fracture mechanics, the particular environment and operating conditions, the knowledge of sizes and geometries of cracks, *etc*. have to be determined first. And there must be useful quantities to connect the related parameters. Generally, the controlling parameter, K, as used in Eq. (1.2) is always adopted for LEFM for evaluating the critical state of a crack, while a path-independent parameter, *J*-integral (Perez, 2004), is defined for handling problems of PFM with even large amounts of yielding. Until now, most of the FCP theories are based on *K*. Considering the plasticity limitations on LEFM, some modifications of *K* have been suggested to solve the small scale yielding problem. The detailed contents of FM will not be discussed in this book while its important concepts will be touched upon to provide background for FCP theory. The concept of *K* and its typical calculation methods, together with other important concepts of fracture mechanics, will be introduced in Chapter 4.

1.4.2.1 The crack growth rate curve

Crack growth in the theory of fatigue is caused by cyclic loading, and the LEFM approach was first introduced by Paris and his co-workers (1961), who equated the fatigue crack growth rate $\frac{\mathrm{d}a}{\mathrm{d}N}$ to the cyclic elastic SIF range, ΔK, at the tip of a long crack subjected to a low value of cyclic stress, as in Eq. (1.2). ΔK can be expressed by a function of the range of cyclic stress $\Delta\sigma$, crack length and other influential factors.

Later, people found that the crack growth rate curve is not linear for all the ranges of ΔK. The general crack growth rates for Mode I cracks in metals are as shown in Fig. 1.2.

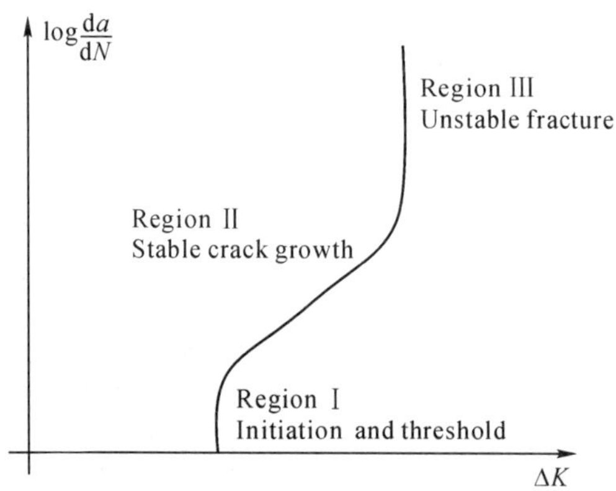

Fig. 1.2 Crack growth rate curve showing the three regions

Fig. 1.2 suggests a subdivision into three regions. At low growth rates in Region I, the curve generally becomes steep and appears to approach a vertical asymptote called initiation and threshold. The crack growth rate goes asymptotically to zero as ΔK approaches a threshold value ΔK_{th} This means that for stress intensities below ΔK_{th}, there is no crack growth, *i.e.* there is a fatigue limit. Behavior in Region I is dependent on microstructural features and it is of considerable importance in service components. Region II is characterized by stable crack growth and can be described by the Paris equation, which has been described as a continuum process not strongly dependent on the microstructure. The rate of crack propagation in Region III is exhibiting a rapidly increasing growth rate towards "infinity", *i.e.* ductile tearing and/or brittle fracture. It corresponds to the onset of unstable and rapid crack growth and is characterized by either the material's fracture toughness or, in the case of ductile materials, by plastic instability. The environment has little effect in this region and deformation mechanisms are similar to those characteristic of monotonic loading (Dowling, 2007).

1.4.2.2 Typical crack growth rate models

The relationship representing the crack growth rate curve can be denoted by crack growth rate models. Up to now, hundreds of different empirical crack growth rate models have been proposed and more progress has been made to establish crack growth rate relations which are able to explain different fatigue phenomena. A review of the development of the models can be referred to (Cui, 2008). Some milestones of this point will be briefly introduced in this section.

In the Paris Eq. (1.2), the range of the Stress Intensity Factor (SIF), ΔK, is utilized as a crack driving force. However, very soon people found that the apparent effectiveness of ΔK is affected by the load ratio R and many other factors such as overload, crack size, environment, microstructure, geometry, temperature will have an influence on fatigue crack growth.

In order to put the threshold value of the SIF range into the model, Donahue, *et al.* (1972) proposed the crack growth relation in the threshold region as

$$\frac{\mathrm{d}a}{\mathrm{d}N} = C\cdot\left[\Delta K - \Delta K_{\mathrm{th}}\right]^m \tag{1.14}$$

Region II crack growth follows a power law, the so-called Paris-Erdogan crack growth law (Paris and Erdogan, 1963) is as given in Eq. (1.2).

Region III crack growth exhibits a rapidly increasing growth rate towards "infinity", *i.e.* ductile tearing and/or brittle fracture. This has led to the relation proposed by Forman, *et al.* (1967),

$$\frac{\mathrm{d}a}{\mathrm{d}N} = \frac{C(\Delta K)^m}{(1-R)K_{\mathrm{C}} - \Delta K} \tag{1.15}$$

where K_{C} is the fracture toughness of the material.

Relations combining the departure from power law behavior at high and low ΔK values also exist (Kanninen and Popelar, 1985; Carpinteri, 1994). The most representative one might be the following, proposed by McEvily and Groeger (1977),

$$\frac{\mathrm{d}a}{\mathrm{d}N} = A(\Delta K - \Delta K_{\mathrm{th}})^2 \left[1 + \frac{\Delta K}{K_{\mathrm{C}} - K_{\max}}\right] \tag{1.16}$$

In order to explain the effects of load ratio R on fatigue crack growth, Elber (1970) introduced the concepts of crack closure and the effective cyclic SIF ΔK_{eff} as the dominant driving force for fatigue,

$$\Delta K_{\mathrm{eff}} = K_{\max} - K_{\mathrm{op}} \tag{1.17}$$

where $K_{\max}$ is the stress intensity calculated for the maximum load, and K_{op} is the stress intensity calculated for the crack opening load. This concept was highly appraised in the 1980s and 1990s but now it is subjected to some challenge (*e.g.* Hertzberg, *et al.*, 1988). In particular, many people have agreed that the physical effects of crack closure have been greatly over-estimated in the past (*e.g.* Vasudevan, *et al.*, 1994). The partial crack closure model (Donald and Paris, 1999; Kujawski, 2001a; 2001b) was proposed, to overcome the difficulty the crack closure model met. The modified effective SIF range is defined as (Kujawski, 2001b),

$$\begin{aligned} \Delta K_{\mathrm{eff}M} &= K_{\max} - K_{\mathrm{op}}[1 + (\frac{2}{\pi} - 1)g] \\ g &= \exp[-(\frac{K_{\max}}{K_{\max\,\mathrm{th}}} - 1)] \end{aligned} \tag{1.18}$$

where $K_{\max\,\mathrm{th}}$ is the maximum SIF at the threshold for a given R-ratio.

Later, Kujawski (2001c; 2001d) further found that without using the crack closure concept, it is possible to explain the stress ratio effect, even better than when using the concept. He defined the following parameter as the fatigue crack driving force,

$$\Delta K^* = (K_{\max})^{\alpha} (\Delta K^+)^{1-\alpha} \tag{1.19}$$

where ΔK^+ is the positive part of the applied SIF range. Using this parameter, the threshold value can be expressed as

$$\Delta K^*_{\mathrm{th}} = \begin{cases} \Delta K^*_{\mathrm{th}\,0} (1-R)^{\alpha} & \text{for } R > 0 \\ \Delta K^*_{\mathrm{th}\,0} (1-R) & \text{for } R \le 0 \end{cases} \tag{1.20}$$

where $\Delta K^*_{\text{th}0}$ is the threshold value corresponding to $R = 0$. From authors' work (Cui, *et al.*, 2011), this equation cannot be regarded as universally valid.

Sadananda, *et al.* (2003) criticized the FCP methods based on the crack closure concept. They pointed out that the drawbacks in the current FLP methods stem from several sources: 1) the assumption of plasticity induced crack closure, 2) the lack of terms in the model that relate to the environmental effects and slip deformation behavior, and 3) several adjustable parameters needed to fit the observed data. In order to overcome these drawbacks and develop a more reliable FLP model, Vasudevan, Sadananda and co-workers have proposed a two-parameter unified approach. This will be discussed in detail in Chapter 3.

1.4.2.3 Physically small crack growth

It is well agreed that for physically small crack growth, Elastic Plastic Fracture Mechanics (EPFM) must be employed. The EPFM approach was first introduced by Tomkins in 1968 (Miller, 1999) who equated $\frac{\mathrm{d}a}{\mathrm{d}N}$ to crack tip decohesion (from knowledge of the cyclic stress-strain curve) and thence to the bulk plastic strain field as occurs, for example, under high strain fatigue. Thus

$$\frac{\mathrm{d}a}{\mathrm{d}N} = B\left(\Delta\varepsilon_p\sqrt{\pi a}\right)^m - \Xi \tag{1.21}$$

where Ξ is the threshold condition in this crack regime.

1.4.2.4 Microstructurally small crack growth

Microstructural Fracture Mechanics (MFM) was proposed to handle the crack propagation at the micro-crack level. The MFM approach was first introduced by Hobson, *et al.* (1986) and followed by Navarro and de los Rios (1988). The crack growth law is expressed as

$$\frac{\mathrm{d}a}{\mathrm{d}N} = C\Delta\gamma^\beta(d - a) \tag{1.22}$$

where d is a microstructural dimension. It should be noted that Eq. (1.22) indicates a zero crack speed when the crack depth, a, is equal to d, and that prior to this state the crack will continuously decelerate until it either stops or continues to propagate according to an overlapping continuum mechanics description.

1.4.2.5 A brief summary of FCP methods

As analysed in Section 1.2, any methods based on the CFD theory cannot be very accurate due to its neglect of several important factors. Now, more and more people have realized that only an FLP method based on FCP theory has the potential to explain various fatigue phenomena observed. However, it is not the case that any type of FCP theory can work. As a matter of fact, from the very fundamental question of fatigue crack driving forces to the more complicated fatigue crack growth rate expression, all need critical examination. For example, crack can be viewed as a macroscopic quantity or a microscopic quantity and different levels of crack propagation will be governed by different parameters. In this book we confine our model to the macroscopic level. No microstructural parameters are involved. The main purpose of this book is to present a UFLP method developed by the authors. The basic idea of the method will be introduced in Chapter 3.

1.5 The Layout of the Book

The book is divided into eight chapters. The first chapter is the introduction, in which the basic definition and the importance of fatigue to marine structures are presented. Then the history of fatigue research is summarized and the general layout of the book is described. Since most of the marine structures are made of metals, in Chapter 2 the latest understanding of fatigue mechanisms of metals is summarized. Chapter 3 explains the meaning of the unified approaches used in this book and the current state-of-the-art of UFLP is introduced. As can be seen from Chapter 3, the basic idea of the unified approach is to use the FCP theory based on fracture mechanics instead of CFD theory, so basic concepts of fracture mechanics are introduced in Chapter 4. In Chapter 5, the general procedure for a UFLP method for marine structures is described and from it one can easily detect the technical problems to be resolved for developing UFLP. Chapter 6 discusses various issues related to fatigue loading, which is the main cause of fatigue failure of marine structures. In Chapter 7, some applications and demonstrations of UFLP are given, which could help the understanding of the philosophy presented in the book and further recognize the problems to be solved for engineering applications. Chapter 8 discusses the relevant issues related to code development based on UFLP for marine structures.

References

ABS (American Bureau of Shipping) (1992). Guide for the Fatigue Strength Assessment of Tankers. Houston, Texas.

Almar-Naess, A. (1985). Fatigue Handbook. Trondheim: Tapir.

Basquin, O. H. (1910). "The exponential law of endurance tests", Proceedings. ASTM, 10(II): 625–630.

Bathias, C. (1999). "There is no infinite fatigue life in metallic materials", Fatigue and Fracture of Engineering Materials and Structures, 22: 559–565.

Baumel, A. Jr. & Seeger, T. (1990). Materials Data for Cyclic Loading (Supplement 1). Amsterdam: Elsevier Science Publishers.

Banvillet, A., Palin-Luc, T. & Lasserre, S. (2003). "A volumetric energy based high cycle multiaxial fatigue criterion", International Journal of Fatigue, 25: 755–769.

Berger, C., Pyttel, B. & Trossmann, T. (2006). "Very high cycle fatigue tests with smooth and notched specimens and screws made of light metal alloys", International Journal of Fatigue, 28: 1640–1646.

BSI (British Standards Institution) (1980). Code of Practice for Fatigue. BS 5400: Part 10. London.

BSI (British Standards Institution) (1993). Code of Practice for Fatigue Design and Assessment of Steel Structures. BS 7608. London.

Carpinteri, A. (ed.) (1994). Handbook of Fatigue Crack Propagation in Metallic Structures (Vols. 1 and 2). Amsterdam: Elsevier Science Publishers.

Cashman, G. T. (2010). "A review of competing modes fatigue behavior", International Journal of Fatigue, 32(3): 492–496.

Castillo, E. & Fernandez-Canteli, A. (2006). "A parametric lifetime model for the prediction of high-cycle fatigue based on stress level and amplitude", Fatigue and Fracture of Engineering Materials and Structures, 29: 1031–1038.

Chaboche, J. L. & Lesne, P. M. (1988). "A non-linear continuous fatigue damage model", Fatigue and Fracture of Engineering Materials and Structures, 11(1): 1–7.

Coffin, L. F. & Tavernelli J. F. (1959). "The cyclic straining and fatigue of metals", Trans of the Metallurgical Society of American Institute of Mechanical Engineers, 215, 794.

Cramer, E., Gran, S., Holtsmark, G., Lotsberg, I., Loseth, R., Olaisen, K. & Valsgard, S. (1994). "Fatigue assessment of ship structures", DNVC Report No. 93–432.

CSR of IACS (2006). "IACS common structural rules for tankers and bulk carriers", the IACS Council, Apr. 1, 2006.

Cui, W. C., Wang, F. & Huang, X. P. (2011). "A unified fatigue life prediction method for marine structures", Marine Structures, 24(2): 153–181.

Cui, W. C. (2002). "A state-of-the-art review on fatigue life prediction methods for metal structures", Journal of Marine Science and Technology, 7: 43–56.

Cui, W. C. (2008). "Mechanisms, mathematical models and preventive measures of fatigue cracking (Chapter 5)", in: Paik, J. K. and Melchers, R. E. (eds.)

Condition Assessment of Aged Structures. Cambridge, UK: Woodhead Publishing Ltd.

Dattoma, V., Giancane, S., Nobile, R. & Panella, F. W. (2006). "Fatigue life prediction under variable loading based on a new non-linear continuum damage mechanics model", International Journal of Fatigue, 28: 89–95.

De Castro, J. T. P., Meggiolaro, M. A. & de Oliveira Miranda, A. C. (2005). "Singular and non-singular approaches for predicting fatigue crack growth behavior", International Journal of Fatigue, 27: 1366–1388.

Donahue, R. J., Clark, H. M., Atanmo, P., Kumble, R. & McEvily, A. J. (1972). "Crack opening displacement and the rate of fatigue crack growth", International Journal of Fracture Mechanics, 8: 209–219.

Donald, K. & Paris, P. C. (1999). "An evaluation of ΔK_{eff} estimation procedure on 6061-T6 and 2024-T3 aluminum alloys", International Journal of Fatigue, 21: S47–57.

Dowling, N. E. (2007). Mechanical Behavior of Materials-Engineering Methods for Deformation, Fracture, and Fatigue (3rd ed.). Upper Saddle River, New Jersey: Pearson-Prentice Hall.

Elber, W. (1970). "Fatigue crack closure under cyclic tension", Engineering Fracture Mechanics, 2: 37–45.

Fatemi, A. & Yang, L. (1998). "Cumulative fatigue damage and life prediction theories: a survey of the state of the art for homogeneous materials", International Journal of Fatigue, 20(1): 9–34.

Forman, R. G., Kearney, V. E. & Engle, R. M. (1967). "Numerical analysis of crack propagation in cyclic-loaded structures", Journal of Basic Engineering, 89: 459–464.

Fricke, W., Cui, W. C., Kierkegaard, H., Kihl, D., Koval, M., Lee, H. L., Mikkola, T., Parmentier, G., Toyosada, M. & Yoon, J. H. (2002). "Comparative fatigue strength assessment of a structural detail in a containership using various approaches of classification societies", Marine Structures, 15(1): 1–13.

Furuya, Y. (2010). "Size effects in gigacycle fatigue of high-strength steel under ultrasonic fatigue testing", Procedia Engineering, 2: 485–490.

Gasiak, G. & Pawliczek, R. (2003). "Application of an energy model for fatigue life prediction of construction steels under bending, torsion and synchronous bending and torsion", International Journal of Fatigue, 25: 1339–1346.

Glen, I. F., Dinovitzer, A., Malik, L., Basu, R. & Yee, R. (2000). "Guide to damage tolerance analysis of marine structures", SSC-409, NTIS#PB2000-108441.

Glinka, G. (1988). "Relations between the strain energy density distribution and elastic-plastic stress-strain field near cracks and notches and fatigue life calculation", Low Cycle Fatigue, ASTM STP 942, 1002–1047.

Hanaki, S., Yamashit, M., Uchida, H. & Zako, M. (2010). "On stochastic evaluation of *S-N* data based on fatigue strength distribution", International Journal of Fatigue, 32: 605–609.

Hertzberg, R. W., Newton, C. H. & Jaccard, R. (1988). "Crack closure: correlation and confusion", In: Mechanics of Fatigue Crack Closure. ASTM STP 982.

Philadelphia, PA: American Society for Testing and Materials, 139–148.
Heuler, P. & Klätschke, H. (2005). "Generation and use of standardised load spectra and load-time histories", International Journal of Fatigue, 27: 974–990.
Hobson, P. D., Brown, M. W. & de los Rios, E. R. (1986). EGF (ESIS) Publication. In: Miller, K. J. & de los Rios, E. R. (eds.) (No.1) The Behavior of Short Fatigue Cracks. London: Mech. Engng, Publications Ltd., 441–459.
ISSC Committee III. 2 (2009). "Fatigue and fracture", in: Jang, C. D. & Hong, H. S. (ed.) Proceedings of the 17th Int. Ship and Offshore Structures Congress, Aug.16–21, 2009, Seoul.
Jahed, H. & Varvani-Farahani, A. (2006). "Upper and lower fatigue life limits model using energy-based fatigue properties", International Journal of Fatigue, 28: 467–473.
Kachanov, L. M. (1986). Introduction to Continuum Damage Mechanics. Dordrecht, the Netherland: Martinus Nijhoff.
Kanninen, M. F. & Popelar, C. H. (1985). Advanced Fracture Mechanics. New York: Oxford University Press.
Kohout, J. & Vechet, S. (1999). "New functions for description of fatigue curves and their advantages", Fatigue'99, 783–788.
Kujawski, D. (2005). "On assumptions associated with ΔK_{eff} and their implications on FCG predictions", International Journal of Fatigue, 27: 1267–1276.
Kujawski, D. (2001a). "Enhanced model of partial crack closure for correlation of R-ratio effects in aluminum alloys", International Journal of Fatigue, 23: 95–102.
Kujawski, D. (2001b). "Correlation of long- and physically short-cracks growth in aluminum alloys", Engineering Fracture Mechanics, 68: 1357–1369.
Kujawski, D. (2001c). "A new $\left(\Delta K^{+} K_{\max}\right)^{0.5}$ driving force parameter for crack growth in aluminum alloys", International Journal of Fatigue, 23: 733–740.
Kujawski, D. (2001d). "A fatigue crack driving force parameter with load ratio effects", International Journal of Fatigue, 23: S239–246.
Lemaitre, J. & Chaboche, J. L. (1990). Mechanics of Solid Materials. Cambridge, UK: Cambridge University Press.
Laue, S., Bomas, H. & Hoffmann, F. (2006). "Influence of surface condition on the fatigue behaviour of specimens made of a SAE 5115 case-hardened steel", Fatigue and Fracture of Engineering Materials and Structures, 29: 229–241.
Lee, B. L., Kim, K. S. & Nam, K. M. (2003). "Fatigue analysis under variable amplitude loading using an energy parameter", International Journal of Fatigue, 25: 621–631.
Lee, K. S. & Song, J. H. (2006). "Estimation methods for strain-life fatigue properties from hardness", International Journal of Fatigue, 28: 386–400.
Li, W., Sakai, T., Li, Q., Lu, L. T. & Wang, P. (2010). "Reliability evaluation on very high cycle fatigue property of GCr15 bearing steel", International Journal of Fatigue, 32: 1096–1107.
Liu, D. & Thayamballi, A. (1995). "Local cracking in ships—causes, consequences,

and control", in: Proceedings of Symposium on the Prevention of Fracture in Ship Structure, Mar. 30–31, 1995, Washington, D.C.

Liu, Y. B., Li, Y. D., Li, S. X., Yang, Z. G., Chen, S. M., Hui, W. J. & Weng, Y. Q. (2010). "Prediction of the *S-N* curves of high-strength steels in the very high cycle fatigue regime", International Journal of Fatigue, 32: 1351–1357.

Liu, Y. M. & Mahadevan, S. (2007). "A unified multiaxial fatigue damage model for isotropic and anisotropic materials", International Journal of Fatigue, 29: 347–359.

LR (Lloyd's Register) (1992). HTS—Something for Nothing? 100A1, Issue 3. London, UK.

LR (Lloyd's Register) (1996). Ship Right—Fatigue Design Assessment Procedure. London, UK.

Manson, S. S. (1965). "Fatigue: a complex subject—some simple approximations", Experimental Mechanics—Journal of the Society for Experimental Stress Analysis, 5(7): 193–226.

Manson, S. S. & Hirschberg, M. H. (1964). Fatigue: an Interdisciplinary Approach. Syracuse, New York: Syracuse University Press.

Marines, I., Bin, X. & Bathias, C. (2003). "An understanding of very high cycle fatigue of metals", International Journal of Fatigue, 25: 1101–1107.

Maymon, G. (2005a). "A 'unified' and a $\left(\Delta K^{+} K_{\max}\right)^{1/2}$ crack growth models for aluminum 2024-T351", Internal Journal of Fatigue, 27: 629–638.

Maymon, G. (2005b). "Probabilistic crack growth behavior of aluminum 2024-T351 alloy using the 'unified' approach", International Journal of Fatigue, 27: 828–834.

McEvily, A. J. & Groeger, J. (1977). "On the threshold for fatigue-crack growth", 4th International Conference on Fracture, Waterloo: University of Waterloo Press, 2: 1293–1298.

McEvily, A. J. & Ishihara, S. (2001). "On the dependence of the rate of fatigue crack growth on the $\sigma_a^n(2a)$ parameter", International Journal of Fatigue, 23: 115–120.

McEvily, A. J., Bao, H. & Ishihara, S. (1999). "A modified constitutive relation for fatigue crack growth", Fatigue'99, 329–336.

Meggiolaro, M. A. & Castro, J. T. P. (2004). "Statistical evaluation of strain-life fatigue crack initiation predictions", International Journal of Fatigue, 26: 463–476.

Memon, I. R., Zhang, X. & Cui, D. Y. (2002). "Fatigue life prediction of 3D problems by damage mechanics with two-block loading", International Journal of Fatigue, 24: 29–37.

Miller, K. J. (1999). "A historical perspective of the important parameters of metal fatigue and problems for the next century", in: Wu, X. R. & Wang, Z. G. (ed.) Proceedings of the seventh International Fatigue Congress (Fatigue'99). Beijing: Higher Education Press, 15–39.

Miner, M. A. (1945). "Cumulative damage in fatigue", Journal of Applied

Mechanics, Trans. ASME, Mech. 67, A159–164.

Mitchell, M. R., Socie, D. F. & Caulfield, E. M. (1977). "Fundamentals of modern fatigue analysis", Fracture Control Program Report No. 26, University of Illinois, 385–410.

Muralidharan, U. & Manson, S. S. (1988). "A modified universal slopes equation for estimation of fatigue characteristics of metals", Journal of Engineering Materials and Technology—Transactions of the ASME, 110: 55–58.

Nakajima, M., Tokaji, K., Itoga, H. & Shimizu, T. (2010). "Effect of loading condition on very high cycle fatigue behavior in a high strength steel", International Journal of Fatigue, 32: 475–480.

Navarro, A. & de los Rios, E. R. (1988). "A microstructurally short fatigue crack growth equation", Fatigue and Fracture of Engineering Materials and Structures, 11(5), 383–396.

Newman, Jr., J. C., Irving, P. E., Lin, J. & Le, D. D. (2006). "Crack growth predictions in a complex helicopter component under spectrum loading", Fatigue and Fracture of Engineering Materials and Structures, 29: 949–958.

Noroozi, A. H., Glinka, G. & Lambert, S. (2005). "A two parameter driving force for fatigue crack growth analysis", International Journal of Fatigue, 27: 1277–1296.

Ong, J. H. (1993a). "An evaluation of existing methods for the prediction of axial fatigue life from tensile data", International Journal of Fatigue, 15(1): 13–19.

Ong, J. H. (1993b). "An improved technique for the prediction of axial fatigue life from tensile data", International Journal of Fatigue, 15(3): 213–219.

Paik, J. K. & Thayamballi, A. K. (2003). Ultimate Limit State Design of Steel-Plated Structures. Chichester: John Wiley & Sons.

Paik, J. K., Brennan, F. P., Carlsen, C. A., Daley, C., Garbatov, Y., Ivanov, L., Rizzo, C. M., Simonsen, B. C., Yamamoto, N. & Zhuang, H. Z. (2006). "Report of specialist committee V.6: condition assessment of aged ships", in: Frieze, P. A. & Shenoi, R. A. (eds.) Proceedings of the 16th International Ship and Offshore Structures Congress, Aug. 20–25, Southampton, UK, 2: 269–320.

Palin-Luc, T. & Lasserre, S. (1998). "An energy based criterion for high cycle multiaxial fatigue", European Journal of Mechanis, A/Solids, 17(2): 237–251.

Palmgren, A. (1924). "Die Lebensdauer von Kugellagern (Durability of ball bearings)", ZDVDI, 68(14): 339.

Pan, W. F., Hung, C. Y. & Chen, L. L. (1999). "Fatigue life estimation under multiaxial loadings", International Journal of Fatigue, 21: 3–10.

Paris, P. C. & Erdogan, F. (1963). "A critical analysis of crack propagation laws", Journal of Basic Engineering, 85: 528–534.

Paris, P. C. Gomez, M. P. & Anderson, W. P. (1961). "A rational analytical theory of fatigue", The Trend in Engineering, 13: 9–14.

Park, J. H. & Song, J. H. (1995). "Detailed evaluation of methods for estimation of fatigue properties", International Journal of Fatigue, 17(5): 365–373.

Perez, N. (2004). Fracture Mechanics. Boston: Kluwer Academic Publishers.

Pyttel, B., Schwerdt, D. & Berger, C. (2011). "Very high cycle fatigue—is there a fatigue limit?" International Journal of Fatigue, 33: 49–58.
Pyttel, B., Schwerdt, D. & Berger, C. (2010). "Fatigue strength and failure mechanisms in the VHCF-region for quenched and tempered steel 42CrMoS4 and consequences to fatigue design", Procedia Engineering, 2: 1327–1336.
Qian, G. A., Hong, Y. S. & Zhou, C. G. (2010). "Investigation of high cycle and Very-High-Cycle Fatigue behaviors for a structural steel with smooth and notched specimens", Engineering Failure Analysis, 17: 1517–1525.
Qian, Y. & Cui, W. C. (2010). "An overview on experimental investigation on variable amplitude fatigue crack growth rule", Journal of Ship Mechanics, 14(5): 556–565.
Ravi Chandran, K. S., Chang, P. & Cashman, G. T. (2010). "Competing failure modes and complex *S-N* curves in fatigue of structural materials", International Journal of Fatigue, 32(3): 482–491.
Roessle, M. L. & Fatemi, A. (2000). "Strain-controlled fatigue properties of steels and some simple approximations", International Journal of Fatigue, 22: 495–511.
Sadananda, K., Vasudevan, A. K. & Kang, I. W. (2003). "Effect of superimposed monotonic fracture modes on the ΔK and $\Delta K_{\max}$ parameters of fatigue crack propagation", Acta Materialia, 51: 3399–3414.
Sakai, T., Takeda, M., Shiosawa, K., Ochi, Y., Nakajima, M., Nakamura, T. & Oguma, N. (1999). "Experimental evidence of duplex *S-N* characteristics in wide life region for high strength steels", Fatigue'99, 573–578.
Schütz, W. (1996). "A history of fatigue", Engineering Fracture Mechanics, 54(2): 263–300.
Skaar, K. T., Valsgard, S., Kohler, P. E. & Murer, C. (1987). "How low can steel weight go with safety and economy?" In: Proceedings 3rd International Conference on Practical Design of Ships and Mobile Offshore Units, PRADS'87, Jun. 22–26, 1987, Trondheim, Norway.
Skorupa, M. (1998). "Load interaction effects during fatigue crack growth under variable amplitude loading—a literature review (Part I: Empirical trends)", Fatigue and Fracture of Engineering Materials and Structures, 21(8): 987–1006.
Skorupa, M. (1999). "Load interaction effects during fatigue crack growth under variable amplitude loading—a literature review (Part II: Qualitative interpretation)", Fatigue and Fracture of Engineering Materials and Structures, 22(10): 905–926.
Smith, K. N., Watson, P. & Topper, T. H. (1970). "A stress-strain function for the fatigue of metals", J. Mater., 5(4): 767–778.
Stoychev, S. & Kujawski, D. (2005). "Analysis of crack propagation using ΔK and $K_{\max}$", International Journal of Fatigue, 27: 1425–1431.
UKDE (1984). U. K. Department of Energy, "Offshore installations: guidance on design and construction", HMSO, Apr. 1984, London.
Vasudevan, A. K., Sadananda, K. & Glinka, G. (2001). "Critical parameters for fatigue damage", International Journal of Fatigue, 23: S39–53.
Vasudevan, A. K., Sadananda, K. & Louat, N. (1994). "A review of crack closure,

fatigue crack threshold and related phenomena", Materials Science and Engineering, A188: 1–22.

Wang, X. S., Kawagoishi, N., Yu, S. W., Nisitani, H. & Goto, M. (1999). "Prediction of fatigue life of carbon steel using only the tensile strength", Fatigue'99, 845–850.

Wang, Q. Y., Li, T. & Zeng, X. G. (2010). "Gigacycle fatigue behavior of high strength aluminum alloys", Procedia Engineering, 2: 65–70.

Wästberg, S., Brennan, D., Chen, R., Hodgson, T., Kihl, D., Litonov, O., Mahérault-Mougin, S., Osawa, N., Paetzold, H., Shin, C.-Ho, Tedeschi, R. & Wang, X. (2006). "Fatigue and fracture", TC III.2 Report. In: Frieze, P. A. & Shenoi, R. A. (eds.), Proceedings of 16th International Ship and Offshore Structures Congress, Aug. 20–25, Southampton, UK, 1, 459–541.

Weibull W. (1951). "A statistical distribution function of wide applicability", Journal of Applied Mechanics, 293–297.

Yang, L. & Fatemi, A. (1998). "Cumulative fatigue damage mechanisms and quantifying parameters: a literature review", Journal of Testing and Evaluation, 26(2): 89–100.

Ye, D. Y., Matsuoka, S., Suzuki, N. & Maeda, Y. (2004). "Further investigation of Neuber's rule and the equivalent strain energy density method", International Journal of Fatigue, 26: 447–455.

Zhang, J. Z., He, X. D. & Du, S. Y. (2005). "Analyses of the fatigue crack propagation process and stress ratio effects using the two parameter method", International Journal of Fatigue, 27: 1314–1318.

2

Current Understanding of Fatigue Mechanisms of Metals

2.1 Introduction

Ships and marine structures are very large welded metal structures operating in a marine environment. Fatigue failure is well recognized to be the main structural failure mode by surveyors from various classification societies. Since high strength steel is more and more widely used in ship structures, which makes the yield strength and ultimate strength increase, the traditional strength criteria of structures are usually satisfied. But fatigue strength is almost unconnected to the yield strength and ultimate strength of the material (Baumel and Seeger, 1990). It is noted that high strength steel can guarantee that the structure can withstand higher stress but at the same time it increases the stress amplitude of the response, so that fatigue failure is more likely to occur in those ship structures made of high strength steel. Fatigue failure is the main reason for damage to ship structures, especially for large ships and ships built with high strength steel (Meggiolaro and Castro, 2003; Lee and Song, 2006; Dowling, 2007).

Materials science reveals (Fuchs and Stephens, 1980) that there are inevitable defects in the manufacturing process. Micro-cracks are always present in materials, especially in welds. The fatigue crack nucleus can be initiated on a microscopically small scale. These micro-cracks grow and coalesce to form a macroscopic crack, and the macro-crack propagates to the critical state of structural failure.

At present, the fatigue research of structures and materials has shown great and growing vitality. Every year thousands of articles on fatigue are published. Human beings have accumulated a large amount of data on fatigue, have a more profound understanding of fatigue problems and have some methods to prevent fatigue fracture. However, because of the complexity of fatigue problems and too many factors which influence fatigue, the mechanism of fatigue failure is not fully understood so far. In many cases, there are still big differences between theoretical analysis and practical situations, which cannot provide effective guidance on engineering. The fatigue life of materials and structures is still one of the most important issues in engineering, mechanics, metallurgy, materials and other fields, especially the morphology of presence slip bands, dislocation structure and formation mechanisms, and their role in fatigue crack initiation are the major subjects of basic research in fatigue.

Understanding the fatigue mechanism is essential for considering various technical conditions which affect fatigue life and fatigue crack growth, such as the material properties, surface quality, residual stress, and environmental influence. This knowledge is essential for the analysis of fatigue properties of an engineering structure. Fatigue prediction methods can only be established if the fatigue mechanism is reasonably understood. For that reason, the present chapter is a prerequisite for most chapters of this book.

The fatigue mechanism is different for different materials. In this chapter, it is confined to metal fatigue only. For fatigue mechanism of composite materials, readers are advised to refer to Carlson and Kardomateas (1996) and Mortensen (2006). The purpose of this chapter is to introduce the current state of the art about the understanding of fatigue mechanisms. The chapter contains seven sections. Section 2.1 is an introduction. Section 2.2 gives a brief description of the different phases of fatigue life. Sections 2.3 and 2.4 discuss current understanding of fatigue crack initiation and crack growth respectively and Section 2.5 considers some special issues about fatigue cracking while Section 2.6 further discusses the fatigue crack growth mechanism of small defects. The last section is a brief summary of this chapter.

2.2 Different Phases of the Fatigue Life

Metal fatigue cracks are often initiated from defects, such as inclusions. Other areas are subjected to stress concentration. For a metal material that does not contain a big enough internal stress concentration source, fatigue cracks in metals are usually initiated from the surface of a component, where fatigue damage is initiated as shear cracks on crystallographic slip planes. The surface shows the slip planes as intrusions and extrusions (Schijve, 2003). After a fatigue crack is initiated in a slip, first of all cracks propagate along the maximum shear stress acting on the plane, with some characteristics of crystallography. For a uniaxial loading condition, the angles between the micro-cracks and the loading direction are roughly 45°. These micro-cracks further propagate or mutually connect under cyclic loading, and most of the micro-cracks stop expanding quickly. Only a few micro-cracks can reach a few tens of microns in length. This is Stage I crack growth. Then the crack growth direction gradually deviates from the original direction, after a transient period, forming the main crack perpendicular to the maximum tensile stress. Crack growth shifts from Stage I to Stage II of crack growth. Finally the crack becomes unstable and fracture occurs. An unstable fracture is the state of accumulation damage relative to the critical value, *i.e.* the results of a crack propagating to the critical crack size, or the SIF of the crack being equal to the fracture toughness of the material. The mechanism of an unstable fracture is the same as that of a brittle fracture under monotonic tensile loading. Therefore, the following sections focus on describing the mechanism of crack initiation and propagation. Fig. 2.1 shows a schematic representation of the two stages in the fatigue crack initiation, growth and fracture.

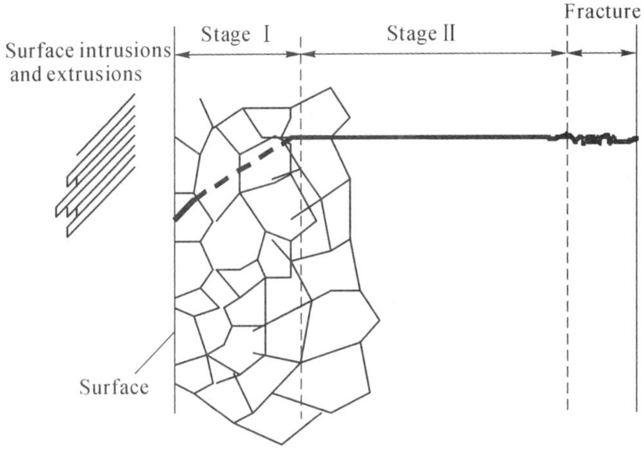

Fig. 2.1 Schematic representation of crack initiation and growth in polycrystalline metals (Cui, 2002b)

This two-stage process was first recognized by Forsyth (1962) and it was one of the most important achievements in metal fatigue in the 20th century (Miller, 1999). However, that division may be too vague for further discussion. So in some other sources, this process may be further divided. Schijve (1967) divided it into four phases: crack nucleation, micro-crack growth, macro-crack growth and failure. Shang, *et al.* (1998) divided it into five stages: 1) early cyclic formation and damage; 2) micro-crack nucleation; 3) short crack propagation; 4) macro-crack propagation; 5) final fracture. Miller (1987a; 1987b) divided the crack length into three regimes: microstructurally small crack; physically small crack and long crack. A slightly different classification of small cracks has also been proposed by Ritchie (1986) as: microstructurally small crack of critical microstructural dimension, *e.g.* grain size; physically small crack of the order of less than 1 mm; mechanically small crack of the order of plastic zone length (several millimeters); chemically small crack of the order up to 10 mm.

Based on these different divisions and applications, Cui (2002) recommended that the fatigue failure process should be divided into five parts: 1) crack nucleation ($a<a_m$); 2) microstructure short crack growth ($a_m \le a<a_p$); 3) physical short crack growth ($a_p \le a<a_1$); 4) long crack growth ($a_1 \le a<a_c$); 5) final fracture. Here, a is the characteristic dimension of an equivalent crack in a component, a_m is the smallest crack length detectable by current technology which is about 0.1 μm; a_p is the smallest crack length for physically small cracks which is about 10 μm, a_1 is the smallest crack length for long cracks which is about 1 mm, a_c is the critical crack length at which component fracture occurs. Obviously, these boundary divisions

are subjective and depend on the ability of crack measuring systems. Traditionally, the process before long crack propagation is named "fatigue crack initiation", while long crack propagation is called "fatigue crack propagation". This division is also too vague and could not be expected to provide an accurate prediction of fatigue life. However, it may be worth pointing out here that, depending on the initial crack length in a component, some of the early stages may be skipped. These five stages of the fatigue process only exist in "defect free" metal components.

It is now well recognized that small pre-existing defects are an inherent feature of engineering components and structures. They may be formed as a result of material forming and fabrication techniques, or may be an adventitious result of careless transportation or handling. The size of such defects may range from the order of microns, for metallurgical in homogeneities such as nonmetallic inclusions, to several millimeters in the case of welding defects such as slag inclusions or heat-affected zone cracks (Hussain, 1997). For welded metal structural components, the crack nucleation phases and microstructure short crack growth phases need not be considered, but physically small crack growth, long crack growth and final fracture are included.

2.3 Crack Initiation Mechanisms for Different Metals

2.3.1 Definition of a Crack Initiation

The initiation of fatigue cracks is an event whose very definition is strongly linked to the size and scale of observation. For example, materials scientists are likely to consider the nucleation of flaws along Persistent Slip Bands (PSBs) as the initiation stage of fatigue failure, whilst a mechanical engineer may associate the resolution of crack detection with the threshold for crack nucleation. The differences in the approaches to fatigue crack initiation constitute the fundamental distinction between fatigue design philosophies currently practiced in industry (Suresh, 2001). Typically, crack initiation is a complex process that can lead to a crack that is extremely small (<1 μm) or one that "pops in" (initiates and rapidly grows) at up to 100 μm in length. To the engineer, initiation is likely to mean the smallest crack or an engineering-sized crack that can be determined by a reliable non-destructive evaluation (NDE) technique, which would, to the scientist, probably mean that the growth of small cracks is included in this definition (Chan, 2010). In our five stages definition, a_m is the scientific definition of crack initiation and the term is taken to mean the process of forming a fatigue crack of a length that is in the order of a grain size or less while a_1 is the engineering definition of crack initiation.

2.3.2 Fatigue Crack Initiation in Slip Bands

Forsyth (1960) first reported the surface roughness phenomena induced by fatigue slip. The interface between the PSB and the matrix are the plane of discontinuity across which there are abrupt gradients in the density and distribution of dislocation (Suresh, 2001). Direct experimental evidence of crack initiation at the interface has been obtained by Hunsche and Neumann (1986) and Ma and Laird (1989a). There are several models to explain the formation mechanism of PSB that have been proposed by some authors (Cottrell and Hull, 1957; Mott, 1958; May, 1960; Kennedy, 1963; Eassmann, *et al.*, 1981; Mughrabi and Endo, 1986). Because there is not a widely accepted model yet, the model proposed by Cottrell and Hull, which is a geometric demonstration model, is adopted in the section. Cottrell and Hull (1957) presented the formation mechanism of slip bands based on the cross-slip model shown in Fig. 2.2. Under cyclic loading at low stress, some slip lines are almost parallel and the increment of the dislocation source appears in a local area in the metal surface. Dislocation in slip lines which are blocked forms a dislocation pileup. The dislocation source is in a relatively stable state. With the fatigue continuing, screw dislocation of the parallel slip line changes the slip plane and cross slip occurs. Opposite dislocations meet in the cross slip plane and cancel each other. While the dislocation source on the original slip plane is re-activated, dislocations increase with slip lines increasing. Occurrence and development of many slip lines constitute the slip bands. The slip band is an irreversible plastic deformation, which shows a raised and recessed groove of extrusion ridges and invasive ditch on the surface of the specimen, as shown in Fig. 2.2. Stress concentration is relatively large at extrusion ridges and invasive ditch of the present slip bands on the material surface. With the development of the PSBs in depth and breadth under the cyclic loading, voids are produced by dislocation increments at recess, and micro cracks are consequently formed.

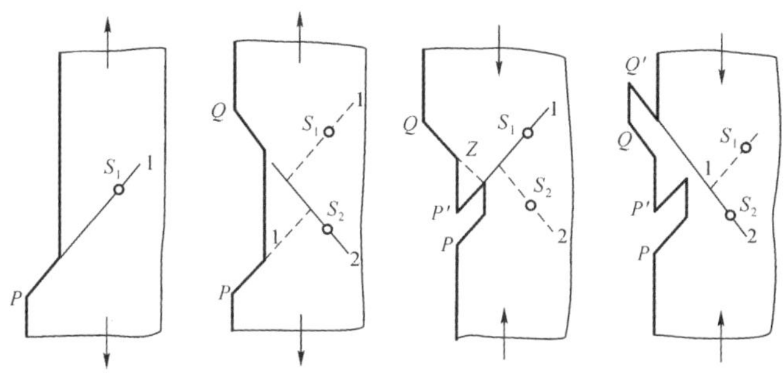

(a) (b) (c) (d)

Fig. 2.2 Model of Cottrel-Hull (Cottrel and Hull, 1957) (with the permission of the Royal Society)

2.3.3 Crack Initiation Along the Grain Boundary (GB)

The nucleation of fatigue cracks at grain boundaries occurs under the influence of embrittling environments and elevated temperature at which GB cavitation and sliding are promoted (Suresh, 2001). The dislocation movement in the grain will be blocked by the GB, which leads to dislocation pileup and stress concentration in the GB. Thus the initiation crack may occur at the GB below the stress level within the slip of the GB. Under a certain stress level, the greater the grain size within materials, the greater the strain on the GB and the more serious the dislocation pileup. GB cracks are more easily triggered. Many GB cracks are present in a corrosive or high temperature environment. In brittle materials residual stress arising from shrinkage between adjacent grains or the existence of the glass phase mismatch, will induce intergranular cracks. For materials with good ductility, such as the GB not containing second phase particles, nor creep deformation and the effect of environmental embrittlement, the probability of GB cracks appearing is relatively small. In general, GB cracking arises from one of two mechanisms during cyclic loading (Suresh, 2001). 1) At low to intermediate plastic strain amplitude, the impingement of PSBs at grain boundaries causes cracking (Figueroa and Laird, 1983). 2) At high plastic strain amplitude, GB cracking occurs as a consequence of surface steps formed at the boundary. This is because under large plastic strain amplitude, the dislocation sub-structure is a cellular structure, cyclic deformation is uniform and not concentrated on the PSB, although extrusion can be observed but does not result in the formation of Tran granular slip bands. However, irreversible sliding creates steps on the surface of the GB and cracks are initiated at the step edges. This GB crack initiation must meet three conditions: 1) the GB separates highly misprinted grains; 2) the active slip system of at least one of the grains is directed at the intersection of the boundary with the specimen surface; 3) the traces of the high angle grain boundaries on the free surface form a large angle (30°–90°) with the tensile stress axis (Kim and Laird, 1978).

It has been recognized that there exists a great difference in the fatigue damage mechanisms between single and polycrystals, which can be mainly attributed to the effects of GBs and the crystallographic orientations. Clear evidence has shown that the interactions of PSBs with various GBs play a decisive role in intergranular fatigue cracking during cyclic deformation. It is suggested that intergranular fatigue cracking strongly depends on the interactions of PSBs with GBs in fatigued crystals, rather than on the GB structure itself (Zhang and Wang, 2008).

2.3.4 Crack Initiation at Inclusions

The effect of inclusions on fatigue crack initiation is often specific to the alloy system. In high strength steels, there are some non-metallic inclusions (such as

MnS, Al_2O_3 and silicate), second phase particles in most metallic materials. Non-metallic inclusions of a microscopic size (10 to 100 μm) have been observed in low-alloy high-strength steels. In aluminum alloys, constituent particles such as S-phase (Al_2CuMg) and β-phase (Al_7Cu_2Fe), typically 1–10 μm in diameter, provide sites for fatigue crack nucleation. In high strength nickel-based super alloys, crack initiation has been identified with the existence of large defects, either pores or nonmetallic inclusions (Suresh, 2001). The inclusions within theses sizes are not considered harmful for the static strength but reduce the ductility of a material due to generating internal voids at large plastic strain and affect the stress distribution on a micro level and thus can contribute to crack nucleation. Fatigue crack nucleation occurs at these inclusions located at the material surface or slightly below the surface. The inclusions act as micro notches in the material which can generate fatigue crack nuclei, especially in high-strength material for higher strength materials have higher notch sensitivity (Schijve, 2009).

The poor adhesion between matrix and aluminum oxide inclusion leads to the formation of an inclusion pit which serves as a simple stress raiser. The fatigue crack originates at the periphery of this inclusion pit at an angle of 45 degrees to the principal stress direction. Metallurgical and micro-fractographical observations revealed that the initiation and early growth of fatigue cracks from non-metallic inclusion are of the shear rather than the tensile mode. The fatigue process is quite localized in the vicinity of inclusion. The mechanism for an initiation and early growth of fatigue cracks is essentially the same as that for ductile materials (Kunio, *et al.*, 1981).

The inclusions and second phase particles may break, separate with the matrix and occur along the interface leading to crack initiation under cyclic loading. Fig. 2.3 is a micrograph of a crack initiated from inclusion showing deboning at the inclusion/matrix boundary corroborated by optical microscopy. Lankford and Kusenberger (1973) proposed a physical model to describe fatigue crack nucleation in certain high strength steels. Initially the inclusions are bonded to the matrix, Fig. 2.4(a). Under repeated stress cycling, a deboned cavity eventually develops at one or other of the tensile poles of the inclusion, Fig. 2.4(b). Subsequently, the crack tip at the inclusion/matrix juncture grows toward the inclusion equator; another cavity may or may not nucleate about the other pole, Fig. 2.4(c). At some critical point determined by the applied stress and the stress field of a partially debonded inclusion of a certain size, point surface defects will be nucleated, as shown in Fig. 2.4(d). It is emphasized that the location of the initial surface defects in general will be some distance, at least several microns, from the inclusion/matrix boundary. As cycling continues, the surface defects coalesce into micro cracks, Fig. 2.4(e), which begin to grow and link up. Further stress cycling leads to transverse fatigue crack growth, and eventually the initiation of micro-cracking, Fig. 2.4(f), at the opposite side of the inclusion. The picture for subsurface nucleation is less clear, but the initial stages of debonding probably are qualitatively similar.

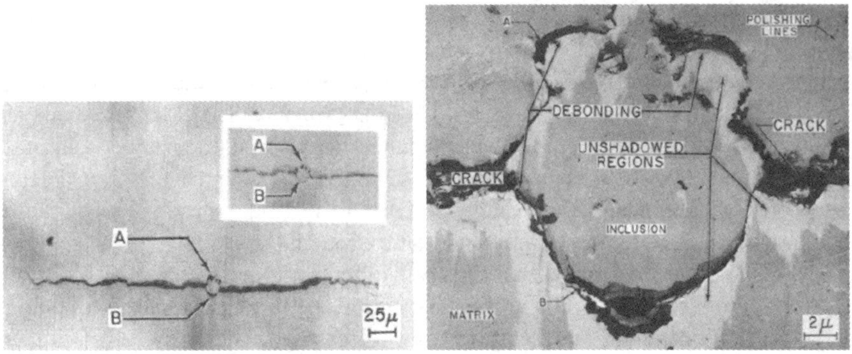

(a) (b)

Fig. 2.3 Electron transmission micrograph showing debonding at inclusion/matrix boundary corroborated by optical microscopy. (a) Optical photomicrograph of replicated specimen surface; 170 ksi stress applied (insert at 0 ksi stress applied); (b) Electron micrograph of debonded inclusion; 170 ksi stress applied (Lankford and Kusenberger, 1973)

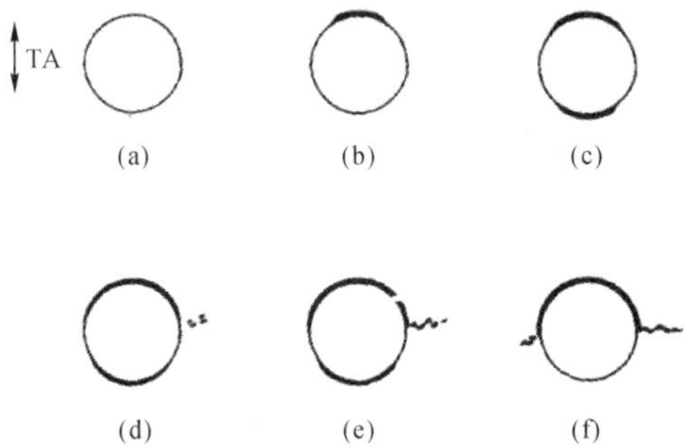

Fig. 2.4 Model of fatigue crack initiation based on debonding of surface inclusions prior to microvoid nucleation within metal matrix. (a) Bonded inclusion; (b) Initial debond at one pole; (c) Growth of initial debond seam; nucleation of debond at opposite pole; (d) Further debond seam growth accompanied by point surface defect nucleation within matrix; (e) Defect growth and coalescence to form fatigue micro-crack; (f) Growth of micro-crack, and nucleation of surface defects at opposite side of inclusion (Lankford and Kusenberger, 1973)

It should be noted that the inclusion mainly promotes crack initiation in slip bands, but is not the decisive factor in the fatigue crack initiation. The impact depends on the nature and size of inclusions, such as the Al_2O_3 and silicate inclusions in steel, the high elastic modulus of which results in stress concentration and the plastic strain concentration at inclusions. Fatigue cracks initiation can be observed at inclusions. The second phase particles of MnS and Fe_3C in steel have a smaller elastic modulus, of which the effect on the fatigue crack initiation is not found. The larger the inclusions scale, the more uneven the distribution, the more serious the stress concentration and the greater the impact of crack initiation.

2.3.5 Slip Band and Dislocation in Single Crystal Metal

Li, *et al.* (2011) have systematically summarized the cyclic deformation behavior of different kinds of face-centered cubic (fcc) single crystals, including copper, nickel, silver, as well as copper-aluminium, copper-zinc alloys, from a historical perspective of the developments over the last several decades. Combined with plenty of previous research results, the influencing factors on cyclic deformation behavior are listed as follows: orientations, Stacking Fault Energy (SFE), Short-Range Order (SRO) and friction stress, or more generally, the ease of cross slip. Among them, the effect of orientations is mainly reflected in the formation of complex dislocation patterns, which depends on activating the secondary slip system. According to the effect of slip mode, the materials can be divided into two types: pure metals and alloys. For pure fcc metals, the effect of SFE is decisive. Due to the easy cross slip of screw dislocations, regular dislocation arrangements, *e.g.* veins, PSBs, labyrinth and cell patterns, will always form. With an increase in alloying elements, antiphase boundary energy gradually replaces SFE to become a new decisive factor affecting the cyclic deformation behavior of fcc alloy single crystals. The corresponding dislocation arrangements consist of dipole array and Stacking Faults (SFs) under the influence of planar slip. The relationship between several factors is well explained. Fig. 2.5 shows a new criterion for the formation of regular dislocation arrangements, or not, in different kinds of fcc metals or alloys by taking into account the effect of the slip mode and the energy difference.

2.3.6 Slip Band and Dislocation in Polycrystal Metal

The fundamental fatigue damage mechanisms and cyclic deformation behavior for single crystals are well studied and the main influencing factor is the crystallographic orientation. However, people observed that there are big differences between single crystals and polycrystals, mainly because of the presence of different GBs in the latter. GBs are very important with respect to the properties of polycrystals. The understanding of the role of these GBs is the key to establishing the connection between well-known features which are typical of the deformation of single crystals and polycrystalline materials. For that purpose, Zhang and Wang (2008) have carried out a systematic study mainly on copper bicrystals with various GBs and different component crystals, including the macroscopic cyclic stress-strain responses and fatigue damage mechanisms on the micro-scale. From their study, the following conclusions have been drawn:

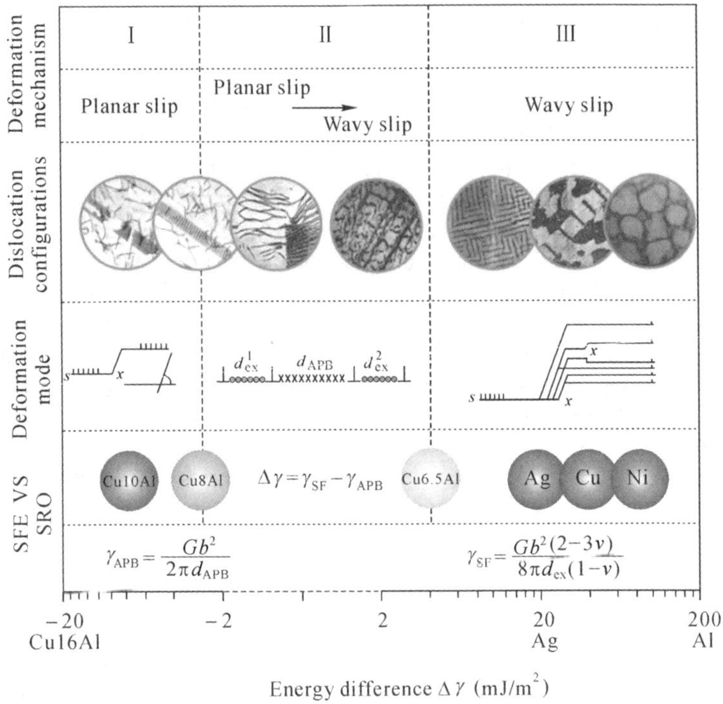

Fig. 2.5 New criterion for the formation of regular dislocation arrangements, or not, in different kinds of fcc metals or alloys by taking into account the effect of the slip mode and the energy difference (Li, *et al.*, 2011) (with the permission of Pergamon)

(1) Under uniaxial and cyclic loadings, GBs often have a strengthening effect on polycrystals. In bicrystals with a GB parallel to the loading direction, the direct evidence for strengthening can be attributed to the surface observations of a GB Affected Zone (GBAZ) with strong secondary slip bands and some deformation bands near the GB. This makes it possible to formulate similar strengthening mechanisms in the case of polycrystals.

(2) When large-angle GBs are perpendicular to the loading direction, the copper bicrystals displayed quite different Cyclic Stress-Strain (CSS) curves. In this case, the component crystal orientations have a decisive effect on the CSS curves of those bicrystals. Therefore, for such bicrystals, the GB strengthening effect can be neglected during cyclic deformation. In combination with the cyclic deformation behavior of copper single crystals oriented for single slip, it is suggested that the occurrence of the plateau or pseudo-plateau of their CSS curves results mainly from the cyclic saturation of the single slip oriented component crystals in the region B of their CSS curves.

(3) In the case of large-angle GBs in copper bicrystals, fatigue cracks always prefer to nucleate at the GBs, rather than along PSBs. Aside from details, the GB cracking is independent of the interaction angles between the GB plane and the stress axis. The intergranular fatigue cracking can be well explained by the

PSB-GB damage (or piling-up of dislocations) mechanism. Therefore, the large angle GBs have a weakening effect on the bicrystals or polycrystals during cyclic deformation although they often strengthen polycrystals under uniaxial loading conditions.

(4) It is suggested that intergranular fatigue cracking strongly depends on the interactions of PSBs with GBs, rather than on the GB structure itself. The nucleation of fatigue cracks along a PSB or a GB is a competitive process, depending on the dislocation glide processes in PSBs. The fatigue damage of ductile fcc crystals originates from the dislocation processes in slip bands. When the dislocations can be freely transported to the surface by slip bands, the gradual surface roughening will become the origin of fatigue cracks and will result in most cases in PSB cracking. When dislocations carried by PSBs cannot transfer through a GB, they will pile-up at the GB, resulting in intergranular fatigue cracking caused by the accumulation of the residual dislocations.

(5) In the newly developed ultrafine-grained or nanocrystalline materials, GBs often play a more and more important role in the mechanical properties. For a better understanding of the mechanical properties and the strengthening mechanisms of these materials it is very important to understand in detail the GB effects on a micro-scale or nano-scale level. This will be a new challenge for materials scientists in the future.

2.3.7 Fatigue Mechanism of Ultrafine-Grained Materials

Fatigue failure in the high-cycle and ultra high-cycle regimes is often dominated by the crack initiation processes, which are strongly influenced by the salient features and defects in the microstructure. Competing fatigue mechanisms involving crack initiation at PSBs, grain boundaries, pores, and non-metallic inclusions or particles, have been reported to occur at surface sites in the high-cycle fatigue regime (10^6–10^7 cycles), but shift to interior sites in the ultra high-cycle fatigue regime (10^9–10^{10} cycles). The changes in the fatigue mechanism and crack initiation site result in large variations in fatigue life (Chan, 2010). Grain refinement has been practiced as an effective means to enhance the strength of crystalline materials for many decades. With the development of a number of so-called Severe Plastic Deformation (SPD) techniques, most notably Equal Channel Angular Pressing (ECAP), it has almost become routine to produce bulk materials with submicron grain sizes and with dimensions of some cm (Valiev, 2000; 2009). In the conventional definition, these materials, with typical grain sizes in the range 100 nm$<D<$1 μm are referred to as ultrafine-grained (UFG) materials, as opposed to truly nanocrystalline (NC) materials with grain sizes $D<$ 100 nm. The latter NC materials are usually produced by techniques such as Inert Gas Condensation (IGC) or an electrodeposition technique and are not available as readily as the UFG materials in bulk form suitable for standard mechanical testing.

Hence, most of the studies performed to date have been made on UFG materials prepared by ECAP processing (Mughrabi and Höppel, 2010).

Most of the early studies of the fatigue properties of UFG single-phase metals produced by ECAP focused mainly on copper as the classical model material and subsequently also on nickel, titanium and aluminium. Subsequently, attention was turned to other materials such as, in particular, some aluminium alloys and also some steels. The main results obtained in the early observations on cyclically deformed UFG copper and partly also for UFG nickel can be summarized as follows (Mughrabi and Höppel, 2010):

(1) The cyclic stresses in strain-controlled tests were about 2–4 times larger than in similar tests on CG material.

(2) More or less severe cyclic softening was observed in strain controlled tests, whereas there was almost no cyclic softening in stress-controlled tests.

(3) On the surface, macroscopic shear bands, extending over much larger distances than the UFG grain size, were observed. Fatigue cracks were initiated in these shear bands.

(4) Inside the fatigued specimens, with a few exceptions, it was difficult to see marked changes in the microstructure related to the shear bands. However, there were other indications that the microstructure of the severely deformed ECAP-processed material was cyclically unstable, as evidenced in the form of patches of severely coarsened grains next to regions with the original UFG microstructure. In other words, a bimodal grain size distribution had evolved, in particular during strain-controlled fatigue.

(5) The fatigue lives of the UFG specimens were generally found to be larger than those of CG specimens in all Wöhler/*S-N* plots, whereas the opposite was true when the data were plotted in a Coffin-Manson diagram. This finding is in broad agreement with expectations according to the total strain fatigue life diagram, UFG material exhibiting larger lives than CG material in the High Cycle Fatigue (HCF) regime, and smaller in the LCF regime.

2.4 FCP Mechanisms

The mechanisms go into the propagation stage after fatigue crack initiation. According to the crack propagation rate, crack growth direction and the mechanisms in the propagation process, the fatigue crack growth can be divided into two stages as shown in Fig. 2.6 (Schijve, 2009). The microscopic mode of fatigue crack growth is strongly affected by the slip characteristics of the material, characteristic microstructural dimensions, applied stress level and the extent of near-tip plasticity. In ductile solids, cyclic crack growth is envisioned as a process of intense localized deformation in macroscopic slip bands near the crack tip. When the crack and the zone of plastic deformation surrounding the crack tip are confined to within a few grain diameters, crack growth occurs predominantly by single shear, in the direction of the primary slip system. This single slip mechanism leading to a zig-zag crack path has been termed Stage I crack growth by Forsyth (1969).

In Stage II, the cracks typically grow in a duplex slip mechanism along the direction perpendicular to the maximum tensile stress until unstable crack propagation occurs.

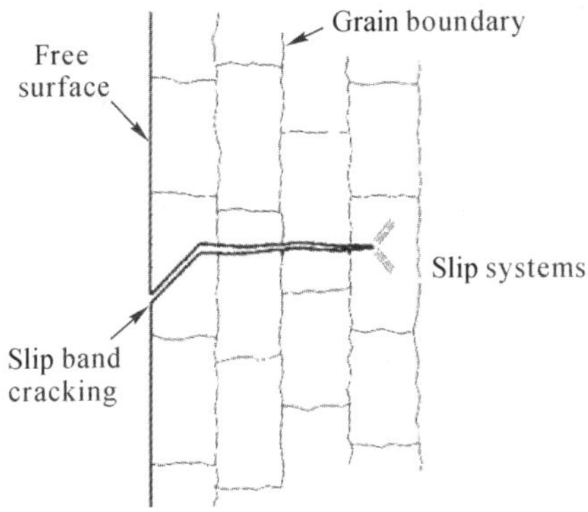

Fig. 2.6 Cross section of micro-crack (Schijve, 2009)

2.4.1 *Stage I FCP*

Stage I crack growth in a single shear mechanism moves in specific slip plane directions which can grow at high rates at low stress ranges, below the threshold value for long crack propagation. The difference in growth rate can be explained by the differences in growth mechanisms. As long as the size of the micro-crack is still in the order of a single grain, the micro-crack is obviously present in an elastically anisotropic material with a crystalline structure and a number of different slip systems. The micro-crack contributes to an inhomogeneous stress distribution on a micro level, with a stress concentration at the tip of the micro-crack. As a result, more than one slip system may be activated. Moreover, if the crack is growing into the material in some adjacent grains, the constraint on slip displacements will increase due to the presence of the neighbouring grains. Similarly, it will become increasingly difficult to accommodate the slip displacements by slip on one slip plane only. It should occur on more slip planes. The micro-crack growth direction will then deviate from the initial slip band orientation. In general, there is a tendency to grow perpendicular to the loading direction (Fig. 2.6). Because micro-crack growth is dependent on cyclic plasticity, barriers to slip can imply a threshold for crack growth. Illustrative results are presented in Fig. 2.7. The crack growth rate measured as the crack length increment per cycle decreased when the crack tip approached the first GB. After penetrating through the GB the crack growth rate increased during growth into the next grain, but it decreased again when approaching the second GB. After passing that GB, the micro-crack continued to grow at a steadily increasing rate (Schijve, 2009). The initially inhomogeneous micro-crack growth, which starts with a relatively high crack growth rate and then slows down or even stops due to material structural barriers, will be discussed as short crack growth in the following section. The fracture surfaces created by Stage I fatigue crack growth exhibit a serrated or faceted profile (Suresh, 2001).

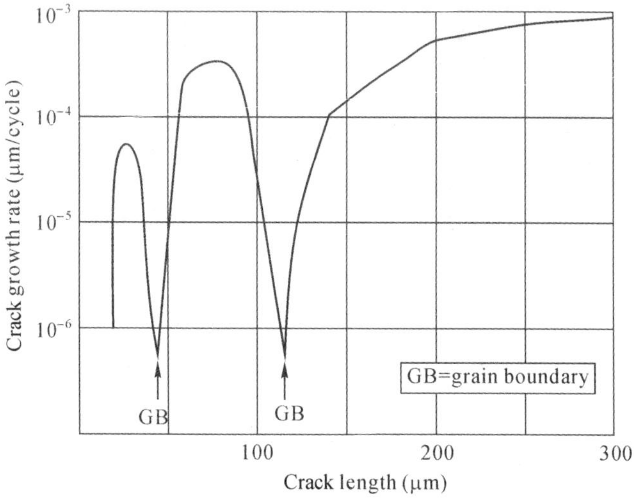

Fig. 2.7 GB effect on crack growth in an Al-alloy (Blom, *et al.*, 1986)

2.4.2 *Stage II Crack Growth and Fatigue Striation*

Stage II is reached when the crack propagation direction is perpendicular to the direction of maximum tensile stress. Stage II cracks typically grow in a duplex slip mechanism perpendicular to the applied load direction. The crack growth rate and the striation spacing in Stage II are much greater than that in Stage I. The propagation under each cycle reaches micron level. So investigations of the mechanism of crack propagation are much more than in Stage I. In Stage II, the crack propagation is mainly transgranular when subjected to normal stress in a noncorrosive ambient environment but at high temperature or in corrosive media the crack growth was along the GB. Stage II fatigue crack growth in many engineering alloys leads to the formation of fatigue "striations". These striations, first observed by Zappfe and Worden (1951), are ripples on the fracture surface (Suresh, 2001). Forsyth and Ryder (1960) and Forsyth (1962) thought that the fatigue striations can be divided into ductility striation (A type) and brittle striation (B type). Ductile striations are formed from light and shade on the irregular non-crystalline high platform, while brittle striations form on the crystalline surface (Zhao and Wang, 1997).

Richards and Lindley (1972) summed up four mechanisms of the striation mechanism, micro-cleavage, pore connectivity and intergranular separation in Stage II. The striation mechanism and the pore connectivity are two plastic propagation mechanisms. They will be presented below.

2.4.2.1 Striation formation mechanism of ductile crack growth

A number of conceptual models have been proposed over the years to rationalize the formation of fatigue striations and the planar growth of Stage II fatigue. Laird and Smith (1962), Laird and Thomas (1967) proposed an appealing physical concept "plastic blunting idealization" to describe the crack propagation in Stage II, based on the direct observation of crack tip geometric change in the ductile metal, as shown in Fig. 2.5. In this model, the increment of crack extension per fatigue cycle is envisioned as occurring due to the plastic blunting of the crack tip. A crack tip amenable to duplex slip plastically blunts as shown in Fig. 2.5 upon the application of a tensile stress. The blunting process also effectively results in the extension of the crack by a distance of the order of the crack tip opening displacement (Suresh, 2001). More detailed processes of the model are as follows. 1) Zero loading: the crack is closed and the crack tip stays in a sharp state (Fig. 2.8(a)); 2) subjected to the tensile load, under the promotion of stress concentration at the crack tip, the local material at the crack tip slips in a direction of 45° (Fig. 2.8(b)); 3) with the load increasing, the slip area expands, the crack tip expands into a semi-circular shape and the stress concentration decreases sharply. The growth of the crack stops with the slide stop and then the plastic blunting occurs (Fig. 2.8(c)); 4) under reverse loading, the reverse slip occurs on the material, the gap between the crack faces are pressed close and the crack tip is bent into a pair of "ear-shaped" gaps (Fig. 2.8(d)) which is the source of stress concentration for the slip near the crack tip in the next cycle; 5) when the reversing loading reaches the maximum compressive stress, the crack surface is pressed together, the blunt crack tip becomes sharp and a new pair of sharp corners appear. Since the closure of the crack during compression cannot fully negate the blunting and the attendant extension of the crack during the preceding tension load, net crack growth occurs during a fatigue cycle leading to the formation of striations. The fatigue crack propagates constantly under cyclic loading and unloading.

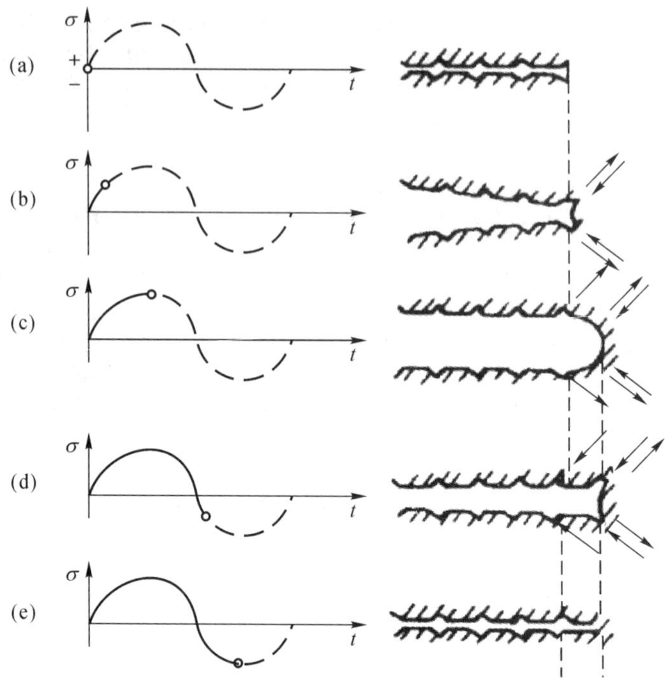

Fig. 2.8 Model of Laird's FCP. (a) Zero load; (b) Small tensile load; (c) Peak tensile load; (d) Onset of load reversal; (e) Peak compressive load (Laird and Thomas, 1967)

2.4.2.2 Micro void coalescence mechanism of ductile crack propagation

The extension of the crack by a distance of the order of the crack tip opening displacement in Laird's model implies that the fatigue crack growth rate decreases with the material yield strength increasing, which has not been confirmed by experiments (Kikukawa, *et al.*, 1979).

The microvoid coalescence mechanism is also known as the cumulative damage mechanism (Weertman, 1978). In this model, crack propagation is similar to the ductile fracture process. Fatigue crack growth is thought to be due to the formation of micropores in the stress regions around the main crack front, coarsening and connections, and the growth of a small crack face. Zheng and Hirt (1983) pointed out that the pore connectivity model can be used to explain the experimental results of Laird's model, insuring that the fatigue crack growth rate has nothing to do with the material yield strength, but is relevant to resistance to the fracture toughness of the material. However, aluminum alloy test results show that the fatigue crack growth mechanism is consistent with the Laird propagation mechanism, rather than the cumulative damage mechanism (Smith, 1984).

2.5 Some Important Issues in Crack Growth

It should be mentioned that current fracture mechanics based fatigue crack growth life prediction methods have difficulty in predicting crack growth in small near micron sized flaws. For example, a crack can be viewed as a macroscopic quantity or a microscopic quantity and different levels of crack propagation will be governed by different parameters. A brief list of the characteristics of the three regimes of fatigue crack growth can be found in *Fatigue of Materials* by Suresh (2001), shown in Table 2.1. Cui, *et al.* (2011) have grouped the influence factors into four types. As mentioned in Chapter 1, it is impossible to consider all the factors which affect the fatigue behavior in the FLP method, but we include several of the most significant factors. The basic idea of our unified method is that initial defects always exist in engineering marine structures and fatigue is purely a crack propagation problem. However, for most welded structures, a large part of their total life is spent in crack propagation around the threshold. This could accurately account for many loading effects such as mean stress effect, load sequence effect and structural defects such as notch effect and size effect.

Table 2.1 Characteristics of the three regimes of fatigue crack growth (Suresh, 2001)

Regime	A	B	C
Terminology	Slow-growth rate (near threshold)	Mid-growth rate (Paris regime)	High-growth rate
Microscopic failure model	Stage I, single shear	Stage II, striation and duplex slip	Additional static model
Fracture surface feature	Faceted or serrated	Planar, with ripples	Additional cleavage or microvoid coalescence
Crack closure level	High	Low	—
Microstructural effects	Large	Small	Large
Load ratio effects	Large	Small	Large
Environmental effects	Large	*	Small
Stress state effects	—	Large	Large
Near-tip plasticity#	$r_c \le d_g$	$r_c \ge d_g$	$r_c >> d_g$

* Large influence on crack growth for certain combinations of environment, load ratio and frequency;

r_c and d_g refer to the cyclic plastic zone size and the grain size respectively

2.5.1 Short Crack

Pearson (1975) first reported the so-called "anomalous" behavior of small fatigue cracks. It has been well established that the growth rates of small cracks can exceed those of long cracks at theoretically equivalent SIF ranges. The observations of Pearson have drawn attention to the fact that the fatigue crack growth behavior of short cracks differs in a non-conservative manner from expectations based upon long crack behavior. This behavior is illustrated in Fig. 2.9, in which the crack-growth rate $\frac{da}{dN}$ or $\frac{dc}{dN}$, is plotted against the linear-elastic SIF range, ΔK. These results show that small cracks grow at ΔK levels even below the long-crack threshold and also grow faster than long cracks at the same ΔK level above the threshold. There is crack retardation phenomenon in short crack growth. This short crack retardation generally occurs when the crack size is comparable to, or is a multiple of, the grain size. The depth of the retardation also depends on the crystallographic orientations of the neighboring grain (Lankford, 1982).

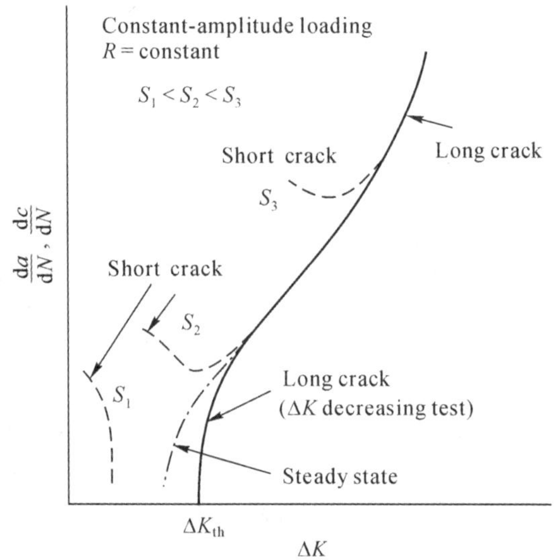

Fig. 2.9 Typical fatigue crack growth behavior for short and long cracks (Ritchie and Lankford, 1986) (with the permission of Elsevier)

The difference in growth rate can be explained by the differences in growth mechanisms. Long cracks, typically, grow in a duplex slip mechanism perpendicular to the applied load direction, whereas short cracks grow in a single shear mechanism along specific slip plane directions. According to the different crack size and crack propagation mechanism, fatigue small cracks are generally divided into four categories, microstructural small cracks, mechanical small cracks, physical small cracks and chemical small cracks (Suresh and Ritchie, 1984).

McEvily and his co-worker (McEvily and Ishihara, 2001; Ishihara and McEvily, 2002) thought that it is impossible to explain the small crack growth behavior through the use of a traditional LEFM approach for the following reasons:

(1) Ongoing from the micro-crack growth range to the macro-crack growth range, fatigue crack growth behavior undergoes a transition from stress control to stress intensity factor control;

(2) Because of the large plastic zone size to crack length ratio in the small crack growth regime, crack growth is elastic-plastic in nature rather than linear-elastic;

(3) The crack closure level increases from zero up to the level associated with a macroscopic fatigue crack as the fatigue crack increases in length in the small crack regime.

The presence of a GB in the neighborhood of a short crack will have a significant influence on the crack growth rates. There are several different mechanisms for dislocation slip propagation movements across grain boundaries (Hansson, 2009). In a study by Shen, *et al.* (1988), a number of different mechanisms were identified and it was found that a GB could act both as dislocation obstacle and as a source for dislocation generation. Bjerkén and Melin (2004) have performed an investigation on how a GB affects the growth behaviour of a short edge crack. In the model, it is assumed that the GB acts as an obstacle, which the dislocations cannot pass. When the stresses in the neighbouring grain are sufficiently high, nucleation of dislocations will occur in that grain, resulting in a spread of the plasticity.

The influence on the crack growth rate for a short edge crack under fatigue loading due to changes in crack length, grain size, load range and GB configuration is investigated using a discrete dislocation technique by Hansson (2009) and the conclusions are cited here. It was found that when the plasticity was restricted to one grain only, the crack growth rate might both increase and decrease for larger distances between the crack tip and the GB, depending on the chosen load range. The closure stress increases with an increase in distance between the crack tip and the GB, and decreases with an increase in crack length. The crack growth rate increases linearly with the SIF range, contrary to long cracks which follow Paris' law with an exponent two to four, and good agreement was found when comparing the growth rates for different load ranges, changing either the crack length or the maximum load. However, some discrepancies were obtained when comparing the increase in crack growth rate due to an increase in crack length and due to an increase in maximum load. In the presence of a low angle grain boundary it was found that a pile-up at this first GB was created before the plasticity could spread into the second grain. Both the crack growth rate and the size of the pile-up before breakthrough depended on the sign of the dislocations of the low angle GB, the distance between the dislocations constituting it and the distance between the slip plane and the closest dislocation in the low angel GB. It was also found that positive dislocations forming the GB resulted in the boundary repelling dislocations, resulting in higher growth rates than when using negative dislocations, which had an attractive influence on the dislocations.

It is convenient to distinguish the short crack and the long crack in predicting the fatigue crack growth life of a component with an initial small crack. But, until now, there is no broadly accepted definition of the critical size of the short crack a_n. Toyler and Knott (1981) thought that a_n should be taken as 10 times the average grain size; Tokaji proposed that a_m be taken to be 9 times the grain size when the average grain size is less than 30 microns, and a_m taken to be 260 microns when the average grain size is larger than 30 microns (Tokaji and Ogawa, 1988); Suresh (2001) pointed out that a_n can be determined by the fatigue limits range and the long crack fatigue threshold through tests. This issue will be discussed in Section 2.6.

2.5.2 Crack Closure

Elber (1970) suggested that crack growth rates are not only influenced by the conditions ahead of the crack tip, but also by the nature of crack face contact behind the crack tip. Crack closure is a phenomenon which is strongly dictated by microstructural and environmental factors, and mechanical loading parameters. The crack closure mechanism has extended from the plasticity induced crack closure mechanism proposed by Elber to several types: plasticity induced crack closure, roughness induced crack closure and oxide induced crack closure, viscous fluids induced crack closure and transformation induced crack closure, crack deflection, crack-bridging by fibers and crack-bridging by particles. A schematic illustration of the mechanisms is shown in Fig. 2.10. Among them, plasticity induced crack closure, oxide induced crack closure and roughness induced crack closure have been studied intensively but only plasticity induced crack closure is well understood and modeled for application, while the others are far from being applied because of their complexity. The mechanisms of plasticity induced closure, oxide induced closure and roughness induced closure are briefly summarized based on "Fatigue of Materials" by Suresh (2001). More details can be found in that book.

(1) Plasticity induced crack closure

Ebler thought that there was a residual deformation behind the fatigue crack tip zone, which made the crack opening displacement decrease and in early contacts between the crack surfaces the fatigue crack growth driving force is less than the nominal value. In Fig. 2.10(a), during one cycle of crack growth, residual tensile strains are left in the material behind the advancing crack front, as only elastic recovery occurs after the creation of the fracture surface. With an increase in the SIF and the size of the plastic zone due to crack advance, the material which has previously been deformed permanently within the plastic zone forms an envelope of plastic zones in the wake of the crack front Suresh (2001). Since the conditions extant in the wake of the crack tip are a result of such factors as the history of loading, the length of the crack and the stress state, Elber's work also brought to light the very dependence of fatigue crack growth rates on the prior history.

(2) Oxide induced crack closure

When a fatigue crack propagates in a moist environment, oxide will form on the freshly formed fracture surface and the oxide is deposited on the surface near the crack tip. The oxide layers on fracture surfaces, whose thickness is much larger than the crack tip opening displacement near threshold, work as a wedge between the crack surface to reduce the unloading sufficiently. Under cyclic loading, the shear displacement, plasticity and roughness make the oxide film break and the fresh metal surface re-oxidises rapidly. Schematically illustration of the mechanisms is shown in Fig. 2.10(b). In such a continuous process of breaking and reforming of the oxide, the corrosive products accumulate in the gap between cracks, the thickness of which is 20–40 times that of the oxide film of metal exposed to the air. The process occurs mainly in the lower range of the SIF, and the degree of closure relates to the size and crack length. Oxide induced crack closure is promoted by a moisture containing environment, elevated temperature, low load ratios, low SIF levels, high cyclic frequencies, and lower strength and coarser grained microstructures where fretting and rubbing along the fracture surface asperities become favorable.

(3) Roughness induced crack closure

Roughness induced crack closure, as shown in Fig. 2.10(c), has come to be recognized as one of the mechanisms by which some of the apparent microstructural effects on fatigue crack growth can be rationalized. This phenomenon provides an explanation for many apparently anomalous effects of microstructure on fatigue crack growth, especially in the near threshold regime, which is counter to conventional interpretation. Experimental observations in a wide range of ductile materials reveal that crack growth in the near threshold fatigue regime occurs by means of single slip mechanics. This Stage I growth mechanism leads to a highly serrated fracture morphology and elevation of the crack closure stress. It is also known that permanent plastic deformation ahead of the crack tip, as well as the possibility of slip irreversibility during unloading from the peak stress, causes a mismatch between the fracture surface asperities. The tortuous crack path promoted by the crystallographic growth mechanisms, in conjunction with the occurrence of mixed model sliding of crack faces and the mismatch between the crack face asperities, immediately provides a mechanism for enhanced crack closure. The phenomenon occurs mainly at a low SIF, when the largest crack tip plastic zone size is smaller than the scale of the microstructure.

From the above description, we can draw a primary conclusion that the crack closure induced by oxide or roughness mainly occurs around the crack growth at low SIF and near the threshold.

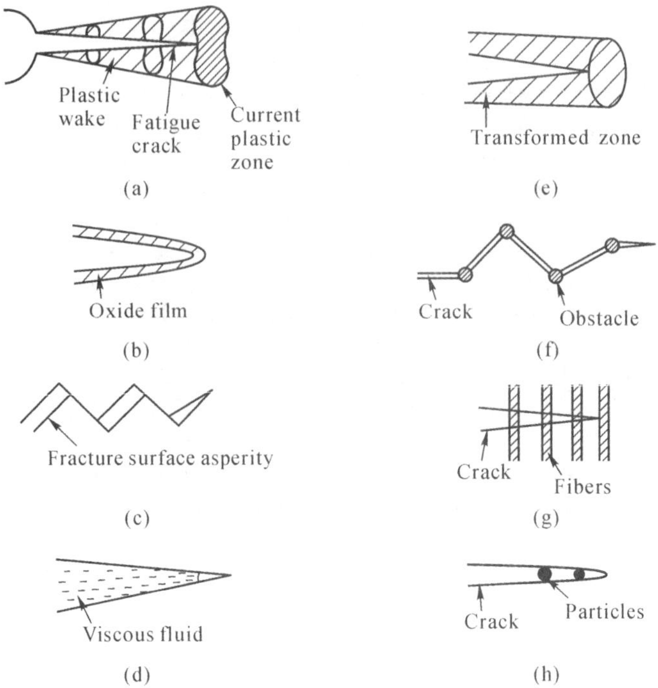

Fig. 2.10 A schematic illustration of the mechanism which promotes retardation of fatigue crack growth in constant amplitude fatigue. (a) Plasticity induced closure; (b) Oxide induced crack closure; (c) Roughness induced crack closure; (d) Fluid induced crack closure; (e) transformation induced crack closure; (f) Crack deflection; (g) Crack-bridging by fibers; (h) Crack-bridging by particles (Suresh, 2001) (with the permission of Cambridge University Press)

2.5.3 Effect of Loading Sequence

Crack growth in structures depends on the stress amplitude, stress ratio, and sequence of the load. Due to the random nature of variable loading, it is difficult to model all these influential parameters correctly. Overloads are known to retard crack growth, while underloads accelerate crack growth relative to the background rate. These interactions, which are highly dependent upon the loading sequence, make the prediction of fatigue life under variable amplitude loading more complex than that under constant amplitude loading. Schijve and Brock (1962) found "overload retardation effect" in 2024-T3 aluminum alloy under constant amplitude cyclic loading. Since then, many scholars have focused their attention on the effect of the loading sequence and tried to describe the phenomenon in crack growth under complex fatigue loading. Many models have been developed to predict the fatigue life of components subjected to variable amplitude loading (Paris, *et al.*, 1999; Huang, *et al.*, 2008; Makabe, *et al.*, 2004; Rushton and Taheri, 2003; Silva,

2007). The earliest of these are based on calculations of the yield zone size ahead of the crack tip and are still widely used. The Wheeler model (Wheeler, 1972) and Willenborg model (Willenborg, *et al.*, 1971), for example, both fall into this category. Another category of models based on the crack closure approach, which considers plastic deformation and crack face interaction in the wake of the crack, was subsequently proposed by Elber (1970), and has been used to model crack growth rates under variable amplitude loads (Newman, 1984; Schijve, 1981). More recent proposals include combinations of the Wheeler model with the Newman crack closure model (Huang, 2005). It is well known that the effect of the loading sequence on FCP is very complex. Many kinds of factors are included, such as the characteristics of random loading itself, specimen geometry characteristics, material properties and microstructure and the surrounding environment (Skorupa, 1998). Silva (2007) has summarized four kinds of retardation mechanisms for overload (underload) induced retardation (acceleration). They are: 1) the residual stress influence; 2) crack tip blunting; 3) crack front irregularity; 4) plastic induction crack closure. Ray and Patanker (2001) have summarized the phenomena of the loading sequence in fatigue crack growth under different loading conditions as follows:

(1) An overload may introduce significant crack growth retardation. Up to certain limits, the tenure of crack retardation effects is increased by the following: larger magnitudes of the overload excitation; periodic repetition of the overload during the crack propagation life; application of short blocks of overload instead of isolated single-cycle overload.

(2) Crack retardation may not always immediately follow the application of an overload. There could be a short delay before the crack growth rate starts decreasing. Under some circumstances, a small initial acceleration in crack growth has been observed. The delayed retardation in crack growth due to overload was clearly verified by observation of striation spacing.

(3) The instantaneous crack growth caused by an overload itself is larger than that expected from a constant amplitude load equal to the amplitude of the overload. This observation has been confirmed by fractography.

(4) An underload has smaller effects on crack growth than an overload of the same magnitude. However, an underload applied immediately after an overload may significantly compensate for the effects of crack growth retardation due to the overload. If underload precedes the overload, the compensation is much smaller due to a sequence effect of the overload cycles.

(5) In step loading, a high-low sequence produces qualitatively similar results as overload cycles including delayed retardation. Interaction effects after a high-low sequence are barely detectable in the macroscopic sense.

(6) Duration of crack growth retardation depends upon ductility of the material. If ductility of an alloy is modified by heat treatment, lower (higher) yield strength corresponds to a longer (shorter) retardation period. Moreover, the specimen geometry also affects the retardation period. A reduced retardation period was observed with an increase in thickness.

(7) Rest periods at zero stress following a tensile peak overload have no

significant influence on subsequent fatigue crack retardation for ductile alloys at room temperature.

The mechanism of the loading sequence in FCP is still not fully understood yet. Due to the complexity of the mechanisms involved in this problem, no universal model exists yet. The phenomena listed above should be considered in FLP of components under random fatigue loading.

2.5.4 Surface Effects

Fatigue in the crack initiation period is a surface phenomenon. As a consequence, various kinds of surface effects can be of great importance for the fatigue life. Surface effects include all conditions which can change the crack initiation period. In other words, they cover the phenomena which enhance the crack initiation mechanism. Surface roughness and surface damage imply that the free surface is no longer perfectly flat. As a consequence, small sized stress concentrations along the material surface occur. Although the stress concentration will rapidly fade away from the surface, it is still significant for promoting cyclic slip and crack nucleation at the material surface. The most detrimental consequence of an unfavorable surface effect is the large reduction in the fatigue limit. This is especially important for structural components designed for an infinite life, *i.e.* with all amplitudes in service below the fatigue limit. Unintentional surface damage, such as nicks and dents, can then be very harmful. Although surface damage can accelerate crack initiation, the high stress amplitude cycles can also generate crack nuclei early in the fatigue life, and the assistance of surface damage is less important for the initiation process. However, if the design life is large in numbers of cycles, the significance of adverse surface effects should be recognized.

The defect, roughness and the residual stress are the three important factors of fatigue. Surface roughness has an important influence on the fatigue life, and even plays a decisive role. The rougher the surface finish, the lower the fatigue limit. Moreover, the higher the material strength, the more significant the impact of surface roughness on the fatigue limits. The magnitude and distribution of residual stress on the surface is another important factor which will increase or decrease the initial fatigue life of the component.

High sensitivity to surface effects at low stress amplitudes and relatively low sensitivity to surface effects at high stress amplitudes can lead to more scatter of the fatigue life at low amplitudes and less scatter at high amplitudes. This trend is generally observed in fatigue experiments (Schijve, 2009).

With plastic deformation of the metal on the surface inducing the strain hardening layer and the residual stress, the fatigue strength increases as the degree of strain hardening increases. As a result the increase in the plastic deformation region caused by cyclic loading is bound to increase the thickness of the hardened layer and reverse the bending fatigue limit. The residual stress consists of self-balanced internal forces, with compressive stress on the surface of metal components, the tensile stress inside. All of the above improve the bending fatigue strength of components and surface fatigue strength of components subjected to tensile stress concentration, but in terms of smooth specimens the tensile fatigue strength will be reduced.

In mechanically processed metal components, tensile residual stress will be induced on the surface of components, while compressive residual stress will be generated inside. Hence the components of tension and compression, bending and torsional fatigue strength, are reduced. If surface defects are produced on the surface of the processed parts, this will easily lead to stress concentration, so that defects become the origin of the fatigue crack. Obviously, for high strength metal materials, we should pay more attention to the surface processing, to minimize the defects and micro cracks in surface processing.

Shot peening is used as a remedy if fatigue problems are anticipated. Shot peening introduces plastic deformation in the surface layer of the material. As a result of the plastic deformation, residual compressive stresses are left in a thin surface layer. Because residual stresses do not affect cyclic shear stresses, cyclic slip may still occur. It even can lead to small micro-cracks, but crack growth is difficult. The residual compressive stresses reduce or prevent crack opening of the micro-cracks. As a result, the stress concentration at the crack tip is much lower and crack growth will be impeded. It may well be stopped completely. The residual compressive stress zone serves as a barrier to micro-crack growth. There are many other ways to introduce residual stresses in the material of structural components.

2.5.5 Environmental Effects

The fatigue tests of specimens are usually performed at room temperature, in an atmospheric environment under cyclic tension and compression or bending loading. The service environment and the loading of structures may be quite different from that of performed fatigue tests. In practical design, the environmental factors should be taken into account. If the design is based on fatigue data performed at room temperature, necessary amendments should be made to counter the difference. Low temperature brittleness should be taken into account for low temperature equipment. An increase in fatigue crack growth and high temperature corrosion should be considered at high temperature, corrosion fatigue in a corrosive environment.

Corrosion fatigue has the following characteristics:

(1) Corrosion fatigue may occur in any metal in corrosive media, and does not need any specific combination of materials and media. Stress corrosion requires a

certain degree of stress conditions, materials—environmental conditions and electrochemical conditions, compared to which corrosion fatigue is more general.

(2) There is no infinite fatigue life in corrosion fatigue, *i.e.* no fatigue limit.

(3) There is no clear relationship between the corrosion fatigue limit and static strength. Improving the static strength of materials contributes little to the fatigue resistance in corrosive media.

(4) Most of the corrosive fatigue fractured surface is covered with corrosive products. Corrosion fatigue fracture from both the micro and macro viewpoint has many similarities to general fatigue. In the general fatigue process, when the crack front opens under tensile stress, slip deformation and dislocation appear in a favorable direction and dislocation pile-up can prevent the slip; in corrosive media, corrosion can reduce the pile-up of dislocations, resulting in the slip increasing, fatigue striations being wide. Brittle secondary cracks featuring in the fatigue fracture surface can be observed.

Corrosion fatigue can be divided into an active state, passive state and unstable passive state. An active state is typical of corrosion fatigue. Metal surface damage is encountered in the passive film when general or partial dissolution occurs. Semicircular pits form on the surface of the material, under the action of the alternating stress cracks appearing at the bottom of the pit, and gradually develop. Most of these cracks are propagated inside in transgranular manner.

Corrosion fatigue damage is the result of the interaction between the environment, the microstructure of the material and cyclic loading; the mechanism is very complex. Because corrosion fatigue and stress corrosion are both caused by loading and media, corrosion fatigue is the same as stress corrosion. The damage and the fracture process can be thought of as hydrogen embrittlement and anodic dissolution. Based on the understanding of the corrosion process, considering the dynamic effects of stress, the mechanism of the development of corrosion fatigue can be revealed. However, understanding how the media accelerates crack initiation and propagation, and how the cyclic deformation promotes the development of corrosion and differences in the interaction of the medium and deformation of the damage process at various stages, are still significant problems. The four mechanisms for corrosion fatigue are as follows:

(1) Formation of cracking and pitting mechanism: this mechanism was used earlier to explain corrosion fatigue phenomenon. The view is that in the metal surface under the action of corrosive media, pits are formed, a source of stress concentration. Under the cyclic stress, a slip occurs at the corrosion pits, so do slip steps. Corrosion fatigue cracks can be observed in the bottom of semi-circular pits in experiments, and the longer the corrosion time, the more the corrosion fatigue strength decreases, so the crack formation mechanism of pitting is still widely accepted.

(2) The crack formation mechanism of a protective film breaking: the theory is the same as for the stress corrosion theory of protective film damage. Under cyclic stress, a fresh metal surface is induced by the surface protective film breaking due to the plastic slip, the rupture is the anode and film around it is the cathode. Under

the action of the media together with the stress, electrochemical corrosion appears, the anode (rupture) is dissolved, and cracks form.

(3) Selective electrochemical corrosion mechanisms: under an alternating load, concentrated deformation areas generate in components. In the concentrated deformation zone the dislocation or impurity precipitated is different from the body, resulting in anodic dissolution in this zone occurring first. With the repeated slip during the fatigue process, dissolution is continuous, the formation of the corrosive channel from the surface results in surface stress concentration leading to crack initiation.

(4) Medium adsorption mechanisms: active factors in the medium absorbed by the metal surface and wedge effects generated in the micro-cracks form stress concentration and decrease the metal binding capacity under alternating stress, which lead to the formation of a fatigue crack source.

The corrosion fatigue limit of the material concerns the composition and microstructure of the material itself. Frequency of the alternating load, stress ratio, loading, stress cycle waveform, stress concentration factors and properties of the material will affect the corrosion fatigue properties.

2.6 Fatigue Crack Growth Mechanism of Small Defects

In general, structures contain micro structural defects such as porosity, voids, and discontinuities, which can lead to the formation of cracks if the service loading exceeds a certain level. Small pre-existing defects may appear on the surface or inside the component. For a welded structure, the crack is usually initiated from the surface of the weld toe, but for a treated weldment with grounding or with some other weld repair techniques, the crack may also be initiated at the surface or from large enough internal defects. For the components with pre-existing defects, their fatigue life corresponds to the crack propagation life from the initial crack to the final critical crack. Because the fatigue mechanism and its influence on the short crack and long crack is quite different, it is difficult to establish a method for FLP from small initial crack size to final fracture. Over the past decades, several authors have attempted to construct a unified model to explain both the physically short-crack growth and long-crack growth behavior (McEvily, *et al.*, 1999; Sadananda, *et al.*, 2001). There is no simple solution available for engineers faced with developing design guidelines for fatigue critical components containing small flaws (Suresh, 2001).

2.6.1 Engineering Initial Crack Size of Structures

Initial defects always exist in marine engineering structures. For correct estimation of the crack propagation life, the initial crack size a_i is very important. However, initial crack length a_i has no precise definition yet. Physicists, metallurgists and engineers have different research purposes, the accuracy and resolution of the employed detection devices are also different, so that the definitions of the initial crack length by different people are different. From the viewpoint of fatigue mechanisms, the definition should be based on the electron microscopy of the microstructure of the metal, but this definition is not very convenient in engineering practice. In experimental study of initial fatigue life, the initial crack length is generally taken as a_i=0.006–0.5 mm (Hussain, 1997). Some definitions of the initial crack length are summarized here. Usually a crack that is observed with low magnification observation equipment is defined as crack initiation, and the corresponding crack length may be chosen as 0.13–0.25 mm (Schijve, 1981; McEvily, 1977); from the design point of view, the length of crack initiation is defined as 0.25–2.5 mm (Zhou, 1982); the surface crack initiation size is defined as a crack of length of 0.5 mm, depth of 0.15 mm by non-destructive testing (Backlund and Beevers, 1982); in the 1980s–1990s, a_i was taken to be 0.25–0.3 mm by some researchers (*e.g.* Wanhill, *et al.*, 1989; Stephens, 1985; Moan, *et al.*, 2000; Lee, 1985); in the estimation of weldment FCP life, initial crack size was also taken as 0.25 mm (Maddox, 1985). Moan, *et al.* (2000) suggested that the initial crack length taken to be 0.19–0.38 mm is appropriate. Others suggest that engineering the initial crack length could be taken to be 0.8–1 mm (*e.g.* Wells, 1980). Initial crack sizes with 0.10–0.25 mm are recommended for offshore structures in the ABS (2005), and 0.5 mm is in the DNV (2005).

All of these initial cracks with the above crack length are small cracks, but some of them may not be short cracks. This issue is not discussed sufficiently and the fatigue life is calculated with the initial size without considering if the crack is in the effective range of the adopted crack growth law.

2.6.2 Definition of Short Crack and Long Crack

It is very important to be able to predict the crack growth from the initial size to final fracture. It is technically significant to consider the crack initiation and crack growth periods separately because several practical considerations have a large influence on the crack initiation period, but a limited influence or no influence at all on the crack growth period (Schijve, 2009). The difficulty for this two-period division is that there is still no universally accepted definition for the initial crack size. For establishing a unified model to explain the crack growth behavior from pre-existing defects to final failure, the mechanism for fatigue crack growth from pre-existing small defects to final failure should be understood soundly.

The fatigue crack growth behavior of short cracks differs from the long crack. But there is no clear difference between the short crack and the small crack, and the fatigue life of components with a small crack will be predicted using the crack growth law for a long crack in most cases. For correctly dealing with the fatigue crack growth of the short crack and long crack, a clear definition of short crack and long crack should be made. The cracks which cannot be analyzed by linear elastic methods because of large-scale plasticity effects are defined as short cracks (Hussain, 1997). The definition implies that a short crack not only depends on the material properties but also on the applied loading. This definition helps to clear the confusion between the concepts of small/long crack and short/long crack. In fatigue crack growth prediction of a component with a small crack, only the short crack and long crack growth mechanism should be considered, whether the crack is mechanically a small crack or microstructurally a small crack or physically a small crack.

2.6.3 Crack Growth Threshold and Intrinsic Crack Length

The fatigue growth behavior of cracks near the threshold is known to be dependent on the crack length. The threshold range of the SIF ΔK_{th} decreases with decreasing crack length. A similar tendency in the variation in ΔK_{th} with the crack length was also reported in the literature. There are some models that have been proposed for explaining the dependence of ΔK_{th} on the crack length. Smith (1984) suggested that the threshold stress was truncated at the fatigue limit of the smooth specimens. In his model, the threshold condition is $\Delta\sigma_{th}=\Delta\sigma_{w\,0}$ when the crack length is shorter than a certain length a_0, here called the intrinsic crack length, and above this length it is expressed by the constant value of ΔK_{th} equal to that for very long cracks. El Haddad, *et al.* (1979) introduced a concept of the effective crack length which is the sum of the actual crack length and the intrinsic crack length. The situation was modeled on the crack-tip slip band blocked by the GB (BSB model). The BSB model is further extended to incorporate the crack length effect on the threshold by Tanaka, *et al.* (1981). The intrinsic crack length a_0 defined by Smith (1977) and El Haddad, *et al.* (1979) is given from the fatigue limit for a smooth specimen $\Delta\sigma_{w\,0}$ and the crack growth threshold for a long crack $\Delta K_{th\,0}$ as,

$$a_0 = (\Delta K_{th\,0} / \Delta\sigma_{w\,0})^2 / \pi \qquad (2.1)$$

For demonstrating the relationship between threshold stress (or threshold SIF range) and crack length, the ratio of $\Delta\sigma_{th}$ over $\Delta\sigma_{w\,0}$ is plotted against that of a_{eq} to a_0 in Fig. 2.11, and the ratio of ΔK_{th} to $\Delta K_{th\,0}$ is plotted against that of a_{eq} to a_0 in Fig. 2.12. From Figs. 2.11 and 2.12, it can be seen that the fatigue crack growth threshold is nearly independent of crack length when the crack length is bigger than the intrinsic crack length a_0, otherwise the fatigue crack growth threshold decreases with the decrease in the crack length. The threshold versus crack length can be more accurately expressed by Tanaka's equations, *i.e.* Eqs. (2.2) and (2.3) (Tanaka, *et al.*, 1981). For very high strength steels of monotonic yield strength up to 2000 MPa, $a_0 \approx 1\text{–}10$ μm and for very low strength steels with yield strength as low as 200 MPa, $a_0 \approx 1\text{–}1{,}000$ μm. The intrinsic crack length is a sufficient condition for a short crack but not a necessary condition. A crack with a length smaller than a_0 is a short crack and a crack with a length larger than a_0 may be a short crack, which also depends on the loading condition according to the short crack definition.

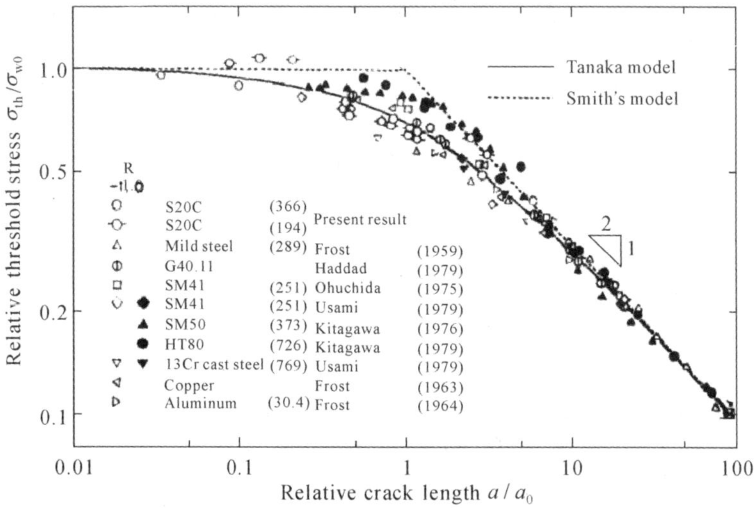

Fig. 2.11 Normalized relationship between threshold stress and crack length for various materials (Tanaka, *et al.*, 1981)

$$\Delta\sigma_{th} = \Delta\sigma_{w0}\sqrt{a_0/(a+a_0)} \tag{2.2}$$

$$\Delta K_{th} = \Delta K_{th0}\sqrt{a/(a+a_0)} \tag{2.3}$$

where a is crack length, a_0 is intrinsic crack length; $\Delta\sigma_{th}$ is the threshold value of the stress; $\Delta\sigma_{w0}$ is fatigue limit of the smooth specimen at very short cracks; ΔK_{th} is crack growth threshold; ΔK_{th0} is crack growth threshold for a long crack.

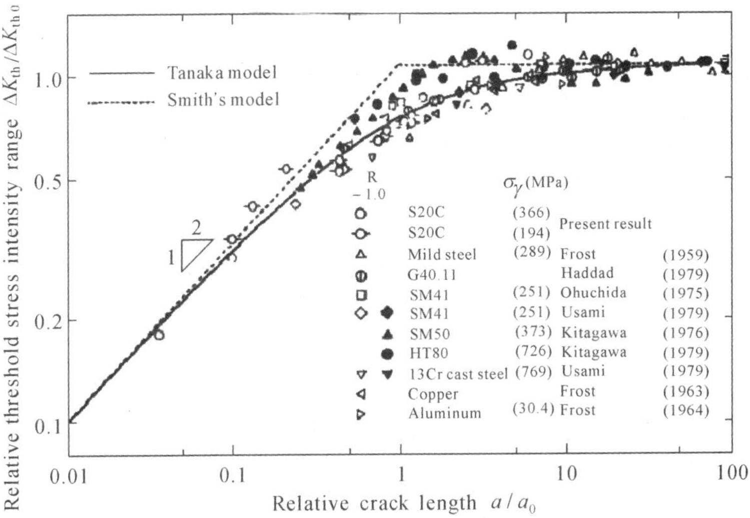

Fig. 2.12 Normalized relationship between threshold SIF range and crack length for various materials (Tanaka, Nakai and Yamashita, 1981)

2.6.4 Equivalent Crack Length for Short and Long Crack

As mentioned above, the intrinsic crack size a_0 can be used as a sufficient condition to judge whether a crack is a short crack or not. For predicting the fatigue life of a component with a small crack, it is more important to establish a method which can deal with the crack growth both for short and long cracks. The intrinsic crack size a_0 has been used as an empirical parameter to account for the differences in crack growth rates between long and short crack fatigue cracks. El Haddad, *et al.* (1979) suggested that when a fictitious or intrinsic crack length a_0, defined in Eq. (2.1), is added to the fatigue crack length a, the resulting characterization based on LEFM provides crack propagation rates that are independent of crack size. They demonstrated that the differences in the fracture mechanics based growth characterization for long and short flaws vanish when the fictitious crack size a_0 is added to the SIF calculation. The major drawback of this empirical approach is that attempts to provide a physical reasoning for the SIF of a crack which include fictitious crack size a_0 have so far remained unconvincing (Suresh, 2001). Newman, *et al.* (1999) have developed a crack growth prediction method which is able to explain small crack growth behavior based on the concept of plasticity induced crack closure and the effective SIF range (ΔK_{eff}) as driving force. The model was used to calculate small and large crack growth rates, and to predict total fatigue lives for notched and un-notched specimens made of two aluminum alloys and a steel under constant-amplitude and spectrum loading. An equivalent initial-flaw-size concept was used to calculate fatigue lives in other cases. Short crack growth behavior has also been examined using the parametric approach developed earlier by Sadananda and Vasudevan (1997). An attempt was also made by Pugno, *et al*. (2007) for a unified treatment of short and long cracks.

McEvily and his co-worker (McEvily and Ishihara, 2001; Ishihara and McEvily, 2002) have proposed a modified expression of fatigue crack growth which is able to explain both short and long crack growth behavior, based on the principles of fracture mechanics for small and large scale yielding. Some efforts in developing a UFLP method based on FCP theory have been made in the past decades. However, most of them are specific to a particular concept and do not represent the most general one. More details about these UFLP methods will be summarized in Chapter 3. For developing a UFLP procedure for predicting the fatigue life of an engineering structural component with a small crack, an equivalent crack length or effective SIF range calculation equation for short and long cracks should include the intrinsic crack length and the cyclic residual stress or plastic strain around the crack tip.

2.7 Summary

Some of the major findings about fatigue crack initiation and growth, and some important influencing factors which are significant for applications in subsequent chapters, are briefly stated.

(1) The fatigue mechanism in metallic materials should basically be associated with cyclic slip (on the surface or around the inclusion) and the conversion into crack initiation and crack growth. Details of the mechanism are dependent on the type of material.

(2) Fatigue in the crack initiation period is a surface phenomenon, which is very sensitive to various surface conditions, such as surface roughness, corrosion pits, *etc*.

(3) In the crack growth period, fatigue is dependent on the crack growth resistance of the material and not on the material surface conditions.

(4) Microstructurally small cracks can be nucleated at stress amplitudes below the fatigue limit. Crack growth is then arrested by microstructural barriers. The fatigue limit as a threshold property is highly sensitive to various surface conditions. At high stress amplitudes, and thus relatively low fatigue lives, the effect of the surface conditions is much smaller.

(5) Most of the reasons for inducing crack closure are mainly concerned with crack growth at low SIF and near the threshold.

(6) It is important to understand the mechanism of short crack growth and the loading sequence in modeling fatigue crack growth of a component with a small crack under service loading conditions.

(7) Fatigue properties and the crack growth mechanism are basically different for the Stage I crack growth period and for the Stage II crack growth period.

(8) Aggressive environments can affect both crack initiation and crack growth. The load frequency and the wave shape are then important variables.

(9) The various characteristics of fatigue fractures can be understood in terms of crack initiation and crack growth mechanisms. These characteristics are essential in fatigue failure analysis, but they are also relevant for understanding the significance of technically important variables of fatigue properties.

References

ABS (American Bureau of Shipping) (2005). Fatigue Assessment of Offshore Structures. Houston, Texas.

Backlund, B., Beevers, C. J. (1982). "Fatigue thresholds", Stockholm: EMAS, 1, 3–10.

Baumel, A. Jr. & Seeger, T. (1990). Materials Data for Cyclic Loading (Supplement 1). Amsterdam: Elsevier Science Publishers.

Bjerkén, C. & Melin, S. (2004). "A study of the influence of grain boundaries on short crack growth during varying load using a dislocation technique", Eng Fract Mech, 71(15): 2215–2227.

Blom, A. E., Hedlund, A., Zhao, W., Fathalla, A., Weiss, B. & Stickler, R. (1986). "Short fatigue crack growth in Al 2024 and Al 7475", Behaviour of Short Fatigue Cracks. Symp., Sept. 1985, Sheffield, EGF 1, MEP, 37–66.

Carlson, R. L. & Kardomateas, G. A. (1996). An Introduction to Fatigue in Metals and Composites. London and New York: Chapman & Hall.

Chan, K. S. (2010). "Roles of microstructure in fatigue crack initiation", International Journal of Fatigue, 32: 1428–1447.

Cottrell, A. H. & Hull, D. (1957). "Extrusion and intrusion by cyclic slip in copper", Proceedings of the Royal Society, 192–203.

Cui, W. C. (2002a). "Relation between crack growth rate curve and *S-N* curve for metal fatigue", Journal of Ship Mechanics, 6(6): 93–106.

Cui, W. C. (2002b). "A state-of-the-art review on fatigue life prediction methods for metal structures", Journal of Marine Science and Technology, 7(1): 43–56.

Cui, W. C., Wang, F., Huang, X. P. (2011). "A unified fatigue life prediction (UFLP) method for marine structures", Marine Structures, 24(2): 153–181.

DNV (Det Norske Veritas) (2005). Fatigue Design of Offshore Steel Structures. Oslo, Norway.

Dowling, N. E. (2007). Mechanical Behavior of Materials-Engineering Methods for Deformation, Fracture, and Fatigue (3rd ed.). Upper Saddle River, New Jersey: Pearson-Prentice Hall.

El Haddad, M. H., Topper, T. H. & Smith, K. N. (1979). "Prediction of non propagating cracks", Engineering Fracture Mechanics, 11: 573–584.

Elber, W. (1970). "Fatigue crack closure under cyclic tension", Engineering Fracture Mechanics, 2: 37–45.

Essmann, U., Gösele, U. & Mughrabi, H. (1981). "A model of extrusions and intrusions in fatigued metals (I: Point-defect production and the growth of extrusions)", Philosophical Magazine A, 44: 405–426.

Figueroa, J. C. & Laird, C. (1983). "Crack initiation mechanisms in copper polycrystals cycled under constant strain amplitudes and in step tests", Materials Science and Engineering, 60: 45–58.

Forsyth, P. J. E. (1962). "A two stage process of fatigue crack growth", in: Crack Propagation, Proceedings of Cranfield Symposium, 76–94, London, Her Majesty's Stationery Office.

Forsyth, P. J. E. & Ryder, D. A. (1960). "Fatigue fracture", Aircraft Engineering, 32: 96–99.

Forsyth, P. J. E. (1969). The Physical Basis of Metal Fatigue. New York: American Elsevier Publishing Co.

Fuchs, H. O. & Stephens, R. I. (1980). Metal Fatigue in Engineering. New York: John-Wiley & Sons Inc.

Goto, M., Han, S. Z., Euh, K., Kang, J.-H., Kim, S. S. & Kawagoishi, N. (2010). "Formation of a high-cycle fatigue fracture surface and a crack growth mechanism of ultrafine-grained copper with different stages of microstructural evolution", Acta Materialia, 58: 6294–6305.

Hansson, P. (2009). "Crack growth rates for short fatigue cracks simulated using a discrete dislocation technique", International Journal of Fatigue, 31: 1346–1355.

Huang, X. P., Moan, T. & Cui, W. C. (2008). "An engineering model of fatigue crack growth under variable amplitude loading", Int. J. Fatigue, 30: 2–10.

Huang, X. P., Zhang, J. B., Cui, W. C. & Leng, J. X. (2005). "Fatigue crack growth with overload under spectrum loading", Theor Appl Fract Mech, 44: 105–15.

Hunsche, A. & Neumann, P. (1986). "Quantitative measurement of persistent slip band profiles and crack initiation", Acta Metallurgica, 34: 207–217.

Hussain, K. (1997). "Short fatigue crack behavior and analytical models: a review", Engineering Fracture Mechanics, 58(4): 327–354.

Ishihara, S. & McEvily, A. J. (1999). "A coaxing effect in the small fatigue crack growth regime", Scripta Mater, 40: 617–622.

Ishihara, S. & McEvily, A. J. (2002). "Analysis of short fatigue crack growth in cast aluminum alloys", International Journal of Fatigue, 24: 1169–1174.

Kennedy, A. J. (1963). Processes of Creep and Fatigue in Metals. New York: Wiley.

Kikukawa, M., Jono, M. & Adachi, M. (1979). "Direct observation and mechanism of fatigue crack propagation", ASTM STP675, 234–253.

Kim, W. H. & Laird, C. (1978). "Crack nucleation and stage I propagation in high strain fatigue", Acta Metall, 26: 777–799.

Kunio, T., Shimizu, M., Yamada, K., Sakura, K. & Yamamoto, T. (1981). "The early stage of fatigue crack growth in martensitic steel", International Journal of Fracture, 17(2): 111–119.

Laird, C. & Thomas, G. (1967). "On fatigue-induced reversion and over aging in dispersion strengthened alloy systems", International Journal of Fracture Mechanics, 3: 81–97.

Laird, C. & Smith, G. C. (1962). "Crack propagation in high stress fatigue", Philosophical Magazine, 8: 847–857.

Lankford, J. (1982). "The growth of small fatigue cracks in 7075-T6 aluminum", Fatigue of Engineering Materials and Structures, 5: 233–248.

Lankford, J. & Kusenberger, F. N. (1973). "Initiation of fatigue cracks in 4340 steel", Metallurgical Transactions, 4: 553–559.

Lee, E. U. (1985). "Corrosion fatigue and stress corrosion cracking of 7475-T7351 aluminum alloy", in: Goel, V. S. (ed.) Corrosion Cracking, Conference Proceedings ASM, Salt Lake City, 123–128.

Lee, K. S. & Song, J. H. (2006). "Estimation methods for strain-life fatigue properties from hardness", International Journal of Fatigue, 28: 386–400.

Li, P., Li, S. X., Wang, Z. G. & Zhang, Z. F. (2011). "Fundamental factors on formation mechanism of dislocation arrangements in cyclically deformed fcc single crystals", Progress in Materials Science, 56: 328–377.

Ma, B. T. & Laird, C. (1989a). "Overview of fatigue behaviour in copper single crystals (I: Surface morphology and stage I crack initiation sites for tests at constant strain amplitude)", Acta Metallurgica, 37: 325–336.

Ma, B. T. & Laird, C. (1989b). "Overview of fatigue behaviour in copper single crystals (II: Population, size, distribution and growth kinetics of stage I cracks for tests at constant strain amplitude)", Acta Metallurgica, 37: 337–348.

Maddox, S. J. (1985). "Fatigue analysis of welded joints using fracture mechanics", in: Goel, V. S. (ed.) Fatigue Life, Conference Proceedings ASM, Dec. 1985, Salt Lake City, 155–166.

Makabe, C. & Kaneshiro, H. (2004). "A method of detecting fatigue crack initiation and an application of an overload using local strains", American Society of Mechanical Engineers, Pressure Vessels and Piping Division (Publication) PVP, 480: 45–51.

May, A. N. (1960a). "A model of metal fatigue", Nature, 185: 303–304.

May, A. N. (1960b). "Random slip model of metal fatigue and Coffin's law", Nature, 188: 573–574.

McEvily, A. J., Bao, H. & Ishihara, S. (1999). "A modified constitutive relation for fatigue crack growth", in: Wu, X. R. and Wang, Z. G. (ed.) Proceedings of the 7th International Fatigue Congress (Fatigue'99). Beijing: Higher Education Press, 329–336.

McEvily, A. J. & Ishihara, S. (2001). "On the dependence of the rate of the fatigue crack growth on the (2a) parameter", International Journal of Fatigue, 23: 115–120.

McEvily, A. J. (1977). "On crack closure in fatigue crack growth", in: Taplin, D. M. R. (ed.) Proceedings ICF4. Oxford: Pergamon Press, 39–42.

Meggiolaro, M. A. & Castro, J. T. P. (2003). "On the dominant role of crack closure on fatigue crack growth modeling", International Journal of Fatigue, 25: 843–854.

Miller, K. J. (1987a). "The behavior of short fatigue cracks and their initiation (Part I: A review of two recent books)", Fatigue and Fracture of Engineering Materials and Structures, 10(1): 75–91.

Miller, K. J. (1987b). "The behavior of short fatigue cracks and their initiation (Part II: A general summary)", Fatigue and Fracture of Engineering Materials and Structures, 10(2): 93–113.

Miller, K. J. (1999). "A historical perspective of the important parameters of metal fatigue and problems for the next century", in: Wu, X. R. & Wang, Z. G. (ed.) 7th Proceedings of the International Fatigue Congress (Fatigue'99). Beijing: Higher Education Press, 15–39.

Miller, K. J. (1993). "The two thresholds of fatigue behavior", Fatigue and Fracture of Engineering Materials and Structures, 16(9): 931–939.

Moan, T., Vårdal, O. T, Hellevig, N. C. & Skjoldli, K. (2000). "Initial crack depth and POD values inferred from in-service observations of cracks in North Sea jackets", Journal of Offshore Mechanics and Arctic Engineering, 122: 157–162.

Mortensen, A. (2006). Concise Encyclopedia of Composite Materials (2nd ed.). Exeter, DEV, UK: Elsevier Science Publishers.

Mott, N. F. (1958). "A theory of the origin of fatigue crack", Acta Metallurgica, 6, 195–197.

Mughrabi, H. & Höppel, H. W. (2010). "Cyclic deformation and fatigue properties of very fine-grained metals and alloys", International Journal of Fatigue, 32: 1413–1427.

Mughrabi, Y. & Endo, M. (1986). "Effects of hardness and crack geometries on ΔK_{th} small cracks emanating from small defects", in: Behaviour of Short Fatigue Cracks, Publication 1 of the European Group on Fracture, 275–293. London: Mechanical Engineering Publications.

Newman J. C., Phillips E. P. & Swain M. H. (1999). "Fatigue-life prediction methodology using small-crack theory", International Journal of Fatigue, 21(2): 109–119.

Newman, Jr. J. C. (1984), "A crack opening stress equation for fatigue crack growth", Int. J. Fract., 24: R131–135.

Newman, Jr. J. C. (1998). "The merging of fatigue and fracture mechanics concepts: a historical perspective", Progress in Aerospace Sciences, 34: 347–390.

Paris, P. C., Tada, H. & Donald, J. K. (1999). "Service load fatigue damage—a historical perspective", Int. J. Fatigue, 21: S35–46.

Pearson, S. (1975). "Initiation of fatigue cracks in commercial aluminum alloys and the subsequent propagation of very short cracks", Engineering Fracture Mechanics, 7: 235–247.

Pugno, N., Cornetti, P. & Carpinteri, A. (2007). "New unified laws in fatigue: from Wöhler's to the Paris' regime", Engineering Fracture Mechanics, 74: 595–601.

Ray, A. & Patanker, R. (2001). "Fatigue crack growth under variable-amplitude loading (Part II: Code development and model validation)", Appl Math Model,

25: 995–1013.

Richards, C. E. & Lindley, T. C. (1972). "The influence of stress intensity and microstructure on fatigue crack propagation in ferritic materials", Engineering Fracture Mechanics, 4: 951–978.

Ritchie, R. O. & Lankford, J. (1986). "Small fatigue cracks: a statement of the problem and potential solutions", Materials Science and Engineering, 84: 11–16.

Rushton, P. A. & Taheri, F. (2003). "Prediction of crack growth in 350WT steel subjected to constant amplitude with over- and under-loads using a modified wheeler approach", Marine Structures, 16(7): 517–539.

Sadananda, K. & Vasudevan, A. K. (1997). "Short crack growth and internal stresses", International Journal of Fatigue, 19(93): S99–108.

Sadananda, K., Vasudevan, A. K. & Holtz, R. L. (2001). "Extension of the unified approach to fatigue crack growth to environmental interactions", International Journal of Fatigue, 23: S277–286.

Sadananda, K., Vasudevan, A. K., Holtz, R. L. & Lee, E. U. (1999). "Analysis of overload effects and related phenomena", International Journal of Fatigue, 21(s1): S233–246.

Schijive, J. (2001). Fatigue of Structures and Materials. Dordrecht, Boston: Kluwer Academic Publishers.

Schijve, J. (1967). "Significance of fatigue cracks in micro-range and macro-range", ASTM STP 415, 415–459.

Schijve, J. (1981). "Some formulas for the crack opening stress level", Engineering Fracture Mechanics, 14: 461–465.

Schijve, J. (2003). "Fatigue of structures and materials in the 20th century and the state of the art", International Journal of Fatigue, 25: 679–702.

Schijve, J. (2009). Fatigue of Structures and Materials (2nd ed. with CD-Rom). Dordrecht, the Netherlands: Springer Science+Business Media, B.V.

Schijve, J. & Brock, D. (1962). "Crack propagation tests based on a gust spectrum with variable amplitude loading", Aircraft Engineering, 34: 314–316.

Shang, D. G., Yao, W. X. & Wang, D. J. (1998). "A new approach to the determination of fatigue crack initiation size", International Journal of Fatigue, 20: 683–687.

Shen, Z., Wagoner, R. H. & Clark, W. A. T. (1988). "Dislocation and grain boundary interactions in metals", Acta Metallurgica, 36(12): 3231–3242.

Silva, F. S. (2007). "Fatigue crack propagation after overloading and underloading at negative stress ratios", International Journal of Fatigue, 29: 1757–1771.

Skorupa, M. (1998). "Load interaction effects during fatigue crack growth under variable amplitude loading—a literature review (Part I: Empirical trends)", Fatigue and Fracture of Engineering Materials and Structures, 21(8): 987–1006.

Smith, R. A. (1984). "Thirty years of fatigue crack growth—a historical review", in: Smith, R. A. (ed.) Fatigue Crack Growth. Oxford: Pergamon Press, 1–16.

Smith, R. A. (1977). "On the short crack limitations of fracture mechanics", International Journal of Fracture, 13: 717–720.

Stephens, R. I. (1985). "Comparison of fatigue resistance of cast steels under different test conditions", in: Goel, V. S. (ed.) Fatigue Life, Conference Proceedings ASM, Salt Lake City, Dec.1985, 69–78.

Suresh, S. (2001). Fatigue of Materials (2nd ed.). Cambridge, UK: Cambridge University Press.

Suresh, S. & Ritchie, R. O. (1984). "Propagation of short fatigue cracks", International Journal of Fracture, 30: 237–259.

Tanaka, K., Nakai, Y. & Yamashita, M. (1981). "Fatigue growth threshold of small cracks", International Journal of Fracture, 17: 519–533.

Tayler, D. & Knott, J. F. (1981). "Fatigue crack propagation behaviour of short cracks: the effect of microstructure", Fatigue and Fracture of Engineering Materials and Structures, 4: 147–155.

Valiev, R. Z., Islamgaliev, R. K. & Alexandrov, I. V. (2000). "Bulk nanostructured materials from severe plastic deformation", Progress in Materials Science, 45: 103–189.

Valiev, R. Z., Langdon, T. G. Alexandrov, I. V., Zhu, Y. T., Estrin, Y. & Kostorz, G. (eds.) (2009). "Proceedings of international symposium on bulk nanostructured materials: from fundamentals to innovation", BNM 2007, Materials Science and Engineering A, 503: 1–208.

Wanhill, R. J. H., De Luccia, J. J. & Russo, M. T. (1989). "Fatigue in aircraft corrosion testing (FACT) programme", AGARD Report No. 713, (AD-A208-359).

Weertman, J. (1978). "Facture mechanics: a unified view for Griffith-Irwin-Orowan cracks", Acta Metallurgica, 26: 1731–1738.

Wells, C. H. (1980). "Analysis of life prediction methods for time-dependent fatigue crack initiation in nickel-base superalloys", Report of the Committee on Fatigue Crack Initiation at Elevated Temperatures, National Materials Advisory Board. Washington, D. C.: National Academy of Sciences.

Wheeler, O. E. (1972). "Spectrum loading and crack growth", J Basic Engng, Trans ASME, Ser D, 94(1): 181–186.

Willenborg, J. D., Engle, Jr. R. M. & Wood, H. A. (1971). "A crack growth retardation model using effective stress concept", FDL-TM-71-1-FBR. January 1971.

Zappfe, C. A. & Worden, C. O. (1951). "Fractographic registrations of fatigue", Transactions of the American Institute of Mining, Metallurgical and Petroleum Engineers, 145: 225–271.

Zhang, Z. F. & Wang, Z. G. (2008). "Grain boundary effects on cyclic deformation and fatigue damage", Progress in Materials Science, 53: 1025–1099.

Zhao, S. B. & Wang, Z. B. (1997). Anti-fatigue Design: Methodology and Data. Beijing: China Machinery Industry Press.

Zheng, X. L. & Hirt, M. A. (1983). "Fatigue crack propagation in steels", Eng, Fract, Mech., 18(5): 965–973.

Zhou, J. E. (1982). "Fatigue properties of a high strength aluminum alloy", Ph. D. thesis. University of Aston, Birmingham.

3

Current State-of-the-Art of UFLP

3.1 Introduction

The unified approach can mean different things to different people. Crupi (2008) has introduced a unifying approach to assess the structural strength in which different failure modes have been unified. The basic concept is that the energy stored and required to produce the structural failure is a material property independent of the loading conditions. Several relationships were used to correlate different mechanical properties and were validated by experimental results. The developed theoretical model was focused on damping, a material property able to correlate vibration parameters with high cycle fatigue parameters and related to the damage of materials and structures. The measure of damping is based on experimental techniques currently used for damage detection, so this unifying approach could be applied also to structural health monitoring. Good predictions were achieved by applying the new approach to structural details of ships. Lo, *et al.* (1996) presented a unified model as a natural extension as well as a generalization of Irwin's concept for mixed mode SIFs ($K_{I\theta}$, $K_{II\theta}$). These unifications are not the main considerations of this book so they are not discussed further.

The unified approach referred to in this book is only concerned with fatigue failure mode. In the fatigue literature, the unified approach has also been used in different contexts and in this chapter these different meanings of unification will be reviewed and the idea for our own unified approach will be introduced.

3.2 Unified Approach for Three Regions of FCP

As introduced in Chapter 1, the earliest theory for predicting the FCP length is the LEFM approach first introduced by Paris and his co-workers (1961; 1963), who equated fatigue crack growth rate to the cyclic elastic SIF range at the tip of a long crack subjected to a low value of cyclic stress. Later, people found that the crack growth rate curve is not linear for all the ranges of ΔK. The sigmoidal shape of the crack growth curve suggests a subdivision into three regions, the initiation and threshold region, the stable crack growth region and the unstable fracture region, see Fig. 1.2. Region II crack growth follows a power law, the so-called Paris-Erdogan crack growth law (Paris and Erdogan, 1963), as given in Eq. (1.2). Region III has led to the relation proposed by Forman, *et al.* (1967). Relations combining the departures from power law behavior at high and low ΔK values also exist (Kanninen and Popelar, 1985; Carpinteri, 1994). The most representative one might be the one proposed by McEvily and Groeger (1977). However, this one cannot explain all the observed fatigue phenomena especially the stress ratio effect and the main challenge in developing a UFLP method is to establish an equation which reflects the actual fatigue failure mechanism and is able to explain all the observed fatigue phenomena.

3.3 Unified Approach to the Stress Ratio Effect or Mean Stress Effect

Some milestones in the development of the crack growth rate model have been briefly introduced in Section 1.4.2.2 of Chapter 1. The concept and expressions of the crack closure model proposed by Elber (1970) and Kujawski (2001b) are mentioned. Furthermore, the parameter ΔK^+ as the fatigue crack driving force introduced by Kujawski (2001c; 2001d) and its threshold value is given. However, this ΔK^+ parameter obviously cannot explain the effect of compressive stresses on fatigue crack growth (Fleck, *et al.*, 1985; Vasudevan and Sadananda, 1999; 2001; Zhang, *et al.*, 2010). Vasudevan and Sadananda have explicitly stated that both ΔK and K_{max} are the fatigue crack driving forces and they developed the two-parameter unified approach (Vasudevan and Sadananda, 1999; Vasudevan, Sadananda and Glinka, 2001; Sadananda, Vasudevan and Holtz, 2001; Sadananda and Vasudevan, 2003; 2004; 2005).

Within the category of partial crack closure models, the modified constitutive relation by McEvily and his co-workers was found to be able to account for various fatigue phenomena observed and valid from a physically small crack to long crack regimes (McEvily, *et al.*, 1999). Their main contributions are:

(1) to explain the elastic-plastic behavior of small cracks;

(2) to explain the variation of the crack closure level;

(3) to explain the transition from the threshold level to the endurance limit as a controlling parameter in the small crack growth regime.

Their modified constitutive relation for fatigue crack growth has been expressed as follows (Ishihara and McEvily, 1999),

$$\frac{\mathrm{d}a}{\mathrm{d}N} = AM^2 \tag{3.1}$$

$$M = K_{\max}(1-R) - \left(1-\mathrm{e}^{-ka}\right)\left(K_{\mathrm{op\,max}} - RK_{\max}\right) - \Delta K_{\mathrm{eff\,th}} \tag{3.2}$$

$$K_{\max} = \sqrt{\pi r_{\mathrm{e}}\left(\sec\frac{\pi}{2}\frac{\sigma_{\max}}{\sigma_Y}+1\right)} \cdot \left(1+Y(a)\sqrt{\frac{a}{2r_{\mathrm{e}}}}\right)\sigma_{\max} \tag{3.3}$$

where r_{e} is the size of an inherent flaw, a parameter whose magnitude is of the order of several microns in length (McEvily, *et al.*, 1999). Newly formed cracks are only significant when their lengths exceed r_{e}. a is the modified crack length which is equal to r_{e} plus the actual crack length. Thus the minimum value of a is r_{e}, which represents a component containing no crack (plain specimen). Eq. (3.11) was derived by combining Irwin's concept of the plastic zone (Irwin, 1960) with Dugdale's result (Dugdale, 1960). This formula was valid for elastic-perfectly-plastic materials and will be modified by us in the next chapter to account for other materials. σ_Y is the yield stress of the material, MPa; $\sigma_{\max}$ is the maximum stress in a loading cycle; R is the stress ratio ($=\sigma_{\min}/\sigma_{\max}$); $\Delta\sigma$ is the stress range applied, MPa. K is a material constant which reflects the rate of crack closure development with crack advance and is similar to that in other partial crack closure models (Kujawski, 2001a). Y is a geometrical factor. $K_{\mathrm{op\,max}}$ is the maximum SIF at the opening level for a macroscopic crack, $\mathrm{MPa}\cdot\mathrm{m}^{\frac{1}{2}}$. Both $K_{\mathrm{op\,max}}$ and $\Delta K_{\mathrm{eff\,th}}$ are functions of R (McEvily, *et al.*, 1999).

This modified constitutive relation was found to be able to account for various fatigue phenomena observed (McEvily, *et al.*, 1999; Ishihara and McEvily, 1999) and valid from physically small crack to long crack regimes (McEvily and Ishihara, 2001).

Their equations are only for near threshold and linear regions and cannot cover the unstable condition. Furthermore, because of the assumption of ideal elastic-plastic material, their definition of $K_{\max}$ implies that the stress will never exceed the yield stress σ_Y even at static failure for "uncracked" plain specimens. These restrictions were removed by Cui and his co-workers in a series of improvements (Cui and Huang, 2003; Wang, *et al.*, 2008; Wang and Cui, 2009), which will be specifically introduced in Chapter 5.

Sadananda and Vasudevan (2003) criticized the FCP methods based on the crack closure concept. They pointed out that the drawbacks in the current FLP methods stem from several sources: 1) the assumption of plasticity induced crack closure, 2) the lack of terms in the model that relate to the environmental effects and slip deformation behavior, and 3) several adjustable parameters needed to fit the observed data. In order to overcome these drawbacks and develop a more reliable FLP model, Vasudevan, Sadananda and their co-workers have proposed a two-parameter unified approach.

The crack growth rate expression for the two parameter model has not been explicitly given in many of their published papers. However, Vasudevan, *et al.* (1994) wrote it as

$$\frac{\mathrm{d}a}{\mathrm{d}N} = A\left(\Delta K - \Delta K^*_{\mathrm{th}}\right)^m \left(K_{\max} - K^*_{\max\,\mathrm{th}}\right)^p \tag{3.4}$$

and Sadananda and Vasudevan (1997) expressed it as

$$\frac{\mathrm{d}a}{\mathrm{d}N} = f\left[\left(\Delta K - \Delta K^*_{\mathrm{th}}\right), \left(K_{\max} - K^*_{\max\,\mathrm{th}}\right)\right] \tag{3.5}$$

These two can only cover near threshold and linear regions but cannot cover the unstable fracture condition.

In order to describe the sigmoidal shape of the crack growth rate curve, Bukkapatnam and Sadananda (2005) proposed the following expression,

$$\begin{aligned} &\frac{\mathrm{d}a}{\mathrm{d}N} = C_1 \Psi(\Delta K, K_{\max}) \\ &\Psi(\Delta K, K_{\max}) = (s_0 - 2\gamma(\Delta K, \Delta K_0, K_{\max}))K^n_{\max} \\ &\gamma(\Delta K, \Delta K_0, K_{\max}) = \frac{\mathrm{e}^{b(\Delta K_0 - \Delta K)^m}}{\mathrm{e}^{b(\Delta K_0 - \Delta K)^m} + \mathrm{e}^{b\left(\Delta\alpha_1 K_{\max} + \alpha_2 \Delta K + \alpha_3 \Delta K_0\right)^m}} \end{aligned} \tag{3.6}$$

Here, b determines the flatness of the cusp of the sigmoidal function, m is a real number between 0 and 1, and n is a real number between 1 and 5, ΔK_0 is the value of ΔK and s_0 is related to the value of $\frac{\mathrm{d}a}{\mathrm{d}N}$ at which the curve reaches the cusp before an accelerated fatigue growth takes place, and α_1, α_2, and α_3 determine the relative contributions of ΔK and $K_{\max}$. These parameters can be adjusted to accurately capture the relative influences of these two drivers. As stated in the foregoing, this structure offers flexibility in capturing the complicated fatigue crack growth profiles prevailing in certain material systems, as well as the multifarious environment-fatigue interactions. Furthermore, the structure can correctly capture the relative contributions of the two drivers ΔK and $K_{\max}$, under various loading conditions and environments.

Although their expression can capture the sigmoidal shape, the physical meanings of the parameters are unclear. We recommend writing the general expression of the two-parameter unified approach following the same idea as McEvily and Groeger (1977) as follows (Cui, *et al.*, 2007),

$$\frac{\mathrm{d}a}{\mathrm{d}N} = \frac{A(\Delta K - \Delta K^*_{\mathrm{th}})^m (K_{\max} - K^*_{\max\,\mathrm{th}})^p}{1 - \left(\dfrac{K_{\max}}{K_{\mathrm{ef}}}\right)^n} \tag{3.7}$$

This expression has been found to be able to explain many fatigue phenomena observed (Bian, *et al.*, 2009; Bian, *et al.*, 2010a; 2010b), especially the stress ratio effect.

The stress ratio effect is sometimes called the mean stress effect, especially in the approaches based on CFD theory. A unified approach to the effect of mean stress on the threshold conditions, *i.e.* fatigue limit and threshold SIF range, is presented by Kujawski and Ellyin (1995). The approach is based on the premise that damage caused by cyclic loading is a function of the strain energy range ΔW, *i.e.* a function of both stress and strain ranges. It is shown that the mean stress effect on the threshold conditions can be treated in a unified manner for members with and without crack-like flaws. The predictions of the proposed method are compared with experimental results and the agreement is found to be fairly good.

Fundamentally speaking, the effect of mean stress is the same as the load ratio effect. Thus, if a fatigue crack driving force, discussed above in this section, can explain the load ratio effect it can also explain the mean stress effect. Thus, such a type of treatment by Kujawski and Ellyin (1995) was an earlier one and now it can be merged in the same category of the load ratio effect. So this problem will not be discussed further.

3.4 Unified Approach for Long- and Physically Short-Crack Growth

The observations of Pearson (1984) and of Kitagawa and Takahashi (1976) have drawn attention to the fact that the fatigue crack growth behavior of short cracks differs in a non-conservative manner from expectations based upon long crack behavior. This behavior is illustrated in Fig. 3.1, in which the crack-growth rate, $\frac{da}{dN}$ or $\frac{dc}{dN}$, is plotted against the linear-elastic SIF range, ΔK. These results show that small cracks grow at ΔK levels even below the large-crack threshold and that they also grow faster than large cracks at the same ΔK level above the threshold. Small-crack effects have been shown to be more prevalent in tests that have compressive loads, such as negative stress ratios (Newman and Edwards, 1988; Edwards and Newman, 1990). Over the past decade, several authors have attempted to construct a unified model to explain both the physically short-crack growth and long-crack growth behavior.

Newman, *et al.* (1999) have shown that the small-crack theory developed by them based on the concept of plasticity induced crack-closure is able to explain this behavior. Crack-tip constraint factors, to account for 3D state-of-stress effects, were selected to correlate large-crack growth rate data as a function of the effective SIF range (ΔK_{eff}) under constant-amplitude loading. Some modifications to the ΔK_{eff} rate relations were needed in the near-threshold regime to fit measured small-crack growth rate behavior and fatigue endurance limits. The model was then used to calculate small and large-crack growth rates, and to predict total fatigue lives for notched and un-notched specimens made of two aluminum alloys and steel under constant-amplitude and spectrum loading. Fatigue lives were calculated using the crack-growth relations and microstructural features like those that initiated cracks for the aluminum alloys and steel for edge-notched specimens. An equivalent initial-flaw-size concept was used to calculate fatigue lives in other cases. Results from the tests and analyses agreed well.

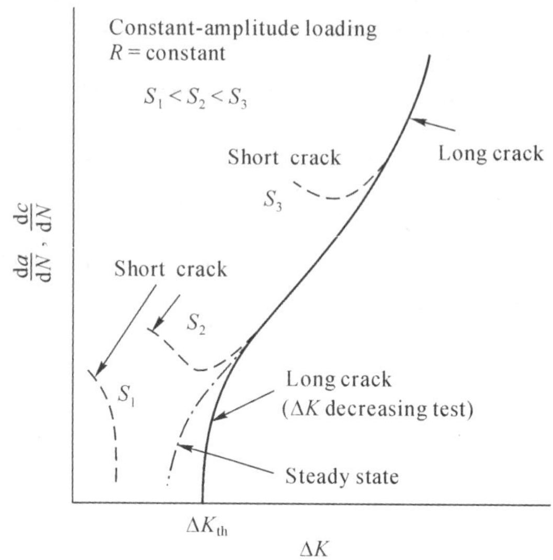

Fig. 3.1 Typical fatigue-crack-growth behavior for small and large cracks (Newman, *et al.*, 1999) (with the permission of IPC Science and Technology Press)

A new approach for correlation of load ratio effects on the long- and physically short-crack growth rate in aluminum alloys is also proposed by Kujawski (2001b). The approach is based on a postulate that an apparent effectiveness of the crack closure phenomenon for long cracks can be different from that for physically short cracks. For long-crack growth behavior the enhanced partial crack closure model is adopted while the conventional Elber's closure model for physically short-crack growth is used. Their approach was shown to provide a fairly good correlation to experimental data taken from the literature for three aluminum alloys under various load ratios ranging from –1 to 0.7.

McEvily and his co-workers thought that it is impossible to explain the small crack growth behavior through the use of a traditional LEFM approach because of the following reasons: 1) On going from the micro-crack growth range to the macro-crack growth range, fatigue crack growth behavior undergoes a transition from stress control to stress intensity factor control. 2) Because of the large plastic zone size to crack length ratio in the small crack growth regime, crack growth is elastic-plastic in nature rather than linear-elastic, and 3) the crack closure level increases from zero up to the level associated with a macroscopic fatigue crack as the fatigue crack increases in length in the small crack regime (McEvily and Ishihara, 2001; Ishihara and McEvily, 2002). They have shown that their modified expression of Eq. (3.9) is able to explain both short and long crack growth behavior.

Modifications are introduced to account for the differences in crack growth behavior of long and short cracks (Ramsamooj and Shugar, 2001; Ramsamooj, 2003) that permit the use of the SIF. These modifications stem from the principles of fracture mechanics for small- and large-scale yielding. Short cracks can grow well below the long crack fatigue threshold range, because the short crack fatigue threshold range is smaller than that for long cracks as it is dependent on the stress level and the plastic constraint factor. Analytical expressions are developed for these relationships and for the fatigue crack growth rates in plane stress and plane strain for short semi-elliptical cracks including those emanating from notches. Microstructural features are not considered. A linear approximation is used for the gradual transition from plane stress to plane strain. The model is formulated using only the readily available material properties. An extensive amount of published fatigue data for a wide range of metals and metal alloys is used to validate the new analytical expression for the crack growth rate. The agreement between the predictions of the new model and the experimental data is fair to excellent.

Short crack growth behavior has also been examined using the two parametric approaches developed earlier by Sadananda and Vasudevan (1997). It is commonly accepted that there is a lack of similitude in the description of the short crack growth behavior. Contrary to this understanding, it is shown that there is no anomaly in the short crack growth behavior. The apparent anomaly arises because of ignoring the second parameter associated with the threshold $K_{\max\ \mathrm{th}}$, and the existence of internal stresses in the crack tip field where the short cracks nucleate and grow. In the case of short cracks nucleating from notches, the internal stresses can be pre-existing or can be generated *in situ* if they form at a free surface. In the latter case, formation of intrusions and extrusions at the PSBs are the precursors for the crack nucleation providing the necessary internal stresses. Thus, short cracks grow under a total force, consisting of both internal and applied stresses, satisfying the same two thresholds for long cracks. Examples are provided from the literature to illustrate the concepts. It is shown that the internal stresses decrease as a short crack grows out of the existing stress field. These internal stresses can be predicted using elastic or elastic-plastic continuum approximations. Application of this concept is extended to understand the role of residual stresses and the transformation induced internal stresses on the crack growth. Based on the analysis, they restated the similitude concept as: *equal crack tip forces result in equal crack growth rates for the same crack growth mechanism, provided all the contributing forces are taken into consideration.*

An attempt is also made by Pugno, *et al.* (2007) for a unified treatment of short and long cracks. The basic idea behind their method is an "intermediate asymptotic matching" of the well-known empirical fatigue laws by Wöhler and Paris, obtained by applying quantized, or finite, fracture mechanics concepts. "Universal" fatigue laws are thus derived and they are introduced in the next section.

3.5 Unified Approach for Initiation and Propagation

From the fatigue mechanisms discussed in Chapter 2, one should be aware that the whole fatigue process can be divided into several stages. These different stages can be roughly grouped as crack initiation and crack propagation.

The main idea of unifying initiation and propagation is that it is assumed that flaws always exist in engineering materials and components and fatigue is purely a crack propagation process. Thus, the problem of unifying initiation and propagation is the same as unifying physically short cracks and long cracks. These methods have been introduced in the previous section.

Traditionally, CFD theory is used to treat the fatigue crack initiation problem using the famous Wöhler's laws while FCP theory is used to treat the FCP problem using the famous Paris' laws. Generalizations and unification of the celebrated Paris' and Wöhler's laws for FCP were derived by Carpinteri and his co-workers (Pugno, *et al.*, 2006a; 2006b; Pugno, *et al.*, 2007; Ciavarella, *et al.*, 2008; Carpinteri and Paggi, 2009; 2010) by applying the recently developed quantized (or finite) fracture mechanics. In particular, three generalized Paris', Wöhler's or unified laws have been proposed and compared, demonstrating their applicability for predicting the life time of structures containing from small (the Wöhler's regime) to large (the Paris' regime) propagating fatigue cracks.

A comparison between the three "universal" laws was given in the paper (Pugno, *et al.*, 2007) as shown in Fig. 3.2. All the three unified laws match asymptotically the Wöhler's (small crack) and Paris' (large crack) regimes. This clearly shows the consistency of the proposed "universal" laws, as "intermediate asymptotic matching" of the well-known empirical fatigue limit laws by Wöhler and Paris. Consequently, their laws must automatically match the empirical observations. Note that in the unified Wöhler's law $\Delta a<<2a_C$ remains a free best-fit parameter that can be fixed to recover the slope of the Paris' law for $a\rightarrow a_C$.

It is clear that the Wöhler's prediction becomes unreasonable for large cracks, whereas the Paris' prediction fails for describing short crack growth, see Fig. 3.2. On the contrary, their universal laws based on quantized or (finite) fracture mechanics represent their unification, useful for treating fatigue growth of cracks, from short to long sizes.

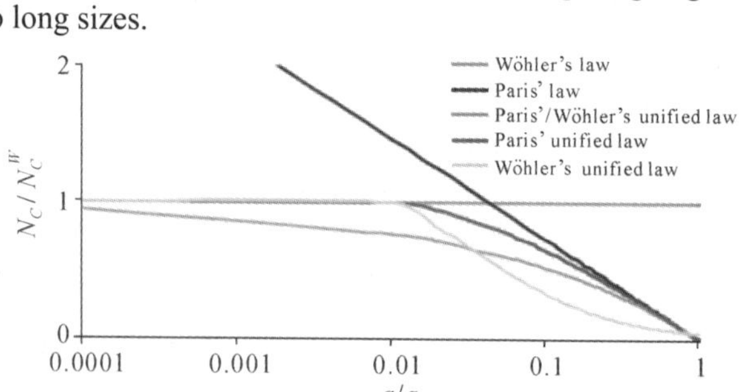

Fig. 3.2 Comparison between different fatigue laws: classical Paris' and Wöhler's laws and the

proposed new universal fatigue laws (Griffith's case, with $C=\bar{C}=\Delta\sigma_2=1$; $m=\bar{m}=2$) (Pugno, *et al.*, 2007) (with the permission of Pergamon)

These advances in understanding the complex phenomenon of fatigue crack growth shed a new light on the possibility to unify the CFD and the FCP approaches, and to solve the challenging task of interpreting the Paris' and Wöhler's power-law regimes within a unified theoretical framework.

3.6 Unified Approach for High and Low Cycle Fatigue

In the past, the low cycle fatigue was often treated with the Manson-Coffin relationship (Manson and Hirschberg, 1964; Coffin and Tavernelli, 1959) and its various modifications (Muralidharan and Manson, 1988; Park and Song, 1995; Roessle and Fatemi, 2000) while high cycle fatigue was treated with *S-N* curved approaches. A method was presented by Ballio and Castiglioni (1995) trying to unify both design and damage assessment methods for high and low cycle fatigue. The basic idea of the method is that by interpreting the stress range $\Delta\sigma$ as the ideal stress range associated to the real strain range $\Delta\varepsilon$ in an ideal perfectly elastic material, high and low cycle fatigue test data can be interpreted by the same Wöhler (*S-N*) lines usually given in recommendations for (high cycle) fatigue design of steel structures. Their conclusions are based on the fact that their scatter is of the same order as usual scatter for the linear cumulative damage theory. However, this scatter is beyond the acceptable range of engineering requirements.

As a matter of fact, most of the unified methods described above can handle both the situation of high cycle fatigue and low cycle fatigue and they can be viewed as the unified approach. The key point for the success of a method is the modification of the stress intensity factor formula by taking account of the material nonlinearity for low cycle fatigue. The McEvily formula, Eq. (3.11), is an example.

3.7 Unified Approach for Fatigue and Creep

For some metals, such as aluminium and titanium, damage accumulation in materials may not only be induced by fatigue but also by creep. The problem becomes more challenging if both types of damage are accounted for.

A general cumulative damage methodology is derived by Christensen (2008) from the basic relation specifying crack growth rate (increment) as a power law function of the SIF. The crack is allowed to grow up to the point at which it becomes unstable, thereby determining the lifetime of the material under the prescribed stress program. The formalism applies for the case of creep to failure under variable stress history, as well as for cyclic fatigue to failure under variable stress amplitude history. The formulation is calibrated by the creep rupture lifetimes at constant stress or the fatigue cycle lifetimes at constant stress amplitude. No empirical (non-physical) parameters are involved in the basic formulation; everything is specified in terms of experimentally determined quantities. Their method is introduced here.

For crack growth under time varying loading, the widely used power law form expressing the crack growth rate as a function of the SIF is

$$\dot{a}(t) = \lambda\left(\sigma(t)\sqrt{a(t)}\right)^{r} \tag{3.8}$$

where crack size $a(t)$ and stress $\sigma(t)$ are functions of time t, r is the power law exponent and λ is a constant.

Separate the variables in Eq. (3.8) and integrate to obtain

$$a^{1-\frac{r}{2}}(t) - a_0^{1-\frac{r}{2}} = \lambda\left(1-\frac{r}{2}\right)\int_0^t \sigma^r(\tau)\mathrm{d}\tau \tag{3.9}$$

where a_0 is the initial crack size and $\sigma(\tau)$ is the given stress history.

Rewrite Eq. (3.9) as

$$\left(\frac{a}{a_0}\right)^{1-\frac{r}{2}} - 1 = \lambda\left(1-\frac{r}{2}\right)a_0^{\frac{r}{2}-1}\int_0^t \sigma^r(\tau)\mathrm{d}\tau \tag{3.10}$$

At this point, let us adopt a more general approach and take the crack size at failure instability as

$$\frac{a}{a_0} = F\left(\frac{\sigma}{\sigma_i}\right), \text{ at failure} \tag{3.11}$$

where σ_i is the instantaneous static strength which characterizes failure for the virgin material with initial crack size a_0. $F(\ )$ is some function yet to be determined.

Combining Eq. (3.10) and Eq. (3.11) gives

$$1 - F^{1-\frac{r}{2}}\left(\frac{\sigma}{\sigma_i}\right) \geq \lambda\left(\frac{r}{2}-1\right)a_0^{\frac{r}{2}-1}\int_0^t \sigma^r(\tau)\mathrm{d}\tau \tag{3.12}$$

The inequality in Eq. (3.12) applies for times that are less than the failure time wherein the crack has not yet reached the critical size for failure. The equality in Eq. (3.12) is at the time then given by t.

It is convenient to express Eq. (3.12) in terms of non-dimensional variables. Let non-dimensional stress be given by

$$\tilde{\sigma} = \frac{\sigma}{\sigma_i} \tag{3.13}$$

and non-dimensional time by

$$\tilde{t} = \frac{t}{t_1} \tag{3.14}$$

where

$$t_1 = \frac{1}{\lambda\left(\frac{r}{2}-1\right)a_0^{\frac{r}{2}-1}\sigma_i^r} \tag{3.15}$$

Hereafter t_1 will be treated as a quantity to be determined from data. On $\log\tilde{\sigma}$ versus $\log\tilde{t}$ scales, different values of σ_i give shifts in the vertical direction while t_1 gives shifts in the horizontal direction.

With Eqs. (3.13)–(3.15) the form Eq. (3.12) becomes

$$1 - F^{1-\frac{r}{2}}(\tilde{\sigma}) \geq \int_0^{\tilde{t}} \tilde{\sigma}(\tau)\mathrm{d}\tau \tag{3.16}$$

Now, let

$$f(\tilde{\sigma}) = 1 - F^{1-\frac{r}{2}}(\tilde{\sigma}) \tag{3.17}$$

giving Eq. (3.16) as

$$\int_0^{\tilde{t}} \sigma^r(\tau)\mathrm{d}\tau \leq f(\tilde{\sigma}(\tilde{t})) \tag{3.18}$$

where the equality applies at the failure event giving the lifetime. Determine $f(\tilde{\sigma})$ from the basic creep rupture behavior at constant stress. Take the given spectrum of creep rupture life times as

$$\tilde{t} = t_c(\tilde{\sigma}), \text{at constant stress} \tag{3.19}$$

$$f(\tilde{\sigma}) = \tilde{\sigma}^r t_c(\tilde{\sigma}) \tag{3.20}$$

As the final step in this procedure, substitute Eq. (3.20) into Eq. (3.18) and write the inequality and equality in separate forms as

$\tilde{t}$ <Lifetime

$$\underbrace{\int_0^{\tilde{t}} \sigma^r(\tau)\mathrm{d}\tau}_{\text{Actual crack size at time } \tilde{t}} < \underbrace{\tilde{\sigma}^r(t)\tilde{t}_c(\sigma(\tilde{t}))}_{\text{Critical crack size for stress at time } \tilde{t}} \tag{3.21}$$

$\tilde{t}$ =Lifetime

$$\int_0^{\tilde{t}} \tilde{\sigma}^r(\tau)\mathrm{d}\tau = \tilde{\sigma}(\tilde{t})\tilde{t}_c(\tilde{\sigma}(\tilde{t})) \tag{3.22}$$

As noted under Eq. (3.21), the actual crack size has not yet reached the critical size. Thereafter, the crack does reach the critical size and the lifetime $\tilde{t}$ is determined from Eq. (3.22).

Next the problem of determining exponent r in Eq. (3.22) is taken up. Rarely does one have direct data on the crack growth rate for a particular material. It would be advantageous to be able to determine r directly from $\tilde{t}_c(\tilde{\sigma})$. It is assumed that exponent r in Eq. (3.22) is determined by the inverse slopes in power law regions, as shown in the creep $\log\tilde{\sigma} \sim \log\tilde{t}$ diagram.

The cumulative damage relation Eq. (3.21) and the lifetime relation Eq. (3.22), along with the determination of r from $\tilde{t}_c(\tilde{\sigma})$, are the main results of this derivation. It is to be expected that the critical crack size at time $\tilde{t}$ in Eq. (3.22) depends upon the stress level at that time. Lower stress levels require larger crack sizes.

There is a special form for creep $\tilde{t}_c(\tilde{\sigma})$ that affords a considerable advantage in applications. This is given by

$$\tilde{t}_c = \frac{\left(1-\tilde{\sigma}^p\right)^q}{\tilde{\sigma}^r} \tag{3.23}$$

where p, q and r are parameters with r being that for the power law range. With Eq. (3.23) the form Eq. (3.22) becomes

$$\frac{1}{\left[1-\tilde{\sigma}^p(\tilde{t})\right]^q}\int_0^{\tilde{t}} \tilde{\sigma}(\tau)\mathrm{d}\tau = 1 \tag{3.24}$$

The previous results for the creep damage type of cumulative effect can be converted to the case of cyclic fatigue by suitable notation and terminology changes.

Let f be the frequency in cycles per unit time and n be the number of cycles.

Then the elapsed time is

$$t = \frac{n}{f} \tag{3.25}$$

Some materials show a frequency dependence, so the present method will be developed for a fixed frequency and fixed form of the variation within one cycle. The starting point is to define and take as given the standard fatigue relation between constant stress $\tilde{\sigma}$ and the number of cycles to fatigue.

Write this as

$$n = N(\tilde{\sigma}), \quad \text{constant stress} \tag{3.26}$$

where n represents the cycles to failure and $\tilde{\sigma}$ is the non-dimensional stress $\tilde{\sigma} = \frac{\sigma}{\sigma_i}$, where σ_i is the instantaneous static strength corresponding to the maximum tensile stress within the cycle of variation. With these notational changes, the previous lifetime damage form Eq. (3.22) converts to the cyclic fatigue case as

$$\int_0^n \tilde{\sigma}(\eta)\mathrm{d}\eta = \tilde{\sigma}^r(n)N(\tilde{\sigma}^r(n)) \tag{3.27}$$

where stress is allowed to have a variable amplitude, so long as the wave form and frequency are unchanged. Variable η in Eq. (3.27) is the past history variable for n. Exponent r in Eq. (3.27) is determined by the same method as in the creep damage case.

The term on the left side of Eq. (3.27) represents the current crack size while the term on the right side is the critical size at the stress existing at n number of cycles. The equality in Eq. (3.27) is for the lifetime n in number of cycles.

Examples:

The creep rupture form given by Eq. (3.23) will be used with parameter values p=15, q=100, r=10. For two stress levels as shown in Fig. 3.3, the present form Eq. (3.24) gives

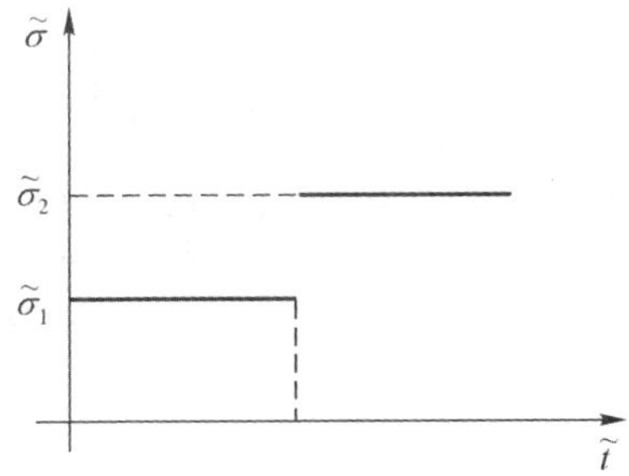

Fig. 3.3 Two step load

$$\tilde{t} = \tilde{t}_1 + \frac{\left(1-\tilde{\sigma}_2^p\right)^q}{\tilde{\sigma}_2^r} - \tilde{t}_1\left(\frac{\tilde{\sigma}_1^r}{\tilde{\sigma}_2^r}\right) \tag{3.28}$$

Example 1:
$\tilde{\sigma}_1 = 0.5$ up to $\tilde{t}_1 = 500$
Then $\tilde{\sigma}_2 = 0.7$ to failure
Lifetimes: $\tilde{t} = 504.7$.
Example 2:
$\tilde{\sigma}_1 = 0.85$ up to $\tilde{t}_1 = 0.0005$
Then $\tilde{\sigma}_2 = 0.5$ to failure
Lifetimes: $\tilde{t} = 1020.8$.

Now let us consider some cases where, at a constant load, the material is taken right up to the point of incipient failure. But just an instant before failure the load is reduced to a specified level. The problem is to determine the remaining lifetime at the reduced load level.

Take $\tilde{\sigma}_1 = 0.6$ up to incipient failure and then reduce the stress levels to $\tilde{\sigma}_2 = 0.5, 0.4$ or 0.3, and find the remaining total lifetimes.

At $\tilde{\sigma}_1 = 0.6$ the creep rupture lifetime is $\tilde{t}_c = 157.8$. Using expression Eq. (3.28) the present lifetime extensions are found as

$$\tilde{\sigma}_2 = 0.5,\ \tilde{t} = \tilde{t}_c + \Delta\tilde{t} = 201.7$$
$$\tilde{\sigma}_2 = 0.4,\ \tilde{t} = \tilde{t}_c + \Delta\tilde{t} = 594.9$$
$$\tilde{\sigma}_2 = 0.3,\ \tilde{t} = \tilde{t}_c + \Delta\tilde{t} = 7938.0$$

In the third case, the lifetime is extended by a factor of 50.

It must be pointed out here that this model is a mixture of cumulative damage theory and FCP theory. In his model, the basic failure property inputs are the creep rupture life times $\tilde{t}_c(\tilde{\sigma})$ in Eq. (3.22) and fatigue lifetimes $N(\tilde{\sigma})$ in Eq. (3.27) both at constant stress amplitude. There are no empirical parameters involved in the theory and final lifetime forms.

In the original paper (Christensen, 2008) several examples are given showing the inadequacy of linear cumulative damage while their nonlinear damage accumulation method overcomes these deficiencies. The general conclusion is that this new cumulative damage formalism is no more difficult to apply than linear cumulative damage, but it has a physical interpretation based upon crack growth and it readily admits generalization to probabilistic conditions.

3.8 Basic Ideas of Our UFLP Method

The basic idea of our unified method is that initial defects always exist in engineering marine structures and fatigue is purely a crack propagation problem. This assumption has been confirmed due to two progresses: 1) FM has been successfully applied at the microstructural level and 2) microstructural level cracks can now be measured. However, for the purpose of easy application, we confine our model to the macroscopic level. No microstructural parameters are involved. In FCP theory, the crack growth rate curve is the most fundamental material property to be defined and many factors will influence the crack growth rate curve. If the method can be regarded as a UFLP method, it must be able to explain all the phenomena observed. However, a complete list of significant phenomena is not available as the fatigue failure mechanism has not been fully understood although it has been studied for more than 170 years. Therefore, some of the phenomena mentioned in the literature may be in conflict with each other. For example, the significance of the crack closure has been debated all the time since its discovery or invention. A UFLP method should be able to explain most of these experimentally observed phenomena. In order to achieve this purpose, the most important factors which affect fatigue behavior must be included in the model. From the description of the fatigue mechanism, it is clear that many factors affect the fatigue life of metal structures. These factors can be categorized into four groups (Cui, 2002): 1) factors related to loading history ($\underline{L}$); 2) factors related to material ($\underline{M}$); 3) factors related to structure ($\underline{S}$); and 4) factors related to the environment ($\underline{E}$). So, in general, the crack growth rate curve can be written as (Cui, *et al.*, 2011),

$$\frac{\mathrm{d}a}{\mathrm{d}N} = f(\underline{L}, \underline{M}, \underline{S}, \underline{E}) \tag{3.29}$$

where $\underline{L}$, $\underline{M}$, $\underline{S}$, $\underline{E}$ represent influence factors due to loading, material, structure and environment respectively. Strictly speaking, if a method can be regarded as a UFLP method, it must be able to explain all the phenomena observed. In order to achieve this purpose, all the factors which affect the fatigue behavior must be included in the crack growth rate model. This is an impossible task. We have to approach it using empirical methods and including several most significant factors. The less the parameters are used, the easier the method is applied. So every empirical crack growth rate curve is a compromise of such an effort. Up to now hundreds of different empirical crack growth rate curves have been proposed and each has its advantages and drawbacks. Since every empirical model only includes a few parameters, it cannot be expected to explain every phenomenon. In that sense every model is not a unified method but the unified approach can be regarded as the long-term target. So the essence of developing a UFLP is to establish a "correct" crack growth rate relation. This will be discussed in Chapter 5.

If the crack growth rate relation is known then, for a given loading history, the accumulated crack length is calculated on a cycle-by-cycle basis. This could accurately account for many loading effects such as mean stress effect, load sequence effect and structural defects such as notch effect and size effect.

By using the unified approach on our terms, the same method should be able to cover both LCF and HCF. The fatigue limit and the ultimate strength are the two extremes of FCP. Various anomalous FCP phenomena must be explained by the method. In this book, the effect of frequency has not been considered but the principle should be the same. If one extends the crack growth rate relation to account for the stress frequency effect, the unified approach can also be extended to dynamic fatigue problems (Plekhov, *et al.*, 2011).

So, in applying the unified procedure for FLP, the basic idea is to calculate the crack size on a cycle-by-cycle basis and the fatigue loading must be given as a loading history rather than spectrum. Fig. 3.4 shows the general flow diagram of the UFLP method.

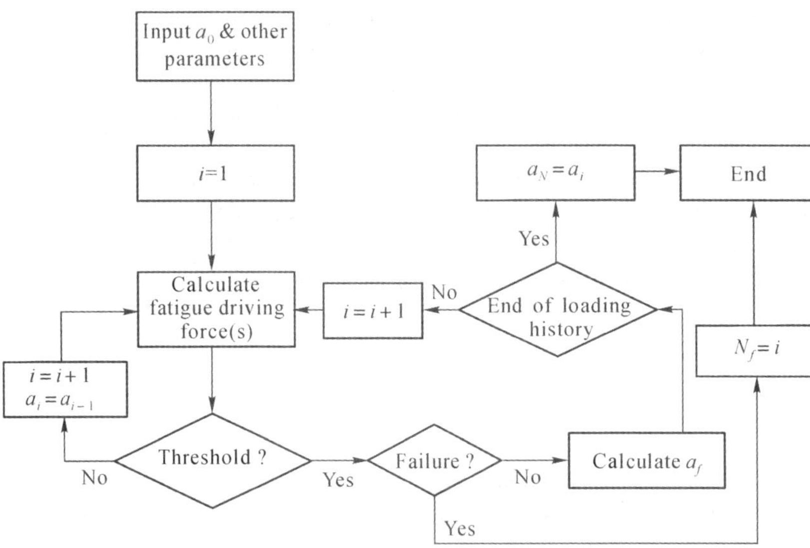

Fig. 3.4 The flow diagram of the UFLP method

3.9 Summary

Now, more and more people have realized that only an FLP method based on FCP theory can overcome the deficiencies existing in CFD theory. FCP theory has the potential to explain various observed fatigue phenomena. However, it is not obvious that any type of FCP theory can work. As a matter of fact, from the very fundamental question of fatigue crack driving forces to the more complicated fatigue crack growth rate expression, all need critical examination. In the past two decades, many authors have made some efforts in developing a UFLP method based on FCP theory. In this chapter, various approaches to the unification have been summarized. However, most of them are specific to a particular concept and do not represent the most general one.

By using the unified approach on our terms, the same method must be able to cover both LCF and HCF and to explain all the observed fatigue phenomena. The fatigue limit and the ultimate strength are the two extremes of FCP. Various anomalous FCP phenomena must be explained by the method. This will be presented in Chapter 5.

References

Ballio, G. & Castiglioni, C. A. (1995). "A unified approach for the design of steel structures under low and/or high cycle fatigue", Journal of Constructional Steel Research, 34: 75–101.

Bian, R., Cui, W. C. & Wan, Z. Q. (2009). "Fatigue crack propagation of the notched specimen under cyclic compression-compression loading based on the two-parameter unified approach", Journal of Ship Mechanics, 13(5): 734–738 (in Chinese).

Bian, R., Cui, W. C., Wan, Z. Q. & Huang, X. P. (2010a). "A method for the evaluation of the two thresholds based on the two-parameter unified approach", Journal of Ship Mechanics, 14(1–2): 94–100 (in Chinese).

Bian, R., Cui, W. C., Wan, Z. Q. & Huang, X. P. (2010b). "Effects of initial cracks and loading sequence on fatigue crack growth of the deepwater structures based on two-parameter unified approach", Journal of Ship Mechanics, 14(5): 516–525 (in Chinese).

Bukkapatnama, S. T. S. & Sadananda, K. (2005). "A genetic algorithm for unified approach-based predictive modeling of fatigue crack growth", International Journal of Fatigue, 27: 1354–1359.

Carpinteri, A. (ed.) (1994). Handbook of Fatigue Crack Propagation in Metallic Structures (Vols. 1 and 2). Amsterdam: Elsevier Science Publishers.

Carpinteri, A. & Paggi, M. (2009). "A unified interpretation of the power laws in fatigue and the analytical correlations between cyclic properties of engineering materials", International Journal of Fatigue, 31: 1524–1531.

Carpinteri, A. & Paggi, M. (2010). "A unified fractal approach for the interpretation of the anomalous scaling laws in fatigue and comparison with existing models", International Journal of Fracture, 161: 41–52.

Christensen, R. M. (2008). "A physically based cumulative damage formalism", International Journal of Fatigue, 30: 595–602.

Ciavarella, M., Paggi, M. & Carpinteri, A. (2008). "One, no one, and one hundred thousand crack propagation laws: a generalized Barenblatt and Botvina dimensional analysis approach to fatigue crack growth", Journal of the Mechanics and Physics of Solids, 56: 3416–3432.

Coffin, L. F. & Tavernelli, J. F. (1959). The cyclic straining and fatigue of metals, Trans of the Metallurgical Society of American Institute of Mechanical Engineers, 215: 794.

Crupi, V. (2008). "A unifying approach to assess the structural strength", International Journal of Fatigue, 30: 1150–1159.

Cui, W. C. (2002). "A state-of-the-art review on fatigue life prediction methods for metal structures", Journal of Marine Science and Technology, 7(1): 43–56.

Cui, W. C. & Huang X. P. (2003). "A general constitutive relation for fatigue crack growth analysis of metal structures", Acta Metallurgica Sinica (English Letters), 16(5): 342–354.

Cui, W. C., Bian, R. G. & Liu, X. C. (2007). "Application of the two-parameter unified approach for fatigue life rediction of marine structures", in: Basu, R. I., Belenky, V., Wang, G. & Yu, Q. (eds.) Proceedings of the 10th International Symposium on Practical Design of Ships and Other Floating Structures, ABS, 2: 1064–1070.

Cui, W. C., Wang, F. & Huang, X. P. (2011). "A unified fatigue life prediction method for marine structures", Marine Structures, 24(2): 153–181.

Donahue, R. J., Clark, H. M., Atanmo, P., Kumble, R. & McEvily, A. J. (1972). "Crack opening displacement and the rate of fatigue crack growth", International Journal of Fracture Mechanics, 8: 209–219.

Donald, K. & Paris, P. C. (1999). "An evaluation of ΔK_{eff} estimation procedure on 6061-T6 and 2024-T3 aluminum alloys", International Journal of Fatigue, 21: S47–57.

Dugdale, D. S. (1960). "Yielding of steel sheets contain slits", Journal of the Mechanics and Physics of Solids, 8: 100–108.

Edwards, P. & Newman, Jr. J. C. (eds.). (1990). "Short-crack growth behaviour in various aircraft materials", AGARD-R-767. Paris: Neuilly-sur-Seine.

Elber, W. (1970). "Fatigue crack closure under cyclic tension", Engineering Fracture Mechanics, 2: 37–45.

Fleck, N. A., Shin, C. S. & Smith, R. A. (1985). "Fatigue crack growth under compressive loading", Engineering Fracture Mechanics, 21(1): 173–185.

Forman, R. G., Kearney, V. E. & Engle, R. M. (1967). "Numerical analysis of crack propagation in cyclic-loaded structures", Journal of Basic Engineering, 89: 459–464.

Hertzberg, R. W., Newton, C. H. & Jaccard, R. (1988). "Crack closure: correlation and confusion", in: Mechanics of Fatigue Crack Closure, ASTM STP 982. Philadelphia, PA: American Society for Testing and Materials, 139–148.

Irwin, G. R. (1960). "Fracture mode transition for a crack traversing a plate", Journal of Basic Engineering, 82: 417–425.

Ishihara, S. & McEvily, A. J. (1999). "A coaxing effect in the small fatigue crack growth regime", Scripta Materialia, 40(5): 617–622.

Ishihara, S. & McEvily, A. J. (2002). "Analysis of short fatigue crack growth in

cast aluminum alloys", International Journal of Fatigue, 24: 1169–1174.
Kanninen, M. F. & Popelar, C. H. (1985). Advanced Fracture Mechanics. New York: Oxford University Press.
Kitagawa, H. & Takahashi, S. (1976). "Applicability of fracture mechanics to very small cracks or cracks in the early stage", in: Proceedings of the Second International Conference on Mechanical Behaviour of Materials, Boston, MA, ASM, 627–631.
Kujawski, D. & Ellyin, F. (1995). "Unified approach to mean stress effect on fatigue threshold conditions", International Journal of Fatigue, 17: 101–106.
Kujawski, D. (2001a). "Enhanced model of partial crack closure for correlation of *R*-ratio effects in aluminum alloys", International Journal of Fatigue, 23: 95–102.
Kujawski, D. (2001b). "Correlation of long- and physically short-cracks growth in aluminum alloys", Engineering Fracture Mechanics, 68: 1357–1369.
Kujawski, D. (2001c). "A new $(\Delta K^{+} K_{max})^{0.5}$ driving force parameter for crack growth in aluminum alloys", International Journal of Fatigue, 23: 733–740.
Kujawski, D. (2001d). "A fatigue crack driving force parameter with load ratio effects", International Journal of Fatigue, 23: S239–246.
Lo, K. W., Tamilselvan, T., Chua, K. H. & Zhao, M. M. (1996). "A unified model for fracture mechanics", Engineering Fracture Mechanics, 54(2): 189–210.
Manson, S. S. & Hirschberg, M. H. (1964). Fatigue: an Interdisciplinary Approach. Syracuse, New York: Syracuse University Press, 133.
McEvily, A. J. & Groeger, J. (1977). "On the threshold for fatigue-crack growth", 4th International Conference on Fracture. Waterloo, Canada: University of Waterloo Press, 2: 1293–1298.
McEvily, A. J. & Ishihara, S. (2001). "On the dependence of the rate of fatigue crack growth on the $\sigma_a^n(2a)$ parameter", International Journal of Fatigue, 23: 115–120.
McEvily, A. J., Bao, H. & Ishihara, S. (1999). "A modified constitutive relation for fatigue crack growth", Fatigue'99, 329–336.
Muralidharan, U. & Manson, S. S. (1988). "A modified universal slopes equation for estimation of fatigue characteristics of metals", Journal of Engineering Materials and Technology—Transactions of the ASME, 110: 55–58.
Newman, Jr. J. C. & Edwards, P. (1988). "Short-crack growth behaviour in an aluminium alloy—an AGARD cooperative test programme", AGARD R-732. Paris: Neuilly-sur-Seine.
Newman, Jr. J. C., Phillips, E. P. & Swain, M. H. (1999). "Fatigue life prediction methodology using small-crack theory", International Journal of Fatigue, 21: 109–119.
Paris, P. C. & Erdogan, F. (1963). "A critical analysis of crack propagation laws", Journal of Basic Engineering, 85: 528–534.
Paris, P. C., Gomez, M. P. & Anderson, W. P. (1961). "A rational analytical theory of fatigue", The trend in Engineering, 13: 9–14.
Park, J. H. & Song, J. H. (1995). "Detailed evaluation of methods for estimation

of fatigue properties", International Journal of Fatigue, 17(5): 365–373.
Pearson, S. (1984). "Fatigue crack propagation in metals", Engineering Fracture Mechanics, 7: 251.
Plekhov, O., Paggi, M., Naimark, O. & Carpinteri, A. (2011). "A dimensional analysis interpretation to grain size and loading frequency dependencies of the Paris and Wöhler curves", International Journal of Fatigue, 33: 477–483.
Pugno, N., Ciavarella, M., Cornetti P. & Carpinteri A. (2006a). "A unified law for fatigue crack growth", Journal of the Mechanics and Physics of Solids, 54: 1333–1349.
Pugno, N., Ciavarella, M., Cornetti, P. & Carpinteri, A. (2006b). "A generalized Paris' law for fatigue crack growth", Journal of the Mechanics and Physics of Solids, 54: 1333–1349.
Pugno, N., Cornetti, P. & Carpinteri, A. (2007). "New unified laws in fatigue: from the Wöhler's to the Paris' regime", Engineering Fracture Mechanics, 74: 595–601.
Ramsamooj, D. V. (2003). "Analytical prediction of short to long fatigue crack growth rate using small- and large-scale yielding fracture mechanics", International Journal of Fatigue, 25: 923–933.
Ramsamooj, D. V. & Shugar, T. A. (2001). "Model prediction of fatigue crack propagation in metal alloys in laboratory air", International Journal of Fatigue, 23: S287–300.
Roessle, M. L., & Fatemi, A. (2000). "Strain-controlled fatigue properties of steels and some simple approximations", International Journal of Fatigue, 22: 495–511.
Sadananda, K. & Vasudevan, A. K. (1997). "Short crack growth and internal stresses", International Journal of Fatigue, 19(93): S99–108.
Sadananda, K., Vasudevan, A. K. & Holtz, R. L. (2001). "Extension of the unified approach to fatigue crack growth to environmental interactions", International Journal of Fatigue, 23: S277–286.
Sadananda, K. & Vasudevan, A. K. (2003). "Fatigue crack growth mechanisms in steels", International Journal of Fatigue, 25: 899–914.
Sadananda, K. & Vasudevan, A. K. (2004). "Crack tip driving forces and crack growth representation under fatigue", International Journal of Fatigue, 26: 39–47.
Sadananda, K. & Vasudevan, A. K. (2005). "Fatigue crack growth behavior of titanium alloys", International Journal of Fatigue, 27: 1255–1266.
Vasudevan, A. K. & Sadananda, K. (1999). "Application of unified fatigue damage approach to compression-tension region", International Journal of Fatigue, 21(s1): S263–273.
Vasudevan, A. K. & Sadananda, K. (2001). "Analysis of fatigue crack growth under compression-compression loading", International Journal of Fatigue, 23: S365–374.
Vasudevan, A. K., Sadananda, K. & Glinka, G. (2001). "Critical parameters for fatigue damage", International Journal of Fatigue, 23: S39–53.
Vasudevan, A. K., Sadananda, K. & Louat, N. (1994). "A review of crack closure,

fatigue crack threshold and related phenomena", Materials Science and Engineering, A188: 1–22.

Wang, F. & Cui, W. C. (2009). "Approximate method to determine the model parameters in a new crack growth rate model", Marine Structures, 22(4): 744–757.

Wang, Y., Cui, W., Wu, X., Wang, F. & Huang, X. (2008). "The extended McEvily model for fatigue crack growth analysis of Metal Structures", International Journal of Fatigue, 30: 1851–1860.

Zhang, J., He, X. D., Sha, Y. & Du, S.Y. (2010). "The compressive stress effect on fatigue crack growth under tension-compression loading", International Journal of Fatigue, 32: 361–367.

4

Basic Concepts of Fracture Mechanics

4.1 Introduction

The mechanical behavior of cracked bodies is typically analyzed using the theory of FM, which came into being as a special methodology when people found that the presence of cracks can obviously weaken the strength of the component and eventually lead to catastrophic failure of the structure. Thus FM deals with the study of how a crack or flaw in a structure propagates under applied loads.

The modern era of fracture mechanics originated with the work of Griffith (1920) who resolved the infinite crack-tip stress dilemma inherent in the use of the theory of elasticity for cracked structures. The second major contribution to the subject was made independently by Irwin (1948) and Orowan (1948) who extended Griffith's approach to metals by including the energy dissipated by local plastic flow. During this same period, Mott (1948) extended Griffith's theory to a rapidly propagating crack. Because of the leading role played in its development, LEFM is also known as Griffith-Irwin fracture mechanics and, somewhat less often, as the Griffith-Irwin-Orowan theory.

Irwin (1956) developed the energy release rate concept related to Griffith's theory. Using the approach of Westergaard (1939) who developed a semi-inverse method to analyze stresses and displacements ahead of a sharp crack, Irwin (1957) showed that the stresses and displacements near the crack tip could be described by a single constant which was related to the energy release rate. This crack tip characterizing parameter is now known as the SIF. During the same period, Williams (1957) also calculated the stress distribution at the crack tip but using a somewhat different technique from the Irwin approach, while both results were identical.

LEFM is generally found to be accurate for brittle materials. Direct application of LEFM to ductile materials is then found to yield overly conservative predictions. Around the 1960s it was realized that LEFM is not applicable when large scale yielding at the crack tip precedes failure. To accommodate the effect of yielding at the crack tip, several researchers proposed some approximate methods, *i.e.* by correcting LEFM (Dugdale, 1960; Wells, 1961; Barenblatt, 1962). While Dugdale (1960) proposed an idealized model based on a narrow strip of yielded material at the crack tip, Wells (1961) suggested the displacement of the crack faces as an alternative fracture criterion when large scale plasticity occurs at the crack tip. The latter parameter is now known as the Crack Tip Opening Displacement (CTOD).

Rice (1968) introduced another parameter to characterize nonlinear material behavior ahead of a crack tip. He generalized the energy release rate to nonlinear materials by idealizing plastic deformation as nonlinear elastic. This parameter is now known as the J-integral. During this same period, Hutchinson (1968) showed that the J-integral could be used to represent the characteristics of crack tip stress fields in nonlinear materials.

To apply fracture mechanics to structural design, a mathematical relationship between material toughness, stress and flaw size must be established. While these relationships for linear elastic problems have been available, Shih and Hutchinson (1976) are the first to provide the theoretical framework to establish such a relationship for nonlinear problems. Shih (1981) also established a relationship between the J-integral and CTOD.

The purpose of this chapter is to give a brief introduction to the basic concepts of fracture mechanics. This is important for readers to understand the rest of the chapters. For an elaborate summary of early fracture mechanics research undertaken during 1913 to 1965, Barsom (1987) is referred to. A comprehensive historical overview on fracture mechanics from 1960 to 1980 is provided by Anderson (1995). The details of fracture mechanics can be found in various textbooks, *e.g.* Machida (1984), Kanninen and Popelar (1985) and Anderson (1995), among others. For practical applications, many handbooks are also available (*e.g.* Sih, 1973; Tada, *et al.*, 1973; Rooke and Cartwright, 1976; Murakami, 1987).

4.2 Types of Cracks

In view of the position of the cracks to the material or structure, they can be divided into through-thickness cracks, surface cracks or part-through-thickness cracks, and embedded cracks, shown in Fig. 4.1. Fatigue cracks in thin sheet material which are not small, will grow through the entire thickness of the material. These cracks are referred to as through-thickness cracks, or simply through cracks. If the fatigue load is cyclic tension only, without bending, the crack front is perpendicular to the material surface. However, in thick material and in general for smaller cracks, the crack front does not pass through the full thickness. The crack front then is generally curved, such as corner or surface cracks, which are also labeled as part-through cracks (Schijve, 2001). And the cracks inside the material are generally labeled as embedded cracks. More consideration for crack form details can be referred to Chapter 8.

Considering the number of the cracks in the material, a single crack and multiple cracks are always referred to and treated separately as shown in Fig. 4.2. Cracks always start in welded positions. If there is a crack at the critical location of welds, some small cracks easily emanate from the main crack under random loading, as shown in Fig. 4.3, which are called kinked cracks (Wang, *et al.*, 2005). Typically, the longest one which will act largely on the structure is called the leading crack or main crack while other minor ones are called disturbing cracks. In actual structures, multiple cracks commonly exist with random length, inclination, number and distribution and even random boundary and loading conditions. Multiple flaws on the same cross-section may lead to an interaction and to more severe effects than single flaws alone. Simple criteria for interaction together with the dimensions of the effective flaws after interaction are always used for assessment of multiple cracks. But from the viewpoint of engineering, if multiple cracks exist, each flaw should be checked for interaction with each of its neighbors using the original flaw dimensions in each case. It is not normally necessary to consider further interaction of effective flaws (BS7910). To determine the predicted failure load of the body with multiple cracks, some modified criteria are proposed, such as establishing an equivalent value of the SIF, which will be introduced specifically in Section 4.4, to consider the effect of multiple cracks. The SIF of the leading crack, including crack interaction effects from multiple cracks, will be used to determine the remotely applied load that will produce an SIF equal to or greater than the measured value called fracture toughness. This criterion is also based on LEFM. The primary assumption is that there exists an effective fracture toughness for the thin sheet which allows more yielding than that under plane strain condition.

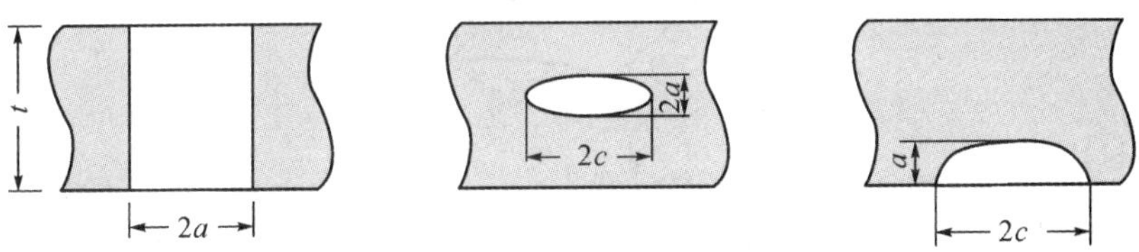

(a) Through-thickness crack (b) Embedded crack (c) Surface crack

Fig. 4.1 Crack forms in view of the position of the crack to the material or structure

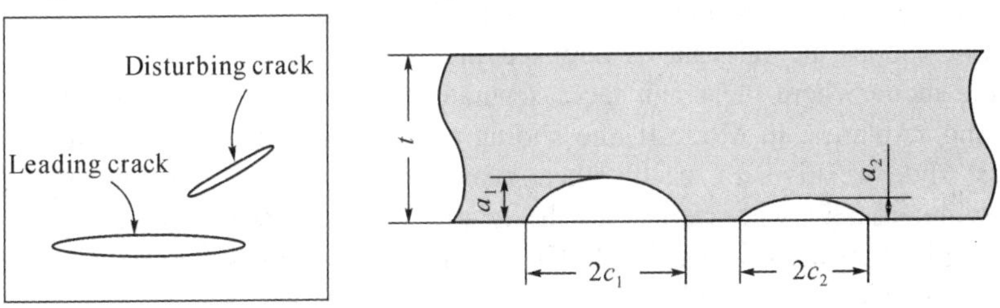

Fig. 4.2 Multiple-cracked body

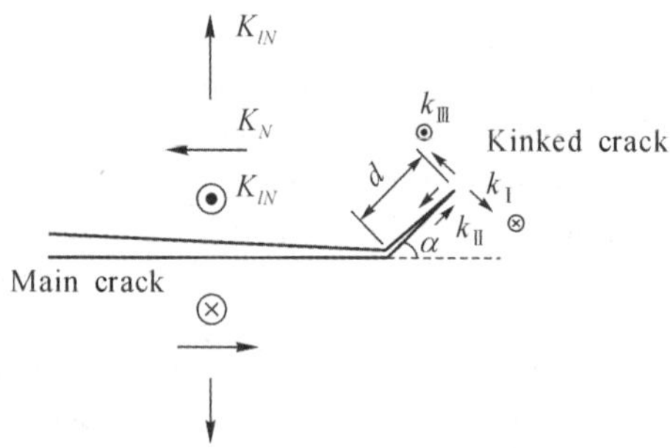

Fig. 4.3 Main crack with a kinked crack (Wang, *et al.*, 2005) (with the permission of Elsevier)

4.3 Types of Opening Modes for a Cracked Body

An investigation of crack-tip stress and displacement fields is important because these fields are typically the ones that govern the fracture process occurring at the crack tip.

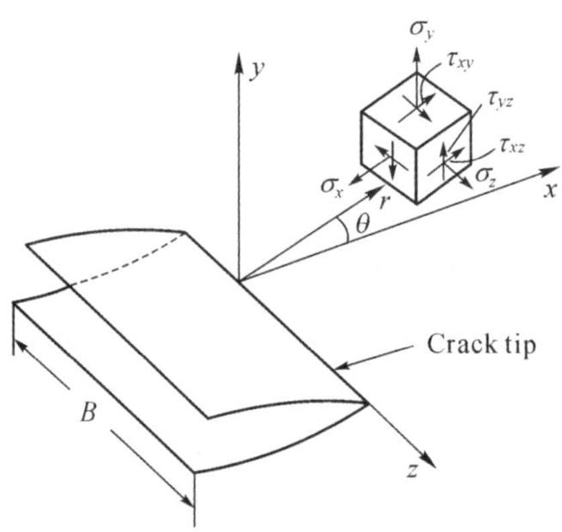

Fig. 4.4 Local coordinate system and the resulting stress components of a cracked body (B= plate thickness)

Considering a cracked body as shown in Fig. 4.4, the crack plane lies in the xz-plane and the crack front is parallel to the y axis. In this case, three basic fracture modes are relevant as depicted in Fig. 4.5. Mode I is the opening or tensile mode where the crack faces separate symmetrically with respect to the xy- and xz-planes. In Mode II, the sliding or in-plane shearing mode, the crack faces slide relative to each other symmetrically about the xy-plane, but antisymmetrically with respect to the xz-plane. In the tearing or antiplane or out-of-plane shearing mode, Mode III, the crack faces also slide relative to each other but antisymmetrically with respect to the xy- and xz-planes. The SIFs associated with the three types of opening modes are denoted as K_{I}, K_{II} and K_{III} respectively.

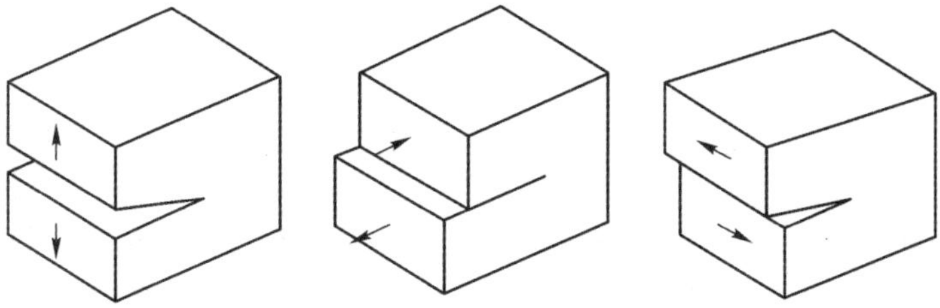

Fig. 4.5 Three basic loading modes for a cracked body. (a) Mode I: opening mode; (b) Mode II: sliding mode (or in-plane shear mode); (c) Mode III: tearing mode (or out-of-plane shear mode) (Kanninen and Popelar, 1985)

4.4 SIFs

4.4.1 Definition

In the practical use of fracture mechanics, the particular environment and operating conditions, the knowledge of sizes and geometries of cracks, *etc*. have to be determined first. And there must be useful quantities to connect the related parameters. Generally, the controlling parameter, K, as used in Eq. (1.2), is always adopted for LEFM for evaluating the critical state of a crack.

According to the coordinate system describing the stresses in the vicinity of a crack, shown in Fig. 4.4, the polar coordinates r and θ are in the xy-plane and normal to the plane of the crack, and the z-direction is parallel to the leading edge of the crack. For any case of Mode I loading, the stresses near the crack tip depend on r and θ as follows:

$$\sigma_x = \frac{K_\mathrm{I}}{\sqrt{2\pi r}}\cos\frac{\theta}{2}\left[1-\sin\frac{\theta}{2}\sin\frac{3\theta}{2}\right]+\cdots \tag{4.1a}$$

$$\sigma_y = \frac{K_\mathrm{I}}{\sqrt{2\pi r}}\cos\frac{\theta}{2}\left[1+\sin\frac{\theta}{2}\sin\frac{3\theta}{2}\right]+\cdots \tag{4.1b}$$

$$\tau_{xy} = \frac{K_\mathrm{I}}{\sqrt{2\pi r}}\cos\frac{\theta}{2}\sin\frac{\theta}{2}\cos\frac{3\theta}{2}+\cdots \tag{4.1c}$$

$$\sigma_z = 0 \qquad \text{(plane stress)} \tag{4.1d}$$

$$\sigma_z = \nu\left(\sigma_x + \sigma_y\right) \qquad \text{(plane strain: } \varepsilon_z = 0\text{)} \tag{4.1e}$$

$$\tau_{yz} = \tau_{zx} = 0 \tag{4.1f}$$

The nonzero stress components in Eqs. (4.1a)–(4.1c) are seen to all approach infinity as r approaches zero—that is, upon approaching the crack tip. Note that this is specially caused by these stresses being proportional to the inverse of $\sqrt{r}$. Thus, a mathematical singularity is said to exist at the crack tip, and no value of stress at the crack tip can be given. Also, all of the nonzero stresses of Eqs. (4.1a)–(4.1c) are proportional to the quantity K_I, and the remaining factors merely give the variation with r and θ. Hence, the magnitude of the stress field near the crack tip can be characterized by giving the value of the factor K_I. On this basis, K_I is a measure of the severity of the crack. Its definition in a formal mathematical sense is (Dowling, 2007)

$$K_I = \lim_{r,\theta \to 0}\left(\sigma_y\sqrt{2\pi r}\right) \tag{4.2}$$

Similarly, the SIFs corresponding to the other two modes can be given as follows (Kanninen and Popelar, 1985),

$$K_{II} = \lim_{r,\theta \to 0}\left(\tau_{xy}\sqrt{2\pi r}\right) \tag{4.3}$$

$$K_{III} = \lim_{r,\theta \to 0}\left(\tau_{yz}\sqrt{2\pi r}\right) \tag{4.4}$$

In relation to LEFM, calculation of the SIF involves most of the work. The SIF is determined as a function of crack size, geometric properties and loading conditions. Specifically, K can be given by the following equation:

$$K = Y\sigma\sqrt{\pi a} \tag{4.5}$$

where a is crack length, σ is nominal stress, where Y is the SIF correction function, which will normally include the effect of factors such as free front surface, finite plate width, crack geometry, non-uniform stress field, presence of geometrical discontinuity and changes in structural restraint (Etube, 2001).

4.4.2 Calculation Methods of SIFs

One of the difficulties in using fracture mechanics is in determining SIFs of cracked structural and mechanical components. The cracks are often subjected to complex stress fields induced by external loads and residual stresses resulting from the surface treatment. Both stress fields are characterized by non-uniform distributions (Wu, *et al.*, 2004).

Generally speaking, analytical approaches and numerical approaches can be used to determine the SIFs. Analytical solutions of the notch SIFs have been found only for a few special cases and SIF solutions under complex conditions with non-uniform stresses are seldom available in databases and handbooks (*e.g.* China Research Institute of Aviation, 1993). Therefore, some new techniques have been developed to solve the problem and extensive studies have been carried out to deal with complex cracked problems. For example, Theocaris and Wu (1990) proposed an Equivalent Through-Crack Model (ETCM) to carry out theoretical analysis of SIFs along the front of semicircular and part-circular surface cracks under mode I loading and this model is also used to study the variation of K_{I}. This method can be used accurately to solve the surface crack problem and, especially, to predict the elastic behavior of the crack on the free surface layer. Wu, *et al.* (2004) presented the generalized weight function technique enabling the SIFs to be calculated for any Mode I loading applied to a planar semi-elliptical surface crack. The calculation is carried out by integrating the product of the stress field and the weight function over the crack area. And the method is particularly suitable for modeling fatigue crack growth in the presence of complex stress fields.

With the development of computer science, FEA and Boundary Element Analysis (BEA) are used for searching for the numerical solutions of SIFs. The solutions vary with the material properties, forms of the crack, geometry of the structure, *etc.* The Finite Element Method (FEM) and Boundary Element Method (BEM) can provide accurate and efficient computation of SIFs with consideration of these influential factors. Typically, the SIFs can be computed through the near tip displacement and stress fields from the conventional FEA. The methods have been included in some commercial software, such as ANSYS. In this software, solving a fracture mechanics problem involves performing a linear elastic or elastic-plastic static analysis and then using specialized post-processing commands or macros to calculate desired fracture parameters. And quadratic elements (singular elements) are with the mid-side nodes placed at the quarter points created for the analysis based on the theoretical foundation that in linear elastic problems, the displacements near the crack tip (or crack front) vary as $\sqrt{r}$, and the stresses and strains are singular at the crack tip, varying as $1/\sqrt{r}$. An example can be seen in Shih and Chen (2002) who built a 3D finite element model of a quarter of the round bar with a semi-elliptic crack, using 20 node brick elements and collapsed singular elements to model the $\sqrt{r}$-stress singularity along the crack front. The SIFs were obtained from the nodal displacements in the vicinity of the crack tip. Collapsed singular elements (with the mid-side nodes in the quarter-point position) were used to model the stress $\sqrt{r}$ -singularity (Toribio, *et al.*, 2009). More discussions on numerical analysis using FEM and BEM to solve SIFs can be found, such as in Wang and Pan (2005); Yuan, *et al.* (2007); Recho, *et al.* (2009).

4.4.3 Typical Examples of SIFs

For steel plates with typical types of cracks under tensile stress, as shown in Fig. 4.6, the SIF value for Mode I is approximately given as follows (Paik and Thayamballi, 2002),

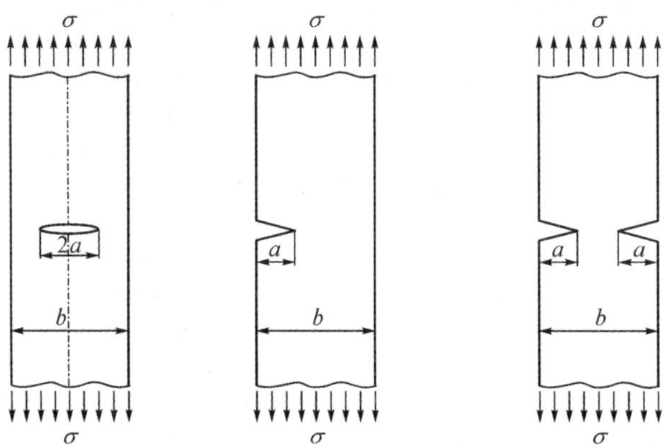

(a) Center crack (b) Crack on one side (c) Crack on both sides

Fig. 4.6 Typical crack locations in a plate under tensile stress

(a) Center crack

$$K_{\mathrm{I}} = F\sigma\sqrt{\pi a} \tag{4.6a}$$

where $F = \left(\sec\dfrac{\pi a}{b}\right)^{1/2}$.

(b) Crack on one side

$$K_{\mathrm{I}} = F\sigma\sqrt{\pi a} \tag{4.6b}$$

where $Q = 30.38\left(\dfrac{a}{b}\right)^4 - 21.71\left(\dfrac{a}{b}\right)^3 + 10.55\left(\dfrac{a}{b}\right)^2 - 0.23\left(\dfrac{a}{b}\right) + 1.12$.

(c) Crack on both sides

$$K_{\mathrm{I}} = F\sigma\sqrt{\pi a} \tag{4.6c}$$

where $F = 15.44\left(\frac{c}{b}\right)^3 - 4.78\left(\frac{a}{b}\right)^2 + 0.43\left(\frac{a}{b}\right) + 1.12$.

If the plate width were infinite, the solution in the first of the above three cases would revert to the classical solution, namely $K_{\mathrm{I}} = \sigma\sqrt{\pi a}$, since F=1 in this case.

The stress concentration magnification factor of semi-elliptical cracks at the toe of fillet welded joints can be expressed as follows,

$$K = M_K K_{\text{plate}} \tag{4.7}$$

where K_{plate} is the SIF in the vicinity of crack tips in a plate and M_K is a magnification factor for welded joints. Bowness and Lee (2000) recommended the empirical expression of M_K of the semi-elliptical crack, and the calculation methods for M_K and the SIFs of surface cracks are introduced in the appendices to Chapter 7.

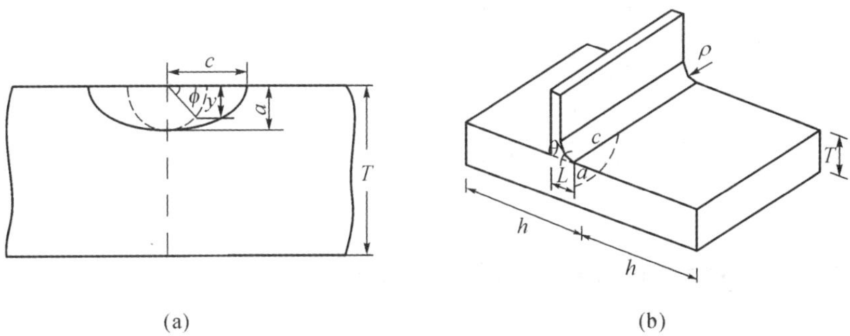

Fig. 4.7 Illustration of surface crack in T-joint. (a) Surface crack sketch; (b) 3D T-joint model

4.4.4 *Plasticity Limitations of the SIFs Based on LEFM*

If the plastic zone is sufficiently small, there will be a region outside of it where the elastic stress field equations (see Eqs. (4.1a)–(4.1f)) still apply, called the region of K-dominance, or the K-field (Dowling, 2007). This is illustrated in Fig. 4.8.

The existence of such a region is necessary for LEFM theory to be applicable. The K-field surrounds and controls the behavior of the plastic zone and crack tip area. As a practical matter, it is necessary that the plastic zone be small compared with the distance from the crack tip to any boundary of the member, such as distances a, b–a, h for a cracked plate, as in Eq(4.8). A distance of $4r_p$ is generally considered to be sufficient. Then the limit on the use of LEFM can be expressed by

$$a,\ (b-a),\ h \geq \frac{\pi}{4}\left(\frac{K}{\sigma_Y}\right)^2 \tag{4.8}$$

This must be satisfied for all three of a, b–a and h. Otherwise, the situation will too closely approach gross yielding with a plastic zone extending to one of the boundaries.

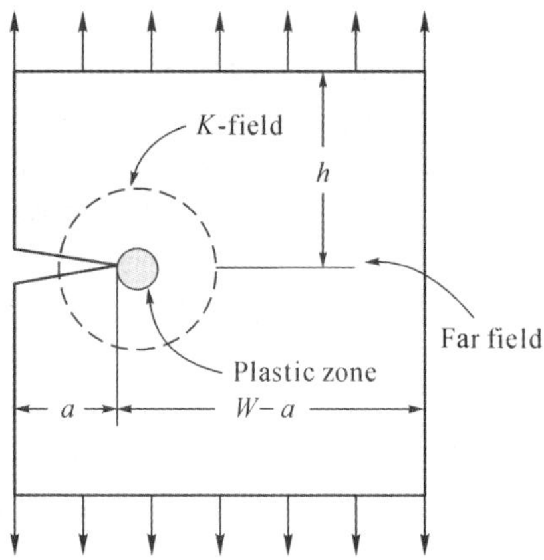

Fig. 4.8 A schematic representation for plasticity limitations on LEFM

4.4.5 Extensions of the SIFs Based on LEFM

In order to reasonably use the SIF in fatigue life calculations, some extensions must be made based on LEFM. McEvily's formula (McEvily, *et al.*, 1999) will be introduced in this section.

Irwin (1960a) has shown that the SIF, K, is related to the stress concentration factor, K_T, by the following equation,

$$K = \lim_{\rho\to\infty} K_T\sigma\sqrt{\frac{\pi\rho}{4}} \tag{4.9}$$

where ρ is the tip radius of the stress concentrator and σ is the remote stress. In order to achieve the desired transition between the threshold level for a fatigue crack and the endurance limit, Eq. (4.9) is modified as follows,

$$K = \lim_{\rho \to \rho_e} K_T \sigma \sqrt{\frac{\pi \rho}{4}} \tag{4.10}$$

where ρ_e, a material constant, is considered to be the effective radius of the fatigue crack, a characterizing parameter. In the case of a centrally cracked panel under Mode I loading, K_T is $\left(1+2\sqrt{\frac{a}{\rho}}\right)$ and Eq. (4.10) becomes

$$K = \left(\sqrt{\frac{\pi \rho_e}{4}} + \sqrt{\pi a}\right)\sigma \tag{4.11}$$

It is noted that an extra term appears in the expression for the SIF, and this modification provides one of the means of dealing with short cracks. The parameter r_e can be expressed as an equivalent length, ρ_e, over which the maximum stress is considered to remain constant. By substituting σ_{max} for $K_T\sigma$ in Eq. (4.10), and then equating this value of K to the value of K at a distance ρ_e ahead of the crack tip, with $\rho = \rho_e$, *i.e.*

$$\sigma_{max} \sqrt{\frac{\pi \rho_e}{4}} = \sigma_y \sqrt{2\pi r_e} \tag{4.12}$$

with σ_{max} set equal to σ_y, r_e is equal to ρ_e /8. In this modified LEFM approach, ρ_e is considered to be the size of an inherent flaw. Newly formed cracks are only significant when their lengths exceed ρ_e, a parameter of the order of a micron in length. The modified SIF in the LEFM range is then expressed as

$$K = \left(\sqrt{2\pi r_e} + Y\sqrt{\pi a}\right)\sigma \tag{4.13}$$

Irwin (1957) proposed that the LEFM approach could be extended to include elastic-plastic behavior by introducing a modified crack length, a_{mod}, which is equal to the actual crack length plus one-half of the crack tip plastic zone size. If the Dugdale expression (1960) (see Eq. (4.24)) for the plastic zone size is used, then

$$a_{mod} = \frac{a}{2}\left(\sec\frac{\pi}{2}\frac{\sigma}{\sigma_Y} + 1\right) \tag{4.14}$$

El Haddad, *et al.* (1979) have also modified the expression for the SIF in the short crack range by adding a constant to the actual crack length, rather than by introducing a separate term as in Eq. (4.14). Then Eq. (4.14) is modified to account for elastic-plastic behavior under cyclic conditions as follows,

$$K_{\mathrm{mod}} = \left(\sqrt{2\pi r_{\mathrm{e\,mod}}} + Y\sqrt{\pi a_{\mathrm{mod}}}\right)\sigma = \left[\sqrt{\pi r_{\mathrm{e}}\left(\sec\frac{\pi}{2}\frac{\sigma_{\max}}{\sigma_Y}+1\right)} + Y\sqrt{\frac{\pi}{2}a\left(\sec\frac{\pi}{2}\frac{\sigma_{\max}}{\sigma_Y}+1\right)}\right]\sigma \tag{4.15}$$

When σ is low enough, the linear and nonlinear equations will be similar, but they will result in a significant difference when the stress level is higher. The comparison between linear and nonlinear equations for K can be shown in Fig. 4.9.

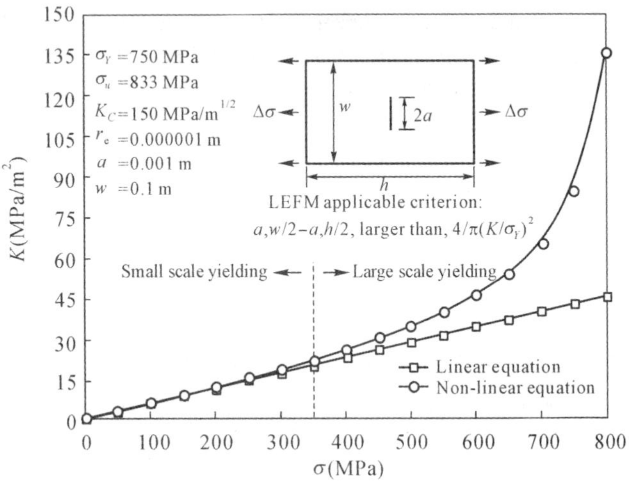

Fig. 4.9 The comparison between linear and nonlinear equations for $K_{\max}$

4.5 Fracture Toughness

4.5.1 Definition

It is considered that fracture takes place if the K value of the structure reaches the critical K value, K_{C}, namely

$$K \geq K_{\mathrm{C}} \tag{4.16}$$

where K_{C} is sometimes called fracture toughness which is typically determined experimentally for a given material, crack and loading situation.

For instance, K_C is determined once the ultimate fracture loads (or failure loads) and the crack sizes are obtained for a mechanical test specimen with the SIF known. In general, the fracture toughness K_C is affected by strain rate, temperature and plate thickness.

4.5.2 Testing

Fracture toughness tests should take account of the orientation of flaws in the structure relative to welded joints, and the constraint, temperature, rate of loading and environment (particularly for materials exposed to hydrogen) experienced in service.

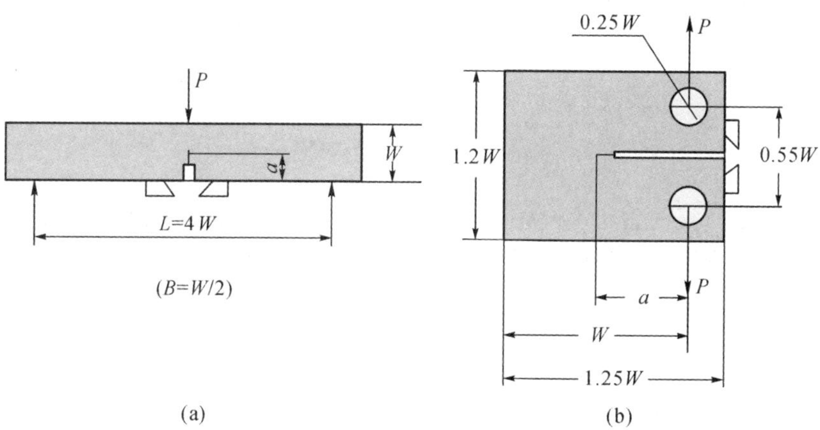

Fig. 4.10 Typical fracture toughness test specimen. (a) Three-point bending specimen; (b) Compact specimen

The SIF for the two specimens can be expressed by the following equations:

(1) Three point bending test

$$K_{\mathrm{I}}=\frac{3PL}{2BW^2}\sqrt{\pi a}\left[1.090-1.735\left(\frac{a}{W}\right)+8.20\left(\frac{a}{W}\right)^2-14.18\left(\frac{a}{W}\right)^3+14.57\left(\frac{a}{W}\right)^4\right] \tag{4.17}$$

(2) Compact test

$$K_{\mathrm{I}} = \frac{P\sqrt{a}}{BW}\left[29.6 - 185.5\left(\frac{a}{W}\right) + 655.7\left(\frac{a}{W}\right)^2 - 1017.0\left(\frac{a}{W}\right)^3 + 638.9\left(\frac{a}{W}\right)^4\right] \tag{4.18}$$

Fracture toughness testing of the metals can be governed by some standards (*e.g.* ASTM E399, 2004). It will be addressed in Section 4.5.3 that fracture toughness decreases with increasing specimen thickness, which results from the plastic zone (see Section 4.6) at the crack tip related to the thickness. It has been addressed in the ASTM standard that the plastic zone should not be too large in order to avoid interaction between the crack and free boundaries of the specimen. Then, the ASTM standard stipulates that the geometrical dimensions should be large enough compared to the crack length and plastic zone size. The requirement can be expressed as follows:

$$a,\, B,\, (W-a) \geq 2.5\left(\frac{K_{\mathrm{I}}}{\sigma_Y}\right)^2 \tag{4.19}$$

In order to accurately measure the value of fracture toughness, the value of K_{I} is always replaced by K_{IC} to express the conservative situation.

Fracture toughness testing at the present time, according to a particular procedure using the pre-cracked specimens with a controlled crack, will present two inconveniences:

- These specimens are regarded as time consuming and expensive.
- For very brittle materials, such as ceramics and high strength steels, it is practically impossible to pre-crack the specimens and the use of a notched specimen is preferred.

Studies on some other determination methods have been discussed (*e.g.* Pardoen, *et al.*, 2000; Margolin, *et al.*, 2006; Kang, *et al.*, 2006; Shyam and Lara-Curzio, 2006; Sarkar and Ray, 2008). For example, the analytical solution to determine the fracture toughness for notch tip distribution is Creager's solution (1967). And the fracture toughness of high strength steel is analyzed using the notch SIF and volumetric approach by El Minor, *et al.* (2003).

4.5.3 Trends

A study of environmental and geometrical effects on the fracture toughness has been reported. However, in this section, the thickness effect will be simply introduced as a foundation of some research content in this book, while environmental effects will be passed over, even though they are equally significant.

As the plate thickness decreases, the K_C value tends to significantly increase. This is because, with a decrease in the plate thickness, it approaches the plane stress state and essentially Modes II or III based on shear fracture are more unlikely to take place than Mode I.

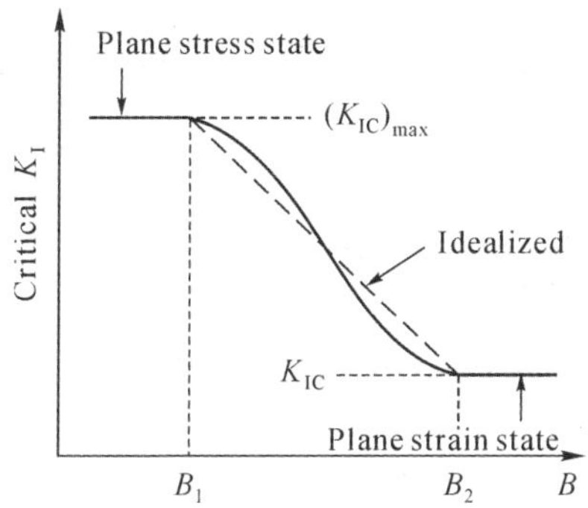

Fig. 4.11 A schematic representation of the critical K_I value versus the plate thickness (B= plate thickness)

For thicker plates, Mode I fracture associated with the plane strain state is more likely to occur. In this case, the fracture toughness is specifically denoted by K_{IC}. Fig. 4.11 shows a schematic representation of the critical K_I value versus the plate thickness. For a given plate thickness, in contrast to through-thickness cracks, surface cracks may sometimes exhibit plane strain behavior because of the related conditions at the crack tip.

It should be noted that the fatigue fracture toughness, which corresponds to the first crack jump and final fracture, can be appreciably lower, *i.e.* up to 50%, than the static fracture toughness under plane strain conditions at the corresponding temperature. An analysis by Troshchenko (1994) has been performed for unstable and stable fatigue crack growth, and a model of unstable crack propagation is proposed that accounts for the observed experimental behavior.

4.6 Crack Tip Plasticity

4.6.1 Plastic Zone for Plane Stress

The region of yielding at crack tips is called the plastic zone. The size of plastic zones should be different for the plane stress situation and the plane strain situation, which will be denoted by r_σ and r_ε respectively. An equation for estimating the plastic zone for plane stress situations can be developed from the elastic stress field equation, Eq. (4.1), with σ_z=0 (Dowling, 2007). With the requirement, in the plane of the crack where θ=0, these simplify to

$$\sigma_x = \sigma_y = \frac{K}{\sqrt{2\pi r}}, \qquad \sigma_z = \tau_{xy} = \tau_{yz} = \tau_{zx} = 0 \tag{4.20}$$

Considering the octahedral shear stress criterion expressed in terms of the uniaxial yield strength,

$$\sigma_Y = \frac{1}{\sqrt{2}}\sqrt{(\sigma_1 - \sigma_2)^2 + (\sigma_2 - \sigma_3)^2 + (\sigma_3 - \sigma_1)^2} \tag{4.21}$$

and σ_x, σ_y, σ_z are principal normal stresses. As all shear stress components are zero, yielding occurs at σ_x=σ_y=σ_Y and then the solution of r_σ can be obtained as follows,

$$r_\sigma = \frac{1}{2\pi}\left(\frac{K}{\sigma_Y}\right)^2 \tag{4.22}$$

This solution is obtained by LEFM theory with elastic stress distribution, which supposes theoretically infinite stresses at the tip of a sharp crack. The actual situation is that the crack tip will become blunted with a region of yielding, as shown in Fig. 4.12. Considering the effect of the plastic zone, large deformation will cause yielding to extend even further than r_σ. The value of the plastic zone size under this situation is sometimes assumed to be equal to r_σ, namely,

$$r_{P\sigma} = 2r_\sigma = \frac{1}{\pi}\left(\frac{K}{\sigma_Y}\right)^2 \tag{4.23}$$

Furthermore, Dugdale (1960) suggested an equation for plastic zone size under plane stress condition,

$$r_{P\sigma} = a\left(\sec\frac{\pi\sigma}{2\sigma_Y} - 1\right) \tag{4.24}$$

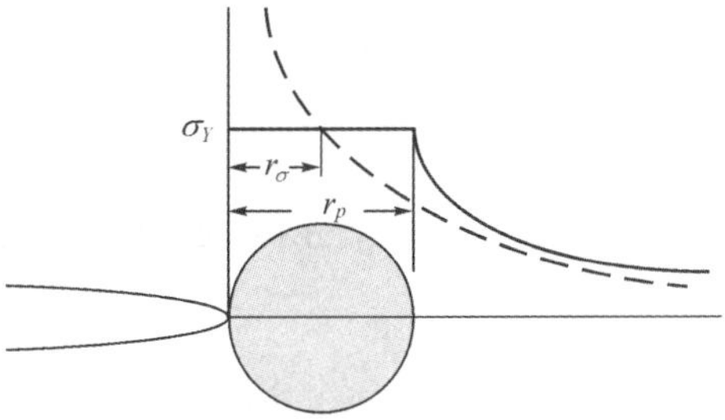

Fig. 4.12 Plastic zone at crack tip

4.6.2 Plastic Zone for Plane Strain

For the cracked body with very large thickness, the deformation of the material inside the plastic zone in the z direction is difficult, namely ε_z is approximately zero, which is called the plane strain condition.

Considering Hooke's law,

$$\varepsilon_x = \frac{1}{E}\left[\sigma_x - \nu\left(\sigma_y + \sigma_z\right)\right] \tag{4.25a}$$

$$\varepsilon_y = \frac{1}{E}\left[\sigma_y - \nu\left(\sigma_x + \sigma_z\right)\right] \tag{4.25b}$$

$$\varepsilon_z = \frac{1}{E}\left[\sigma_z - \nu\left(\sigma_x + \sigma_y\right)\right] \tag{4.25c}$$

Substituting ε_z=0 and σ_x=σ_y into Hooke's law can give a stress in the z direction of ε_z=2$\nu\sigma_y$. Using the octahedral shear stress criterion, we can obtain σ_x=σ_y=σ_Y/(1–2ν). If the typical value of Poisson's ratio of ν=0.3 is adopted, the values of σ_x and σ_y should be 2.5σ_Y. Considering the suggestion by Irwin (1960) that the effect is somewhat smaller, with the yielding around $\sigma_y \approx \sqrt{3}\sigma_Y$, the value of r_ε can be expressed by

$$r_\varepsilon = \frac{1}{6\pi}\left(\frac{K}{\sigma_Y}\right)^2 \tag{4.26}$$

Then, the value of $r_{P\varepsilon}$ should be as follows,

$$r_{P\varepsilon} = 2r_{\varepsilon} = \frac{1}{3\pi}\left(\frac{K}{\sigma_Y}\right)^2 \tag{4.27}$$

Furthermore, Irwin gave an equation to express the plastic zone size under plane strain state,

$$r_{P\varepsilon} = \frac{(1-2\nu)^2}{\pi}\left(\frac{K}{\sigma_Y}\right)^2 \tag{4.28}$$

4.6.3 Plastic Zone Under Real Stress State

Real structures may not always satisfy the conditions of plane stress or plane strain. Then the calculation for plastic zone size should take into account the real stress state at crack tips. Theoretical analysis is difficult to use for this problem. Finite element analysis becomes an effective tool. Actually, there have been some solutions for the plastic zone size considering the effect of the real stress state at crack tips.

Under small scale yielding conditions, Voorwald, *et al.* (1991) suggested that the plastic zone size at the crack tips can be expressed by the following equation,

$$r_P = \alpha\left(\frac{K}{\sigma_Y}\right)^2 \tag{4.29}$$

where the value of α can be an index representing the size of the plastic zone. Voorwald gave the expressions of α as follows,

$$\alpha = \begin{cases} \dfrac{1}{6\pi} & (t \geq 2.5(K_{\max}/\sigma_{ys})^2 \\ \dfrac{1}{\pi} & \left(t \leq \dfrac{1}{\pi}(K_{\max}/\sigma_{ys})^2\right) \\ \dfrac{1}{6\pi} + \dfrac{5}{6\pi}\left(\dfrac{2.5 - t(K_{\max}/\sigma_y)^{-2}}{2.5-\pi^{-1}}\right) & \left(\dfrac{1}{\pi}\left(\dfrac{K_{\max}}{\sigma_y}\right)^2 < t < 2.5\left(\dfrac{K_{\max}}{\sigma_{ys}}\right)^2\right) \end{cases} \tag{4.30}$$

Voorwald has not pointed out the limit of the applicability of the equations above. Moreover, the equations could not consider the effect of Poisson's ratio, ν. In order to consider the effect of ν, Huang, *et al.* (2005) proposed another expression on the basis of Voorwald's one,

$$\alpha=\begin{cases} 1 & \left(t\le\frac{1}{\pi}(K_{\max}/\sigma_{ys})^2\right) \\ \frac{1}{1-2\mu} & t\ge 2.5\left(\frac{K_{\max}}{\sigma_{ys}}\right)^2 \\ 1+\frac{2\mu[t(\sigma_{ys}/K_{\max})^2-1/\pi]}{(1-2\mu)(2.5-1/\pi)} & \left(\frac{1}{\pi}\left(\frac{K_{\max}}{\sigma_y}\right)^2<t<2.5\left(\frac{K_{\max}}{\sigma_{ys}}\right)^2\right) \end{cases} \tag{4.31}$$

However, the equations still have some limitations when expressing the real property of the material, as they do not take into account the effect of strain hardening.

When studying the effect of thickness, Guo (1999) extended the Dugdale model to the 3D condition, and deduced the relationship between the plastic zone size and material thickness together with material properties. He suggested using a weighted average value of the plastic zone size distribution along the z-direction to express the effect of thickness and gave his equation to calculate plastic zone size as

$$\sqrt{\frac{a+r_p}{2}}\int_0^{r_p}\frac{g(R,n)}{f(T_h,k)}\frac{\mathrm{d}r}{\sqrt{(a+r_p)^2-(a+r)^2}}=\sqrt{r_{p0}} \tag{4.32}$$

where the expressions of $g(R, n)$ and $f(T_h, k)$ include the results from finite element analysis and HRR solutions, which are complicated to use in this procedure and iterative calculation brings more difficulties for the application of this equation.

Considering the limitations of current solutions, Liu, *et al.* (2006) further analyzed the problem using the finite element method. In view of the power-law strain hardening model proposed by Hoffmann and Seeger (1985),

$$\varepsilon=\begin{cases} \frac{\sigma}{E} & \sigma\le\sigma_Y \\ \frac{\sigma_Y}{E}\left(\frac{\sigma}{\sigma_Y}\right)^n & \sigma>\sigma_Y \end{cases} \tag{4.33}$$

The analysis results are depicted in Figs. 4.13–4.15.

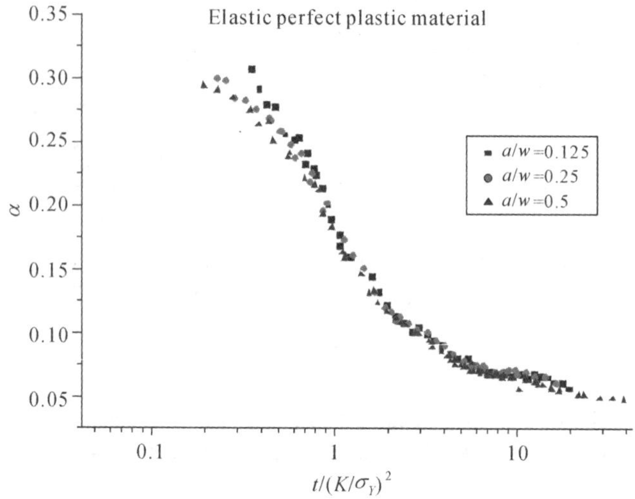

Fig. 4.13 $t/(K/\sigma_Y)^2 - \alpha$ data under different a (Liu, *et al.*, 2006)

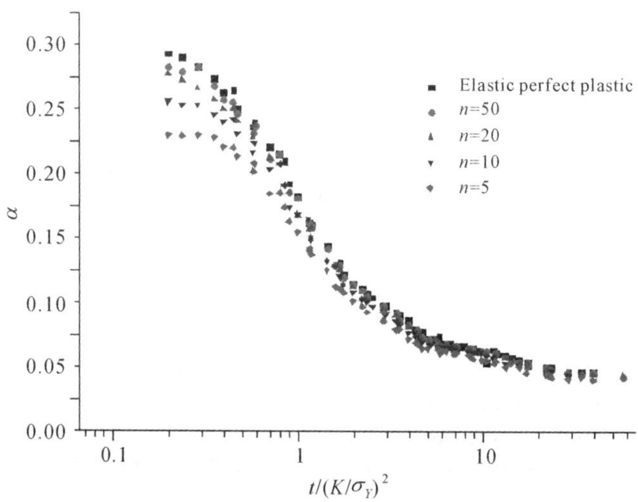

Fig. 4.14 $t/(K/\sigma_Y)^2 - \alpha$ data under different n (Liu, *et al.*, 2006)

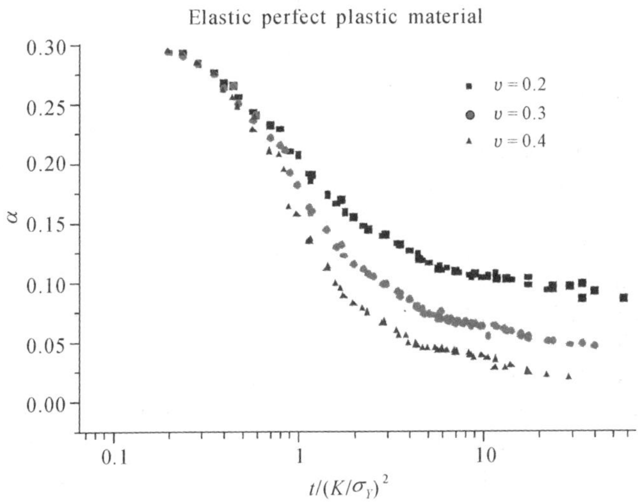

Fig. 4.15 $t/(K/\sigma_Y)^2-\alpha$ data under different ν (Liu, *et al.*, 2006)

Then a continuous equation for α considering the effect of strain hardening and Poisson's ratio is obtained as,

$$\alpha=\frac{(1-1.65\mu)^2}{5}-\frac{1}{20n}((1-1.65\mu)^2)^{\frac{1}{n}}+\frac{\dfrac{1}{\pi}-\dfrac{1}{2.2n}\left(\dfrac{1}{\pi}\right)^{\frac{1}{n}}-\left(\dfrac{(1\text{-}1.65\mu)^2}{5}-\dfrac{1}{20n}((1-1.65\mu)^2)^{\frac{1}{n}}\right)}{\left(1+\dfrac{t/(K/\sigma_{ys})^2}{1+1/n}\right)^{1.6+1/n}} \tag{4.34}$$

The equation can be simplified when an elastic-perfectly-plastic condition is considered, namely $n=\infty$, as follows,

$$\alpha=\frac{(1-1.65\mu)^2}{5}+\frac{\dfrac{1}{\pi}-\dfrac{(1-1.65\mu)^2}{5}}{(1+t/(K/\sigma_{ys})^2)^{1.6}} \tag{4.35}$$

4.7 Summary

The theory of FM is the basis of the UFLP method. It deals with the study of how a crack or flaw in a structure propagates under applied loads. Before investigation of crack propagation, the crack forms and types of fracture modes must be understood. The division of cracks into through-thickness cracks, surface cracks and embedded cracks, or into single cracks and multiple cracks is briefly introduced. The three types of crack opening modes corresponding to different crack-tip stress and displacement fields are illustrated.

The SIF gives an indication of stress severity around the crack tip. It is a concept of LEFM and typically expressed as an equation of crack length, stress level and a correction function accounting for other influential factors. Generally speaking, analytical approaches and numerical approaches can be used to determine the SIFs, and their typical values can be referred to in some handbooks. It is considered that fracture takes place if the SIF of the structure reaches its critical value, *i.e.* fracture toughness which is determined by test. The SIF based on LEFM is applicable only when the plastic zone around the crack tip is sufficiently small. In order to reasonably use it in fatigue life calculations, some extensions must be made and McEvily's formula is introduced in this chapter.

Plastic zone size is another important parameter used for evaluating the characteristics of a cracked body. The deformation inside the plastic zone and the state of stress conditions at the crack tip vary from the plane stress situation to the plane strain condition. The calculation for the plastic zone size should take into account the real stress state at crack tips. Theoretical analysis is difficult to use for this problem. Finite element analysis becomes an effective tool. Some solutions for plastic zone size considering the effect of the real stress state are introduced and will be adopted in Chapter 5.

References

Anderson, T. L. (1995). Fracture Mechanics: Fundamentals and Applications (2nd ed.). London: CRC Press.

ASTM (2004). Annual Book of ASTM Standards. West Conshohocken, PA. ASTM International.

Barenblatt G. I. (1962). "The mathematical theory of equilibrium cracks in brittle fracture", Advances in Applied Mechanics, Academic Press, 7: 55–129.

Barsom, J. M. (ed.) (1987). "Fracture mechanics retrospective: early classic papers (1913–1965)", The American Society of Testing and Materials (RPS 1), Philadelphia, PA.

Bowness, D. & Lee, M. M. K. (2000). "Prediction of weld toe magnification factors for semi-elliptical cracks in T-butt joint", International Journal of Fatigue, 22: 369–387.

Broek, D. (1974). Elementary Engineering Fracture Mechanics. Dordrecht / Boston / Lancaster: Martinus Nijhoff Publshers.

China Research Institute of Aviation (1993). Handbook of Stress Intensity Factors. Beijing: Science Press (in Chinese).

Creager M. & Paris, P. C. (1967). "Elastic field equations for blunt cracks with refrence to stress corrosion cracking", International Journal of Fracture Mechanics, 3: 247–252.

Dowling, N. E. (2007). Mechanical Behavior of Materials-Engineering Methods for Deformation, Fracture, and Fatigue (3rd ed.). Upper Saddle River, New Jersey: Pearson Prentice Hall.

Dugdale, D. S. (1960). "Yielding of steel sheets containing slits", Journal of the Mechanics and Physics of Solids, 8: 100–108.

El Haddad, M. H., Topper, T. H. & Smith, K. N. (1979). "Prediction of non propagating cracks", Engineering Fracture Mechanics, 11(3): 573–584.

El Minor, H., Kifani, A., Louah, M., Azari, Z. & Pluvinage, G. (2003). "Fracture toughness of high strength steel-using the notch stress intensity factor and volumetric approach", Structural Safety, 25: 35–45.

Etube L. S. (2001). Fatigue and Fracture Mechanics of Offshore Structures. London and Bury St Edmunds: Professional Engineering Publishing Ltd.

Griffith, A. A. (1920). "The phenomena of rupture and flow in solids", Philosophical Transactions, Series A, 221: 163–198.

Guo, W. L. (1999). "Three-dimensional analyses of plastic constraint for through-thickness cracked bodies", Engineering Fracture Mechanics, 62: 383–407.

Haddad, M. H., Topper, T. H. & Smith, K. N. (1979). "Prediction of non propagating cracks", Engineering Fracture Mechanics, 11: 573–584.

Hoffmann, M. & Seeger, T. (1985). "Dugdale solutions for strain hardening materials", in: The crack tip opening displacement in elastic-plastic fracture mechanics. Workshop on the CTOD Methodology, Geesthacht, Germany, 57–77.

Huang, X. P., Han, Y., Cui, W. C., Wan, Z. Q. & Bian, R.G. (2005). "Fatigue life prediction of weld-joints under variable amplitude fatigue loads", Journal of Ship Mechanics, 9(1): 89–97 (in Chinese).

Hutchinson, J. W. (1968). "Singular behavior at the end of a tensile crack in a hardening material", Journal of the Mechanics and Physics of Solids, 16: 13–31.

Irwin, G. R. (1948). "Fracture dynamics", Fracturing of Metals. The American Society for Metals, Cleveland, 147–166.

Irwin, G. R. (1956). "Onset of fast crack propagation in high strength steel and aluminum alloys", Sagamore Research Conference Proceedings, 2: 289–305.

Irwin, G. R. (1957). "Analysis of stresses and strains near the end of a crack traversing a plate", Journal of Applied Mechanics, 24: 361–364.

Irwin, G. R. (1960a). "Fracture mechanics", in: Proceedings of 1st Symposium on Naval Structural Mechanics. Oxford: Pergamon Press, 557–594.

Irwin, G. R. (1960b), "Plastic zone near a crack and fracture toughness", in: Proceedings of the 7th Sagamore Research Conference on Mechanics and Metals Behavior of Sheet Material. New York, 4: 463–478.

Kang, J. Y., Choi, B. I. & Lee, H. J. (2006). "Application of artificial neural network for predicting plain strain fracture toughness using tensile test results", Fatigue and Fracture of Engineering Materials and Structures, 29: 321–329.

Kanninen, M. F. & Popelar, C. H. (1985). Advanced Fracture Mechanics. New York: Oxford University Press.

Leis, B. N., Hopper, A. T., Ahmad, J., Broek, D. & Kanninen, M. F. (1986). "Critical review of the fatigue growth of short cracks", Engineering Fracture Mechanics, 23(5): 883–898.

Liu, Q., Wang, F., Huang, X. P. & Cui, W. C. (2006). "Three dimensional FE analysis of the plastic zone size near the crack tip", Journal of Ship Mechanics, 10(5): 90–99 (in Chinese).

Machida, S. (1984). Ductile Fracture Mechanics. Tokyo: Nikkan Kogyo Shimbunsha (Newspaper Co.) (in Japanese).

Margolin, B. Z., Shvetsova, V. A., Gulenko, A. G. & Kostylev, V. I. (2006). "Application of a new cleavage fracture criterion for fracture toughness prediction for RPV steels", Fatigue and Fracture of Engineering Materials and Structures, 29: 697–713.

McEvily, A. J., Bao, H. & Ishihara, S. (1999). "A modified constitutive relation for fatigue crack growth", Fatigue'99, 329–336.

Mott, N. F. (1948). "Fracture of metals: theoretical considerations", Engineering, 165: 16–18.

Murakami, Y. (ed.) (1987). Stress Intensity Factors Handbook. New York: Pergamon Press.

Orowan, E. (1948). "Fracture and strength of solids", Reports on Progress in Physics, XII, 185–232.

Paik, J. K. & Thayamballi, A. K. (2002). Ultimate Limit State Design of Steel Plated Structures. London: John Wiley & Sons.

Pardoen, T., Scibetta, M., Chaouadi, R. & Delannay, F. (2000). "Analysis of the geometry dependence of fracture toughness at cracking initiation by comparison of circumferentially cracked round bars and SENB tests on Copper", International Journal of Fracture, 103: 205–225.

Recho, N. (2009). "Boundary element analysis of generalized stress intensity factors", ICOME2009, Oct. 18, Nanjing, China.

Rice, J. R. (1968). "A path independent integral and the approximate analysis of strain concentrations by notches and cracks", Journal of Applied Mechanics, 35: 379–386.

Rooke, D. R. & Cartwright. D. J. (1976). Compendium of Stress Intensity Factors. Uxbridge: H.M.S.O.

Sarkar, R. & Ray, K. K. (2008). "Estimation of fracture toughness using miniature chevron-notched specimens", Fatigue and Fracture of Engineering Materials and Structures, 31: 340–345.

Schijve, J. (2001). Fatigue of Structures and Materials. Dordrecht, Boston: Kluwer Academic Publishers.

Shih, C. F. (1981). "Relationship between the J-integral and the crack opening displacement for stationary and extending cracks", Journal of the Mechanics and Physics of Solids, 29: 305–326.

Shih, C. F. & Hutchinson, J. W. (1976). "Fully plastic solutions and large scale yielding estimates for plane stress crack problems", Journal of Engineering

Materials and Technology, 98: 289–295.

Shih, Y. S. & Chen, J. J. (2002). “The stress intensity factor study of an elliptical cracked shaft”, Nuclear Engineering and Design, 214: 137–145.

Shyam, A. & Lara-Curzio, E. (2006). “The double-torsion testing technique for determination of fracture toughness and slow crack growth behavior of materials: a review”, Journal of Materials Science, 41: 4093–4104.

Sih, G. C. (1973). Handbook of Stress Intensity Factors. Bethleham, PA: Lehigh University.

Tada, H., Paris, P. C. & Irwin, G. R. (1973). Stress Analysis of Cracks Handbook. Hellertown, PA: Del Research Corporation.

Theocaris, S. & Wu, D. L. (1990). “A theoretical analysis of the stress intensity factors varying along the front of a surface crack”, International Journal of Fracture, 3: 243–250.

Toribio, J., Matos, J. C., González, B. & Escuadra, J. (2009). “An automated procedure for the geometrical modeling of a surface crack front”, Struct Durab Health Monit, 123: 1–16.

Troshchenko, V. T., Pokrovsky, V. V. & Yasniy, P. V. (1994). “Unstable fatigue crack propagation and fatigue fracture toughness of steels”, Fatigue and Fracture of Engineering Materials and Structures, 17(9): 991–1001.

Voorwald, H. J. C. & Torres, M. A. S. (1991). “Modeling of fatigue crack growth following overloads”, International Journal of Fatigue, 13(5): 423–427.

Wang, D. A., Lin, S. H. & Pan, J. (2005). “Stress intensity factors for spot welds and associated kinked cracks in cup specimens”, International Journal of Fatigue, 27: 581–598.

Wang, D. A. & Pan, J. (2005). “A computational study of local stress intensity factor solutions for kinked cracks near spot welds in lap-shear specimens”, International Journal of Solids and Structures, 42(24): 6277–6298.

Wells, A. A. (1961). “Unstable crack propagation in metals: cleavage and fast fracture”, in: Proceedings of the Crack Propagation Symposium, Paper 84, Cranfield, UK.

Westergaard, H. M. (1939). “Bearing pressures and cracks”, Journal of Applied Mechanics, 6: 49–53.

Williams, M. L. (1957). “On the stress distribution at the base of a stationary crack”, Journal of Applied Mechanics, 24: 109–114.

Wu, X., Hu, D., Preuss, M., Withers, P. J. & Loretto, M. H. (2004). “The role of surface condition, residual stress and microstructure on pre-yield cracking in Ti44Al8Nb1B”, Intermetallics, 12(3): 281–287.

Yuan, S., Xu, Y. J. & Williams, F. M. (2007). “Computation of stress intensity factors by the sub-region mixed finite element method of line”, Acta Mechanica Solida Sinica, 20(2): 149–162.

5

Development of a UFLP Method for Marine Structures

5.1 Introduction

We can recall from Chapter 1 and Chapter 3 that currently-used stress-based approaches for FLP of marine structures are subjected to theoretical flaws. FCP theory could potentially overcome these deficiencies. This is the fundamental philosophy for the development of a UFLP method for marine structures. And the general meaning of the UFLP method discussed in this book has been briefly introduced there.

This chapter specifically describes the development of the UFLP method for marine structures from which one can detect the technical problems to be resolved for developing a UFLP method. The contents include the general procedure to develop the fatigue life method based on FCP theory, the development of a general constitutive relationship, the engineering approaches to determine the parameters in the improved model and the analysis of the capabilities of the method. At the same time, future trends for the development of the method are included in the last section.

5.2 A General Procedure for the UFLP Method

5.2.1 The General Function Format of the Fatigue Crack Growth Rate Curve for the UFLP Method

The crack growth rate curve is the fundamental material property under fatigue to be defined. As has been described in Chapter 3, the general crack growth rate curve can be written as a function of $\underline{L}$, $\underline{M}$, $\underline{S}$ and $\underline{E}$. However, within each category the vector (L, M, S or E) contains many variables and they might be further divided into microscale parameters and macroscale parameters according to the scale or mechanistic paramcters and humanistic parameters. More complicated, there are dependencies among the parameters. For example, most of the macroscale parameters are functions of microscale parameters. There are also mutual influences between mechanistic variables and humanistic factors. It is impossible to elicit a set of complete and independent basic factors which influence the fatigue behavior. Thus, from a strictly theoretical point of view, a UFLP method which has included all the influential factors is impossible to achieve. We have to approach it using empirical methods and including several of the most significant factors. From a practical point of view, the less the parameters are used the easier the method is applied. So every empirical crack growth rate curve is a compromise of such an effort. Up to now hundreds of different empirical crack growth rate curves have been proposed and each has its advantages and drawbacks. Some milestones regarding this have been briefly introduced in Section 1.4.2.1 of Chapter 1.

Since every empirical model only includes a few parameters, it cannot be expected to explain every phenomenon. In that sense every model is not a unified method but the unified approach should be our target for long-term endeavors.

Typically, a general fatigue crack growth rate curve can be expressed by the following equation,

$$\frac{\mathrm{d}a}{\mathrm{d}N} = F(\sigma_{\max}, R, a, \cdots) \tag{5.1}$$

where $\sigma_{\max}$ and $\sigma_{\min}$ are maximum and minimum stresses in any given load cycle, R is the stress ratio defined as $R=\sigma_{\min}/\sigma_{\max}$, a is the crack length.

Based on the general format of the fatigue crack growth rate curve, some representative functions have been introduced in Chapter 1 and emphasized again in Chapter 3. And its more precise format will be proposed in the main sections of this chapter as the basis for developing the UFLP method.

5.2.2 Calculation of Fatigue Life

Cycle-by-cycle calculation is the most widely accepted procedure for simulating the real crack propagation. For each loading cycle, a crack growth increment Δa_i is separately calculated by integration and the results are accumulated to predict the fatigue crack length for the whole loading history. According to the improved constitutive model, the total crack length can be mathematically represented in Eq. (5.2) using a cycle-by-cycle integration procedure.

$$a = a_0 + \sum_{i=1}^{n} \Delta a_i = a_0 + \sum_{i=1}^{n} F(\sigma_{\max}, R, a, \cdots) \tag{5.2}$$

For the numerical implementation of the improved constitutive model, a cycle-by-cycle integration program starts with inputting the variables concerning material characteristics and geometric dimensions and the model parameters determined by fitting those data of the crack growth rate under constant amplitude loading. Another important input file is the loading spectrum which lists peak and valley values in the real order of the applied sequence. Fig. 3.4 gives a flow diagram of the UFLP method. The current unified method covers the region from physically small cracks to long cracks. Micro-cracks are not treated. So the basic input includes: 1) an equivalent initial crack size for the component, a_0; 2) the empirical crack growth rate curve in the region: $\frac{da}{dN}=F(\sigma_{\max}, R, a, \cdots)$ where $a_0 \le a < a_f$; 3) a loading history: ($\sigma_{\min}^{(i)}$, $\sigma_{\max}^{(i)}$), i=1, 2, $\cdots$, N. From these input data, we can plot the crack length vs. the number of cycles curve, the S-N curve and other information of interest. This will be done according to the following procedure:

Let the load process be represented by N consecutive load reversals $\{\sigma_{\min}^{(i)}, \sigma_{\max}^{(i)}\}$, i=1, 2, $\cdots$, N, then the corresponding crack lengths $\{a_i\}$ and these crack lengths $\{a_i\}$ can be calculated from the following equation,

$$\int_{a_{i-1}}^{a_i} \frac{1}{F(a, \sigma_{\min}^{(i)}, \sigma_{\max}^{(i)})} \mathrm{d}a = 1 \tag{5.3}$$

Note in the use of Eq. (5.3), the following two conditions must be satisfied,

$$P_1(\sigma_{\max}, R, a) > P_{1\text{th}} \tag{5.4}$$

$$P_2(\sigma_{\max}, R, a) < P_{2c} \tag{5.5}$$

where P_1 and P_2 are some parameters to be defined for the threshold condition and unstable fracture condition. If condition Eq. (5.4) is not satisfied, it means that this load reversal will not lead to crack propagation and it will be ignored. This reflects the existence of a fatigue limit. If condition Eq. (5.5) is violated, it means that this load reversal is large enough to cause unstable fracture. Therefore, using the present model, the characteristics at two extremes of a general S-N curve can also be reflected.

5.3 Development of a Unified Fatigue Crack Growth Rate Model

5.3.1 The Crack Growth Rate Model for Constant Amplitude Loading

As analyzed in sections above, in applying the unified procedure for FLP, the most important problem to solve is to establish a "correct" crack growth rate relation. However, this can only be regarded as a long-term target but cannot be realized in the near future. Therefore, at the moment we have to rely on the empirical methods and, in conjunction with experiments, we can make improvements to existing ones. As far as the application is concerned, we may simply choose the "best one" through some comparisons.

Up to now, many crack growth rate curves are available and, in order to determine the best, the following factors need to be considered:

(1) The function format of the crack growth rate curve;

(2) The quantity (or quantities) to be defined as the crack driving force;

(3) The threshold condition;

(4) The unstable fracture condition.

By comparing the existing ones, the modified constitutive relation developed by McEvily and his co-workers (McEvily, *et al.*, 1999; Ishihara and McEvily, 2002) has been found to be able to explain many phenomena observed. Their equations, Eqs. (3.1)–(3.3) have been introduced in Chapter 3.

But it is well known that a general crack growth rate curve exhibits a sigmoidal shape from near the threshold region to unstable fracture. McEvily's model is certainly only valid for near threshold and linear regions. For a cracked body, the unstable fracture condition is often defined by the condition,

$$K_{\max} = K_{\mathrm{C}} \tag{5.6}$$

where K_{C} is the fracture toughness of the material. It has been pointed out that the use of Eq. (3.3) to define $K_{\max}$ implies that the stress will never exceed the yield stress σ_Y even at static failure for "uncracked" plain specimens. In order to make Eq. (3.3) be also valid for the uncracked plain specimens, an empirical approach is adopted by replacing the yield strength σ_Y by another "virtual strength" (σ_V). The "virtual strength" (σ_V) is defined by the condition that the maximum stress of the uncracked (*i.e.* a=r_{e}) plain specimen is equal to the ultimate strength (σ_u) of the material in an unstable fracture condition. Based on this definition, the virtual strength can be determined from the following equation,

$$\left(1+\frac{Y(r_e)}{\sqrt{2}}\right)\sqrt{\pi\, r_e\left(\sec\frac{\pi}{2}\frac{\sigma_u}{\sigma_V}+1\right)}\sigma_u = K_C \tag{5.7}$$

and the solution of Eq.(5.7) is

$$\frac{\sigma_V}{\sigma_u}=\frac{\pi}{2}\cdot\frac{1}{\cos^{-1}(\frac{1}{\alpha^2-1})}, \qquad \alpha=\frac{K_C}{\sigma_u\sqrt{\pi r_e}\left(1+\frac{Y(r_e)}{\sqrt{2}}\right)}>\sqrt{2} \tag{5.8}$$

From Eq. (5.8) it can be seen that the virtual strength of the material will depend on the crack type (Y) together with three material parameters σ_u, K_C and r_e. Four of them form a new parameter α. Fig. 5.1 shows the influence of the parameter α on the ratio between virtual strength and ultimate strength. As α increases, the ratio approaches unity.

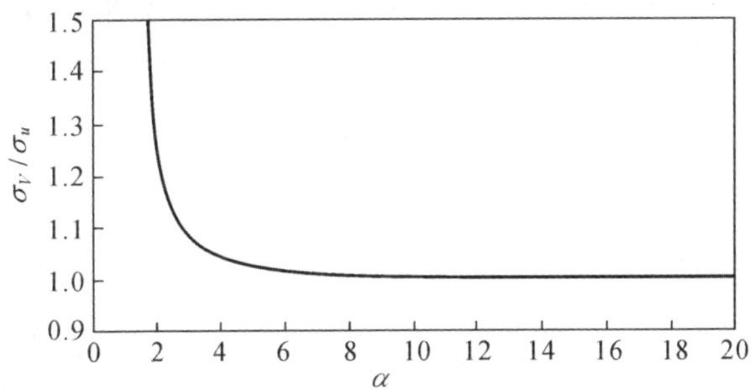

Fig. 5.1 Variation of virtual strength versus α parameter

The virtual strength of the material may represent the material strength at the limit of the “perfect” condition (r_e=0) while the actual ultimate strength of the material (σ_u) represents the strength under the condition that the defect size is equal to the inherent flaw length (r_e>0), a minimum crack size for engineering metals. What it says here is that the virtual strength is only an ideal value when all the inherent flaws have been removed. However, for any existing engineering metal, the inherent flaws exist and its maximum strength is the ultimate strength which will be lower than its virtual strength. The essence of this modification is a small change in the defined length of the plastic zone size. The new modified crack length is expressed as

$$a=\frac{a_{act}}{2}\left(\sec\frac{\pi}{2}\frac{\sigma}{\sigma_V}+1\right) \tag{5.9}$$

which is slightly shorter than the Dugdale expression using σ_Y instead of σ_V. If σ

is less than $0.9\sigma_Y$, then the effect of modification is small but when σ approaches σ_Y, the Dugdale expression will give a crack length of infinity, which is also not very meaningful. The new empirical formula of Eq. (5.9) will give a finite value of the crack length and this may be more reasonable. Obviously, this conclusion needs physical confirmation although the measurement of the plastic zone in the high stress level is a difficult task.

When a body contains a crack its ultimate strength (σ_{ua}) will be lower than the ultimate strength of the material (σ_u) and this ultimate strength can be determined from the following equation,

$$\sqrt{\pi r_e\left(\sec\frac{\pi}{2}\frac{\sigma_{ua}}{\sigma_V}+1\right)}\left(1+Y(a)\sqrt{\frac{a}{2r_e}}\right)\sigma_{ua}=K_C \tag{5.10}$$

It can be seen that σ_{ua} will depend on K_C and a, together with the crack type. As the crack length increases, the ultimate strength will decrease.

Thus the general constitutive relation for fatigue crack growth can be expressed as follows,

$$\frac{da}{dN}=\frac{AM^2}{1-\left(\frac{K_{max}}{K_C}\right)^n} \tag{5.11}$$

$$M=\sqrt{\pi r_e\left(\sec\frac{\pi}{2}\frac{\sigma_{max}}{\sigma_V}+1\right)}\left(1+Y\sqrt{\frac{a}{2r_e}}\right)\sigma_{max}(1-R)-\left(1-e^{-ka}\right)\left(K_{op\,max}-RK_{max}\right)-\Delta K_{eff\,th} \tag{5.12}$$

$$K_{max}=\sqrt{\pi r_e\left(\sec\frac{\pi}{2}\frac{\sigma_{max}}{\sigma_V}+1\right)}\left(1+Y(a)\sqrt{\frac{a}{2r_e}}\right)\sigma_{max} \tag{5.13}$$

Both $\Delta K_{eff\,th}$ and $K_{op\,max}$ are functions of the stress ratio R and let us temporally use the following general format to describe these two quantities,

$$\Delta K_{eff\,th}=\Delta K_{eff\,th\,0}(1-R)^{\gamma} \tag{5.14}$$

$$K_{op\,max}=K_{op\,max\,0}(1-R)^{\gamma} \tag{5.15}$$

where $K_{op\,max\,0}$ and $\Delta K_{eff\,th\,0}$ are corresponding values at the zero stress ratio. Different exponents γ_1 and γ_2 could be used, but for simplicity they are assumed to be the same following the assumption implicitly made by McEvily and his co-workers that these two quantities are close to each other. γ is between 0.5 and 1.0. γ can also take different values for positive R and negative R.

Thus, the independent model parameters for this general constitutive relation are: A, n, σ_u, r_e, K_C, $\Delta K_{\text{eff th 0}}$, $K_{\text{op max 0}}$, k, γ, a total of 9 parameters.

From the condition of M=0, the fatigue limit ($\sigma_{\infty a}$) in terms of the maximum stress for a body containing a crack can be determined from the following equation,

$$\left(1+Y(a)\sqrt{\frac{a}{2r_e}}\right)\sqrt{\pi\, r_e\left(\sec\frac{\pi}{2}\frac{\sigma_{\infty a}}{\sigma_V}+1\right)}\sigma_{\infty a}=\frac{\left[\left(1-e^{-ka}\right)K_{\text{op max 0}}+\Delta K_{\text{eff th 0}}\right](1-R)^{\gamma}}{1-Re^{-ka}} \tag{5.16}$$

$\sigma_{\infty a}$ is a function of σ_u, r_e, K_C, $\Delta K_{\text{eff th 0}}$, $K_{\text{op max 0}}$, k, γ, a and R. If a is set to be equal to r_e, then this equation will give the fatigue limit of material (σ_∞), *i.e.* the plain specimen fatigue limit. As the crack length increases, the fatigue limit will decrease.

The extended constitutive relation can be applied from the LCF down to the static failure to the HCF up to the fatigue limit and from "crack-free" plain specimens to cracked bodies. For a body containing an existing initial crack length a_0, the maximum fatigue stress has to be in the range between the fatigue limit ($\sigma_{\infty a}$) and ultimate strength (σ_{ua}). If the stress is lower than the fatigue limit, then the fatigue crack will not propagate. If the stress approaches the ultimate strength, static failure will occur. Therefore, the sigmoidal shape of the general crack growth rate curve also reflects the general shape of an *S-N* curve.

However, it is shown in Li, *et al.* (2006) that the McEvily model with the fixed slope of 2 is not in agreement with many experimental results. Thus, Wang, *et al.* (2008) proposed a further extended model based on the work of McEvily, *et al.* (1999) and Cui and Huang (2003). In this model, the parameter m is adopted to represent the slope of the fatigue crack growth rate curve for different materials. The extended McEvily model can be described by

$$\frac{\mathrm{d}a}{\mathrm{d}N}=\frac{AM^m}{1-\left(\frac{K_{\max}}{K_C}\right)} \tag{5.17}$$

This model shows a more promising capability to explain various fatigue phenomena. But there are still some shortcomings in the formula. The measurement method of the parameter $K_{\text{op max}}$ is still not expatiated very clearly in the relevant literature. It is not even a material constant, as the assumption of its independence from the SIF range ΔK level may not always be proved valid (Ishihara and McEvily, 2002) and it should be strongly dependent on the stress range (Matos, *et al.*, 2007). In order to reduce the uncertainty resulting from the measurement of this parameter, the item ΔK_{op} defined as the stress intensity factor range at the opening level and $K_{\text{op max}}$ introduced in the modified McEvily model

(Ishihara and McEvily, 2002) will be replaced by Newman's function (Newman, 1984; Newman, *et al.*, 1995) in the improved crack growth rate model. Another shortcoming in Eqs. (5.11) and (5.17) is the expression of the unstable fracture condition. Typically, the unstable fracture condition in the fatigue crack growth model is defined by $K_{max}=K_{Cf}$, where K_{Cf} is the fracture toughness of the material under fatigue condition, which should be strongly dependent on the load sequence. This problem has been specifically pointed out by Schijve (2003) and needs further experimental study. Then, the general constitutive relation in the extended McEvily model will be rewritten in the following form,

$$\frac{\mathrm{d}a}{\mathrm{d}N}=\frac{A(\Delta K_{\mathrm{eff}}-\Delta K_{\mathrm{eff\,th}})^m}{1-(K_{\max}/K_{\mathrm{Cf}})^n} \tag{5.18}$$

where

$$\Delta K_{\mathrm{eff}}=K_{\max}-K_{\mathrm{op}}=K_{\max}\cdot(1-f) \text{ and } f=K_{\mathrm{op}}/K_{\max} \tag{5.19}$$

$$f=f_1(R,\alpha',\sigma_{\max}) \tag{5.20}$$

$$\Delta K_{\mathrm{eff\,th}}=f_2(R) \tag{5.21}$$

$$K_{\mathrm{Cf}}=f_3(\lambda) \tag{5.22}$$

where the parameter α' represents the stress state condition around the crack tip. λ is the coefficient for calculation of the plastic zone size.

5.3.2 The Improved Crack Growth Rate Model Under VA Loading

Based on the improved crack growth rate model introduced above, the prediction results are in good agreement with their corresponding test data for a wide range of metal materials under constant amplitude loading, which will be shown in Section 5.5.2. However, most of engineering structures in real life are subjected to variable amplitude loading. Then it becomes more necessary to accurately predict the fatigue life of engineering structures under variable amplitude loading for both safe and economic design and operation. To attain this objective, the most typical variable amplitude load sequences must be analyzed first. The majority of the work has been carried out on the effect of single peak tensile overload, as it can lead to significant crack growth retardation. However, the inherent mechanisms responsible for the load interaction effects are not fully understood (Borrego, *et al.* 2008).

In metals, the load interaction effects are due to variations in crack closure stress (Skorupa, 1998; Dominguez, *et al.*, 1999; Prawoto, 2009; Irwin, 1960; Guo, 1999; Voorwald and Torres, 1991; Liu, *et al.*, 2006; Dowling, 2007; McEvily, *et al.*, 2004). Taking single overload as an example, it is well known that the fatigue crack growth is retarded after an application of a tensile overload. The crack growth retardation phenomenon can be explained as follows. During a tensile overload, the material ahead of the crack tip is plastically deformed over a distance equal to r_y. During unloading, a large monotonic compressive plastic zone is created as a result of being compressed by the surrounding elastic material. The residual compressive stress can weaken the applied stress and increase the level of the crack opening stress when the crack tip enters the monotonic compressive plastic zone.

As illustrated in Fig. 5.2(a), the crack opening level instantaneously rises to the maximum value, and then gradually recovers to the initial level under constant amplitude loading when the crack penetrates the large plastic zone in the subsequent cycles. On the contrary, a monotonic tensile plastic zone is produced as a result of a compressive underload. The residual tensile stress can decrease the level of crack opening stress and accelerate fatigue crack growth. This process is described in Fig. 5.2(b). To account for the change in the crack opening level, a modified coefficient, Φ, is introduced in the improved crack growth rate model as illustrated in Fig. 5.2, and the variation in the crack opening level is described in Eq. (5.25). And the improved crack growth rate model under variable amplitude loading can be expressed as Eqs. (5.23)–(5.25),

$$\frac{\mathrm{d}a}{\mathrm{d}N} = \frac{AM^m}{1-\left(K_{\max}/K_{\mathrm{C}}\right)^n} \tag{5.23}$$

$$M = K_{\max} - K'_{\mathrm{op}} - \Delta K_{\mathrm{eff\,th}} \tag{5.24}$$

$$K'_{\mathrm{op}} = \Phi K_{\mathrm{op}} = \Phi f_{\mathrm{op}} K_{\max} \tag{5.25}$$

As overload is independently applied in the constant amplitude baseline loading, the modified factor of the crack opening level can be expressed as Eq. (5.26) based on the Wheeler model.

$$\Phi = \begin{cases} \left(\dfrac{a_{\mathrm{OL}} + r_{\mathrm{OL}} - r_{yi}}{a_i}\right)^{\gamma} & a_{\mathrm{OL}} \le a_i \le a_{\mathrm{OL}} + r_{\mathrm{OL}} - r_{yi} \\ 1 & a_i > a_{\mathrm{OL}} + r_{\mathrm{OL}} - r_{yi} \end{cases} \tag{5.26}$$

On the contrary, when the underload is independently encountered in the constant amplitude baseline loading, the modified factor of the crack opening level can be expressed as Eq. (5.27),

$$\Phi = \begin{cases} \left(\dfrac{a_i + r_{yi} - r_{\text{UL}}}{a_{\text{UL}}} \right)^{\gamma} & a_{\text{UL}} \le a_i \le a_{\text{UL}} + r_{\text{UL}} - r_{yi} \\ 1 & a_i > a_{\text{UL}} + r_{\text{UL}} - r_{yi} \end{cases} \tag{5.27}$$

where γ is a shaping exponent determined by fitting the recovering period of the experimental data. a_i is the crack length in the current ith cycle; r_{yi} is the plastic zone in the current ith cycle; r_{OL} is the plastic zone size when the overload cycle occurs; r_{UL} is the plastic zone size when the underload cycle occurs.

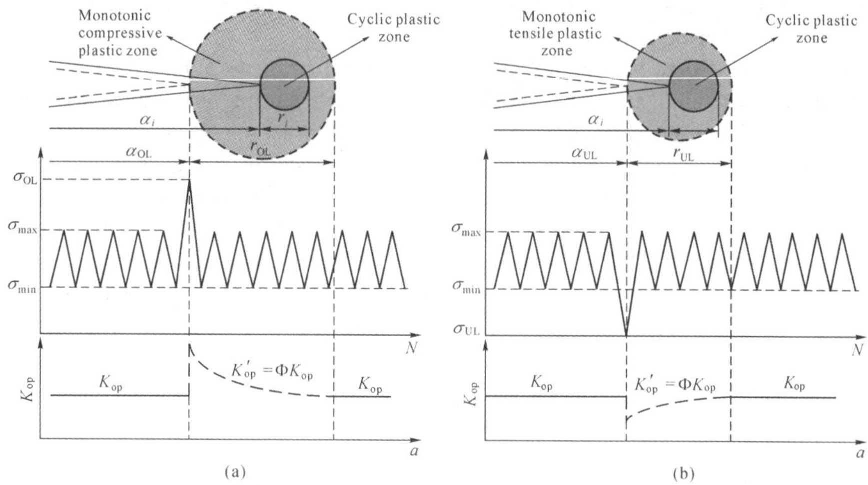

Fig. 5.2 A schematic representation of the assumed change of K_{op} under load sequence with single overload (a) or underload (b) (Chen, *et al.*, 2012) (with the permission of John Wiley and Sons, Inc.)

The modification model introduced above has the capacity for modeling crack growth retardation or acceleration due to the independent application of single and multiple overloads or underloads in an otherwise constant amplitude spectrum. However, this model cannot deal with the effects of the combination of overload and underload. The application of an underload immediately following an overload diminishes the effect of the latter to a greater or lesser extent depending on their relative values (Rama, *et al.*, 2004). It was observed that the retardation effect is lesser due to a tensile overload followed by a subsequent compressive overload, which can even lead to an acceleration of the crack growth (Rama, *et al.*, 2007), while an underload cycle applied prior to overloading has little change in the amount of retardation (Taheri, *et al.*, 2003; Bacila, *et al.*, 2007). Therefore, the crack growth acceleration caused by an underload occurring just before an overload is usually neglected. Herein, the effect of an underload that follows an overload is modeled by subtracting the plastic zone size increment caused by the underload from the plastic zone size of the overload. The schematic representation of modifying plastic zone sizes is illustrated in Fig. 5.3. Accounting for a reduction in the amount of retardation as a result of underload following the modification factor, Φ, can be rewritten as Eq. (5.28).

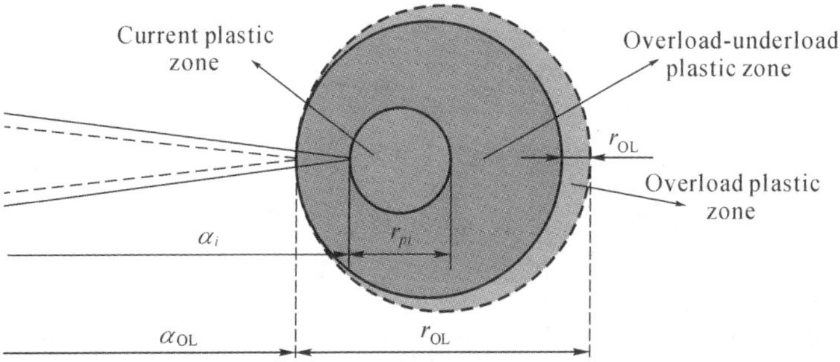

Fig. 5.3 Schematic representation of modifying plastic zone sizes to account for the effect of underload following overloading (Chen, *et al.*, 2012) (with the permission of John Wiley and Sons, Inc.)

$$\Phi = \begin{cases} \left(\dfrac{a_{OL} + r_{OL} - r_{yi} - r_{UL}}{a_i} \right)^{\gamma} & a_{OL} \le a_i \le a_{OL} + r_{OL} - r_{yi} - r_{UL} \\ 1 & a_i > a_{OL} + r_{OL} - r_{yi} - r_{UL} \end{cases} \quad (5.28)$$

- *Plastic zone size*

In considering the effect of load sequence interactions, it is important to estimate the plastic zone size in front of the crack tip as addressed above. In the homogenous materials, the plastic zone size is often calculated by using Irwin's method. Based on von Mises yielding criterion and stress release principle, Irwin (1960) obtained the analytical 2D plastic zone size of a center-cracked plate with finite width as follows,

$$r_y = \lambda \left(\frac{K_{max}}{\sigma_Y} \right)^2, \quad \alpha = \begin{cases} \dfrac{1}{\pi} & \text{for plane stress;} \\ \dfrac{1}{3\pi} & \text{for plane strain} \end{cases} \quad (5.29)$$

Subsequently, some segmented calculation expressions of the plastic zone size have been developed (Guo, 1999; Voorwald, *et al.*, 1991). To make the calculation precise and easy, the plastic zone size ahead of the crack tip is modeled as a continuous function of the maximum applied SIF, yield strength and plate thickness in the reference (Huang, *et al.*, 2008). To further consider the effects of the material Poisson's ratio and extend it to power-hardening material, Liu, *et al.* (2006) proposed the following equation,

$$\lambda = \frac{(1-1.65\nu)^2}{5} - \frac{1}{20n'}((1-1.65\nu)^2)^{\frac{1}{n'}} + \frac{\frac{1}{\pi} - \frac{1}{2.2n'}\left(\frac{1}{\pi}\right)^{\frac{1}{n'}} - \left(\frac{(1-1.65\nu)^2}{5} - \frac{1}{20n'}((1-1.65\nu)^2)^{\frac{1}{n'}}\right)}{\left(1 + \frac{t/(K_{\max}/\sigma_Y)^2}{1+1/n'}\right)^{1.6+1/n'}} \tag{5.30}$$

For the elastic-perfectly-plastic material (y_i), Eq. (5.30) can be simplified to the following form,

$$\lambda = \frac{(1-1.65\nu)^2}{5} + \frac{\frac{1}{\pi} - \frac{(1-1.65\nu)^2}{5}}{(1 + t/(K_{\max}/\sigma_Y)^2)^{1.6}} \tag{5.31}$$

The plastic zone size mentioned above is also called the monotonic plastic zone. As the minimum cyclic load in a cycle is approached, yielding in compression occurs in a region of a smaller size, called the cyclic plastic zone, as expressed in Eq. (5.32) (Dowling, 2007),

$$r_y = \lambda \left(\frac{\Delta K}{2\sigma_Y}\right)^2 \tag{5.32}$$

The tensile plastic zone diameter can be calculated using the above formula. However, evaluation of the compressive plastic zone size will present some uncertainties. Based on the concept of a cyclic plastic zone described by Suresh (1998), Eq. (5.33) was used as an estimate of the compressive plastic zone size,

$$r_{\mathrm{UL}} = \lambda \left(\frac{\Delta K_{\mathrm{UL}}}{2\sigma_{\mathrm{Yld}}}\right)^2 = \lambda \left(\frac{K_{\mathrm{min,CAL}} - K_{\mathrm{min,UL}}}{2\sigma_Y}\right)^2 \tag{5.33}$$

where σ_{yld} is the material's yield stress (assumed equal in tension and compression), and ΔK_{UL} is estimated as $K_{\mathrm{min,CAL}}$–$K_{\mathrm{min,UL}}$.

5.3.3 Establishment of Cycle-by-Cycle Integration Procedure

Based on the model introduced above considering the sequence effect, the loading history must be defined in such a way that the real order of the loading cycles can be reproduced. Under these conditions, cycle-by-cycle calculation is the most widely accepted procedure for simulating the real crack propagation (Dominguez, *et al.*, 1999). For each loading cycle, a crack growth increment Δa_i is separately calculated by integration and the results are accumulated to predict the fatigue crack length for the whole loading history.

$$a = a_0 + \sum_{i=1}^{n} \Delta a_i = a_0 + \sum_{i=1}^{n} f(\Delta K, R, \cdots) \tag{5.34}$$

Because the load sequence is irrelevant for the application of linear damage accumulation, very often the load-time history is counted by a counting method, *e.g.* rainflow cycle counting (Dowling, 1972). In order to take the retardation effect into account, Wheeler introduced a modified factor in front of the equation of the fatigue crack growth rate (Yuen and Taheri, 2006), which is one of the most widely employed but the physical meaning of the factor is ambiguous as it does not reflect the major cause inducing crack growth retardation phenomenon. The improved constitutive model reckons that the crack growth retardation is due to variations in crack closure level so a magnification factor is introduced to the value of crack opening level. According to the improved constitutive model, the total crack length can be mathematically represented in Eq. (5.35) using a cycle-by-cycle integration procedure.

$$a = a_0 + \sum_{i=1}^{n} \Delta a_i = a_0 + \sum_{i=1}^{n} f(\Delta K, R, \Phi, \cdots) \tag{5.35}$$

The integration program has been written using Compaq Visual Fortran software. The program starts with inputting the variables concerning material characteristics and geometric dimensions and the model parameters determined by fitting those data of the crack growth rate under constant amplitude loading. Another important input file is the loading spectrum which lists peak and valley values in the real order of the applied sequence. The flow diagram for a cycle-by-cycle integration procedure of the improved constitutive model is illustrated in Fig. 5.4. Calculation of the fatigue crack growth rate is the core of this program and the correlative expressions are also listed in Fig. 5.4.

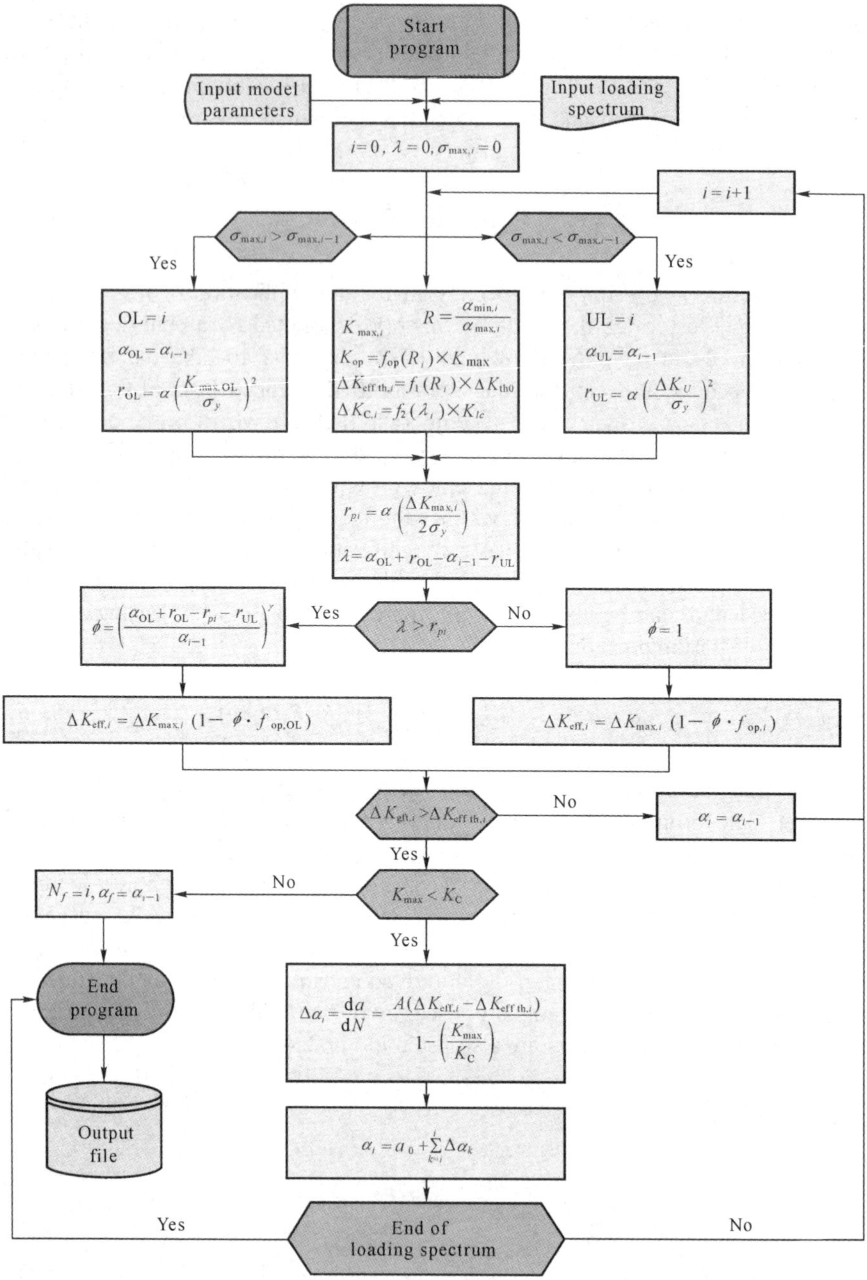

Fig. 5.4 Flow diagram for the cycle-by-cycle integration procedure of the improved crack growth rate model (Chen, *et al.*, 2012) (with the permission of John Wiley and Sons, Inc.)

5.3.4 Discussion of Model Parameters

5.3.4.1 Unstable fracture condition

In view of the lack of this value after costly experiments, the static fracture toughness measured from an easier test is usually adopted as the unstable fracture condition in many models. Typically, the plane stress fracture toughness is adopted for prediction. The difference between fatigue fracture toughness K_{Cf} and static fracture toughness K_C will be left for future study. But, it is well known that fracture toughness is affected by the crack tip stress/strain state, load conditions and environment, *etc*. We will aim at analyzing the effect of the crack tip stress/strain state and applying the results to the UFLP model. The crack tip stress/strain state could be expressed by a constraint factor, which has been pointed out to be a function of plate thickness, t, SIF, K, and yield stress of the material, K_Y, loading conditions and environment *etc*. Generally, the value of $t/(K/\sigma_Y)^2$ is used as a control parameter for the crack tip stress/strain state and consequently for the fracture toughness. That means the fracture toughness should be considered as a variable when $t/(K/\sigma_Y)^2$ is changed during crack propagation, and the value of the fracture toughness is lower than that under the plane stress state but higher than K_{IC} (fracture toughness under the plane strain state). In the current analysis, we will denote the plane stress toughness by K_C, while K_{Cf} is defined as a variable changing with the crack tip stress/strain state.

It is well-known that when $t/(K/\sigma_Y)^2$ increases, the fracture toughness value tends to significantly decrease because the crack tip stress state approaches the plane strain case. On the contrary, when $t/(K/\sigma_Y)^2$ decreases, the fracture toughness value tends to significantly increase because the crack tip stress state approaches the plane stress case. Under the plane stress state and plane strain state, the notations K_C and K_{IC} are used, respectively.

Dong, *et al.* (2000) proposed a relationship between the fracture mechanics parameters K_C and K_{IC} using the energy conservation principle, namely,

$$K_C = \frac{r_{y\sigma}}{r_{y\varepsilon}} \cdot \sqrt{1-\nu^2} K_{IC} \tag{5.36}$$

where $r_{y\sigma}$ and $r_{y\varepsilon}$ stand for the crack tip plastic zone size under the plane stress state and plane strain state, respectively.

Irwin's model (1960) is used here for the calculation of $r_{y\sigma}$ and $r_{y\varepsilon}$, namely,

$$r_{y\sigma} = \frac{1}{\pi}\left(\frac{K}{\sigma_Y}\right)^2 \tag{5.37}$$

$$r_{y\varepsilon} = \frac{(1-2\nu)^2}{\pi}\left(\frac{K}{\sigma_Y}\right)^2 \tag{5.38}$$

Then the relationship between the two fracture mechanics parameters can be obtained. In order to establish the equation of K_C as a function of the stress state parameter $t/(K/\sigma_Y)^2$, we assume that

$$\begin{cases} K_{\mathrm{Cf}} = f_{K_\mathrm{C}} \cdot K_{\mathrm{IC}} \\ f_{K_\mathrm{C}} = C \cdot \dfrac{r_y}{r_{y\varepsilon}} + D \end{cases} \tag{5.39}$$

Here, the parameters of C and D must satisfy the following conditions,

$$\begin{cases} C + D = 1 & \text{when } r_y = r_{y\varepsilon} \\ C \cdot \dfrac{1}{(1-2\nu)^2} + D = \dfrac{\sqrt{1-\nu^2}}{(1-2\nu)^2} & \text{when } r_y = r_{y\sigma} \end{cases} \tag{5.40}$$

The solution of Eq. (5.40) is: $C = \dfrac{(1-2\nu)^2 - \sqrt{1-\nu^2}}{(1-2\nu)^2 - 1}$ and $D = \dfrac{\sqrt{1-\nu^2} - 1}{(1-2\nu)^2 - 1}$.

The value of r_y can be calculated by Eqs. (5.29) and (5.30).

Fig. 5.5 shows a schematic representation of the K_C value versus $t/(K/\sigma_Y)^2$ as an example using Eq. (5.40).

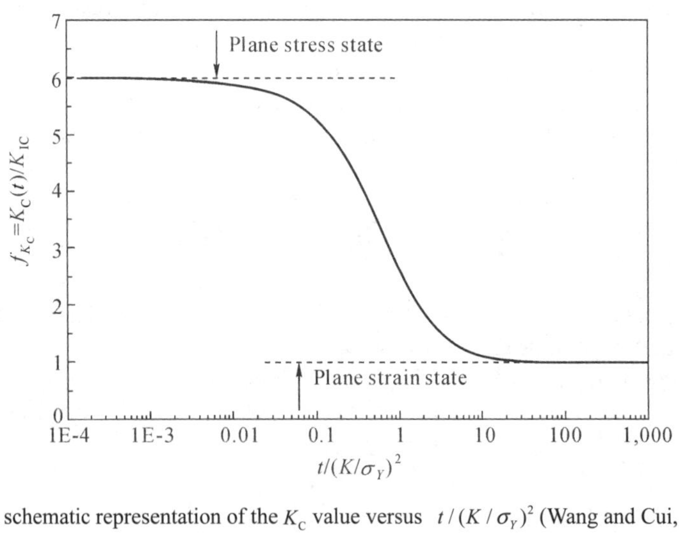

Fig. 5.5 A schematic representation of the K_C value versus $t/(K/\sigma_Y)^2$ (Wang and Cui, 2010a)

5.3.4.2 Crack opening level

In the widely-accepted empirical formulas of Newman's model for calculating the SIF at the opening level used in the prediction model, only two specified conditions are considered, *i.e.* plane stress condition and plane strain condition. In this section, a continuous empirical formula proposed for estimating the value of K_{Cf} during crack propagation will be used in conjunction with the calculation of the SIF at the opening level in the extended McEvily model. In addition, a modified continuous equation for the crack tip stress/strain constraint factor is adopted combined with Newman's model for the SIF at the crack opening level. Then the extended McEvily model will be further improved for application.

Newman and his co-workers (1984; 1995) concluded from finite element calculation that the crack closure will also depend on the maximum stress level σ_{max}. They proposed a crack opening function *f*, defined as the ratio K_{op}/K_{max}. It is expressed as a segmented function of *R*, σ_{max}, the material flow strength σ_{fl} and also a plane stress/strain constraint factor α', namely,

$$f_{op} = \begin{cases} \max\{R, A_0 + A_1 R + A_2 R^2 + A_3 R^3\} & 0 \le R < 1 \\ A_0 + A_1 R & -2 \le R < 0 \end{cases} \tag{5.41}$$

$$\begin{cases} A_0 = (0.825 - 0.34\alpha' + 0.05\alpha'^2)\cdot[\cos(\pi\sigma_{max}/2\sigma_{fl})]^{1/\alpha'} \\ A_1 = (0.415 - 0.071\alpha')\cdot\sigma_{max}/\sigma_{fl} \\ A_2 = 1 - A_0 - A_1 - A_3 \\ A_3 = 2A_0 + A_1 - 1 \\ \sigma_{fl} = (\sigma_Y + \sigma_u)/2 \\ \alpha' = \begin{cases} 1 & \text{for plane stress;} \\ 1/(1-2\nu) & \text{for plane strain} \end{cases} \end{cases} \tag{5.42}$$

In Newman's function, two specified values of α' are adopted for the plane stress state and plane strain state, respectively. In order to extend the equation for more general conditions with the crack increasing, the effect of $t/(K/\sigma_Y)^2$ on the crack tip stress/strain constraint factor should be considered.

According to Liu, *et al.* (2006), the crack tip stress/strain constraint factor could be expressed as a function of $t/(K/\sigma_Y)^2$ and ν, namely,

$$\alpha' = \begin{cases} 1 & t/(K/\sigma_Y)^2 \le \dfrac{1}{\pi} \\ 1 + \dfrac{2\nu[t/(K/\sigma_Y)^2 - 1/\pi]}{(1-2\nu)(2.5 - 1/\pi)} & \dfrac{1}{\pi} < t/(K/\sigma_Y)^2 < 2.5 \\ \dfrac{1}{1-2\nu} & t/(K/\sigma_Y)^2 \ge 2.5 \end{cases} \tag{5.43}$$

As it is inconvenient to use the segmented expression in Eq. (5.43) for quick calculation, we will use a continuous equation here, which is obtained by fitting the data between those under the plane stress state and plane strain state and plotted in Fig. 5.6, namely,

$$\alpha' = \frac{1}{1-2\nu} + \frac{1 - \dfrac{1}{1-2\nu}}{[1 + 0.8861 \cdot (t/(K/\sigma_Y)^2)^{3.2251}]^{0.75952}} \tag{5.44}$$

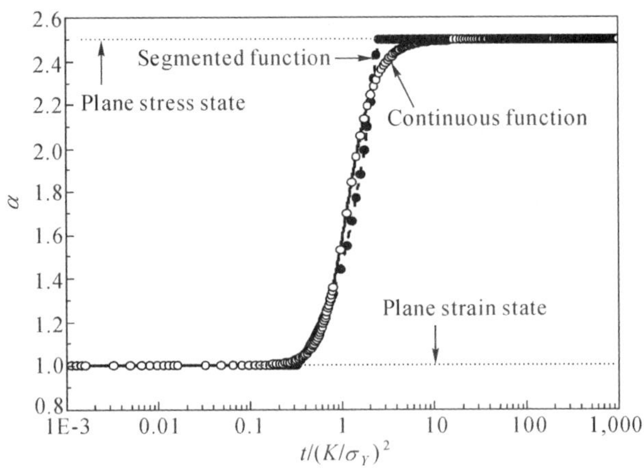

Fig. 5.6 A schematic representation of the α' value versus $t/(K/\sigma_Y)^2$ (Wang and Cui, 2010a)

5.3.4.3 Threshold value of effective SIF range

An improved equation was proposed for prediction of the effect of the load ratio on the fatigue threshold ΔK_{th} (*linear*) from a comprehensive analysis of different kinds of materials (Huang and Moan, 2007). The equation is written in a segmented function for three regions of the load ratio as follows,

$$\Delta K_{\text{th}} / \Delta K_{\text{th}\,0} = \begin{cases} (1-R)^{\beta_1} & -5 \le R < 0 \\ (1-R)^{\beta} & 0 \le R < 0.5 \\ (1.05 - 1.4R + 0.6R^2)^{\beta} & 0.5 \le R < 1 \end{cases} \tag{5.45}$$

where β and β_1 are material constants. And, for quick analysis, the values of the two quantities can be set as the following constants for aluminum and steel respectively,

$$\begin{cases} \beta = 0.7 & \text{for aluminum} \\ \beta_1 = 0.84 & \text{for aluminum} \\ \beta = 0.3 & \text{for steel} \\ \beta_1 = 0.5 & \text{for steel} \end{cases} \tag{5.46}$$

Eq. (5.45) is recommended here to be the function linking the values of ΔK_{th} (*linear*) of a wide range of load ratios.

When ΔK_{th} (*linear*) (R) is obtained, the value of a_{th} (R) can be easily known, and consequently the value of $\Delta K_{\text{eff th}}$ (R) will be calculated by the following equation,

$$\Delta K_{\text{eff th}}(R) = K_{\max}(a_{\text{th}}) \cdot \left[1 - f_{\text{op}}(a_{\text{th}})\right] \tag{5.47}$$

5.3.4.4 Effect of specimen thickness

An example will be used here for discussion of the effect of specimen thickness using the model introduced above. During the simulation, we used a multivariable nonlinear regression method to determine the parameters in the model first according to the test data for the thick plate and obtained the values of A, m, n, r_e which are equal to $1.5144\times10^{-6}(\text{MPa})^{-m} \cdot \text{m}^{1-m/2}$, 1.36204, 6 and 3.15×10^{-7}m respectively. The detailed approaches to estimate parameters will be introduced in Section 5.4. These values for the thick plate will be used for the calculation of the crack growth rate for the thin plate. The simulation results and their comparison with the test data (Bao and McEvily, 1998) are illustrated in Fig. 5.7. It can be seen that they agree with each other very well. The comparison proves the validity of the method proposed above.

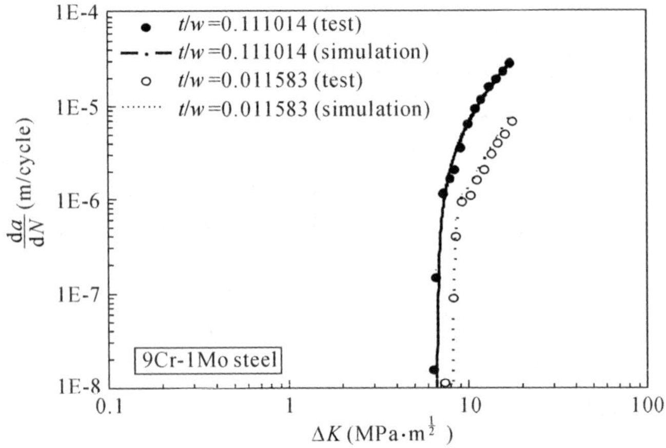

Fig. 5.7 Comparison between test data (Bao, 1998) and simulation results of the effect of plate thickness on the curve of crack growth rate versus SIF range (Wang and Cui, 2010a)

However, Eqs. (5.18)–(5.22) are mostly used in the cases of 1D or 2D problems. In order to extend the concepts in the model to surface cracks, the concept of "thickness" at each point along the crack front should be re-defined. The concept of equivalent thickness (Zhang and Guo, 2005) will be applied to the analysis here. Fig. 5.8 shows the schematic illustration for calculation of equivalent thickness at each point. If there is a semi-elliptical crack A_i-O-C_i in a plate, then the general equation of the semi-elliptical shape is

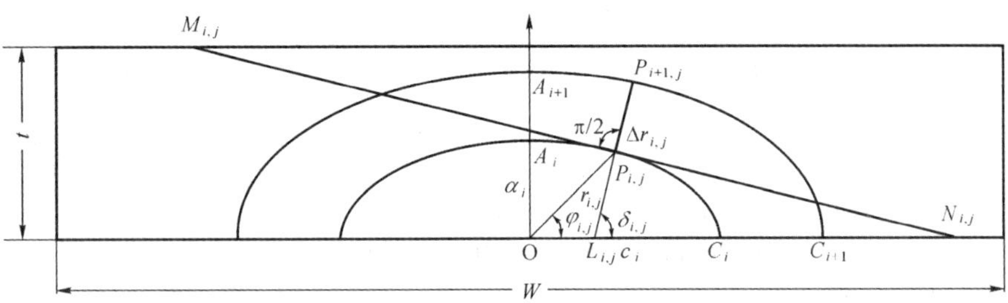

Fig. 5.8 Schematic illustration for calculation of equivalent thickness at each point

$$\frac{x^2}{c_i^2}+\frac{y^2}{a_i^2}=1 \qquad (y \geq 0) \tag{5.48}$$

It is assumed that incremental growth of the crack at points on the crack front is perpendicular to the crack front (Zhang and Guo, 2005). And the equivalent thickness of a certain point, $P_{i,j}$, is defined to be twice the minimum value of $|P_{i,j}N_{i,j}|$ and $|P_{i,j}M_{i,j}|$, where $N_{i,j}$ and $M_{i,j}$ are respectively the intersection point of the tangent line across point $P_{i,j}$ and the boundary lines OC (y=0) and OA (y=t). For a certain point $P_{i,j}$:($r_{i,j}\cdot\cos\varphi_{i,j}$,$r_{i,j}\cdot\sin\varphi_{i,j}$) in the crack front,

$$r_{i,j}^2 = \frac{1}{\cos^2\varphi_{i,j}/c_i^2 + \sin^2\varphi_{i,j}/a_i^2} \tag{5.49}$$

The equation of the tangent line can be written as

$$\frac{r_{i,j}\cos\varphi_{i,j} \cdot x}{c_i^2} + \frac{r_{i,j}\sin\varphi_{i,j} \cdot y}{a_i^2} = 1 \quad (y \geq 0) \tag{5.50}$$

When y=0, $x_{N_{i,j}} = \dfrac{c^2}{r\cos\theta}$

When y=t, $x_{M_{i,j}} = \dfrac{c^2}{r\cos\theta}\left(1 - \dfrac{r\sin\theta \cdot t}{b^2}\right)$

$N_{i,j}:\left(x_{N_{i,j}}, 0\right)$

$M_{i,j}:\left(x_{M_{i,j}}, t\right)$

$$t_{eq_{i,j}} = \min\left(\left|P_{i,j}N_{i,j}\right|, \left|P_{i,j}M_{i,j}\right|\right) \cdot 2 \tag{5.51}$$

Then the growth rate of point $P_{i,j}$ can be calculated according to Eqs. (5.48)–(5.51). Under a certain cycle, the point $P_{i,j}$ will reach point $P_{i+1,j}$ supposing that the initial crack length is $r_{i,j}$ and $\Delta r_{i,j} = \dfrac{dr}{dN}(r_{i,j,\cdots}) \cdot \Delta N$.

The equation of the perpendicular line can be written as

$$y - r_{i,j}\sin\varphi_{i,j} = \frac{c_i^2 \sin\varphi_{i,j}}{a_i^2 \cos\varphi_{i,j}}\left(x - r_{i,j}\cos\varphi_{i,j}\right) \tag{5.52}$$

When y=0, $x_{L_{i,j}} = r_{i,j}\cos\theta_{i,j}\left(1 - \dfrac{a_i^2}{c_i^2}\right)$

$L_{i,j}:\left(x_{L_{i,j}}, 0\right)$

Then the coordinates of $P_{i+1,j}$ can be obtained.

$$\begin{cases} x_{P_{i,j}} = \left(\left|P_{i,j}L_{i,j}\right| + \Delta r_{i,j}\right) \cdot \cos\delta_{i,j} + x_{Li,j} \\ y_{P_{i,j}} = \left(\left|P_{i,j}L_{i,j}\right| + \Delta r_{i,j}\right) \cdot \sin\delta_{i,j} \end{cases} \tag{5.53}$$

The propagated crack shape can be depicted by calculating the coordinates of enough points along the new crack front. Typically, the shape of the new crack front will not remain semi-elliptical. Then the function of the SIF coefficient Y should be calculated by finite element analysis at each step. To save calculation time, it is assumed that the new front is still semi-elliptical but the ellipse should be obtained by fitting the calculated points. The parameters of the new semi-elliptical crack will be used for the next step of the calculation and then Eqs. (5.18) to (5.22) can be used reasonably. Accordingly, the crack shape ratio evolution can be obtained through "cycle-by-cycle" calculation.

Putra and Schijve (1992) conducted tests on a series of 7075-T6 plates with a semi-elliptical crack and Zhang and Guo (2005) gave predicted shape ratios using Paris' law. They are depicted together with the current predicted results in Fig. 5.9. When the shape ratio *a/t* is small, the results from Paris' law and the present model are similar and agree well with the test data. But when the shape ratio is larger with crack propagation, the results from Paris' law show a discrepancy compared to the test. On the contrary, the present model can give satisfactory evaluation results by considering the nonlinear effect when calculating the stress intensity factors. It should be noted that the values of *a* and *c* for each step in Fig. 5.9 are the fitted ones. Fig. 5.10 gives the current predicted crack profile of the semi-elliptical crack with a_0/t=0.2, a_0/c_0=1.0 in Putra and Schijve (1992). Apparent growth retardation can be observed in the predicted fronts.

A similar conclusion can be obtained by comparing the current predicted results with the test data on 7075-T6(51) (Kim and Song, 1992) and the predicted results (Zhang and Guo, 2005) based on Paris' law depicted in Figs. 5.11 and 5.12. It can be seen that the present model can give more precise results compared to Paris' law. But a different shape evolution tendency is observed compared to Fig. 5.9. The difference results from the values of the initial crack shape ratios. A larger initial *a/c* will result in an ascending trend with crack propagation while the opposite trend is for a small initial *a/c*.

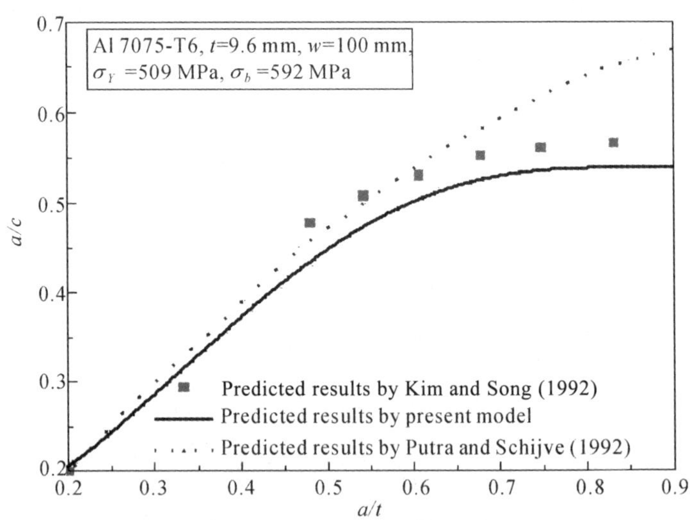

Fig. 5.9 Predicted shape ratios comparing Putra and Schijve (1992) with Zhang and Guo (2005)

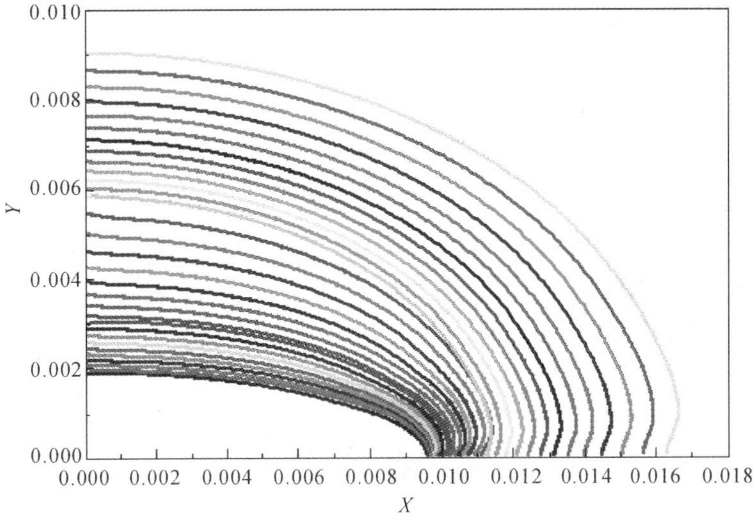

Fig. 5.10 Predicted crack profile of the semi-elliptical crack with a_0/t=0.2, a_0/c_0=1.0

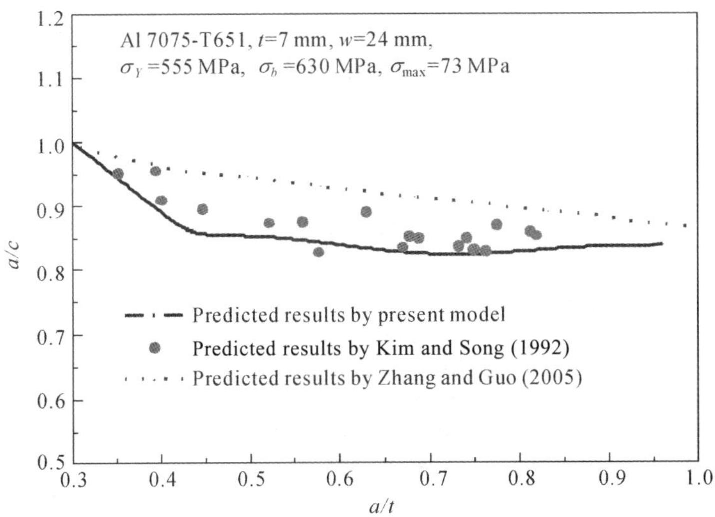

Fig. 5.11 Predicted shape ratios comparing Kim and Song (1992) with Zhang and Guo (2005)

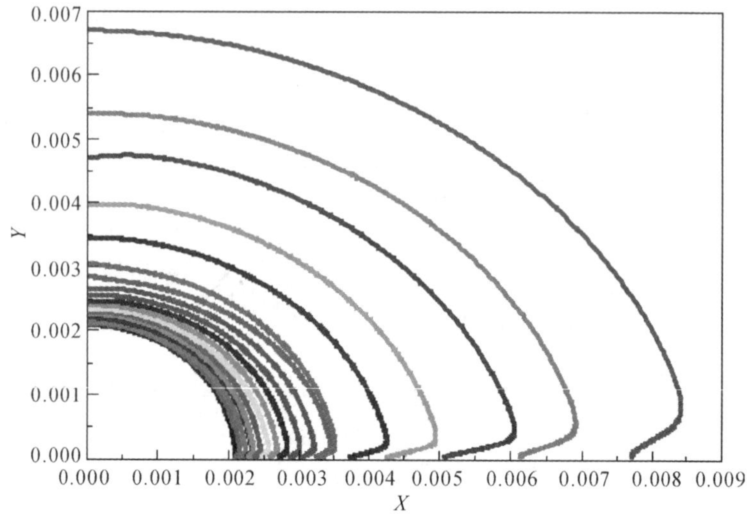

Fig. 5.12 Predicted crack profile of the semi-elliptical crack with a_0/t=0.2, a_0/c_0=0.2

5.4 Engineering Approaches to Determine the Parameters in the Improved Model

5.4.1 General Methods to Estimate the Model Parameters

Costly fatigue tests for determining so many model parameters will certainly be an obstacle for its engineering application, as one may realize that fatigue failure is the consequence governed by these model parameters if this model really reflects the correct fatigue crack growth behavior. Thus, we may be able to estimate the model parameters from any set of fatigue test data. In this section, the approaches to estimate the model parameters will be introduced.

If the model parameters are regarded as deterministic, then determination of the model parameters from available test data is basically an optimization problem. The general form of the model can be expressed as follows,

$$y_i = f(\underline{X}^{(i)};\underline{\theta}) + e_i \qquad i = 1,\ 2,\cdots,N \qquad (5.54)$$

where the observed values of the y_i constitute the responses or values of the dependent variables $\underline{X}$; the known $\underline{X}^{(i)} = (x_1^{(i)}, x_2^{(i)}, \cdots, x_K^{(i)})$ are the vectors of values of the independent K variables; f is a known function of an unknown parameter vector $\underline{\theta}$=(θ_1,θ_2, ⋯ ,θ_M); and the values of e_i are independently distributed normal errors.

The residuals for the model are

$$e_i(\underline{\theta}) = y_i - f(\underline{X}^{(i)}; \underline{\theta}) \qquad i = 1, 2, \cdots, N \tag{5.55}$$

A value of $\underline{\theta}$ can be determined by minimizing the following error function,

$$Error = \sum_{i=1}^{N} \left[y_i - f(\underline{X}^{(i)}; \underline{\theta}) \right]^2 \tag{5.56}$$

5.4.2 Estimation Method from Crack Growth Rate Data

Theoretically speaking, all model parameters can be regarded as variables to be determined but this involves very complicated mathematical manipulation. For simplicity, we generally treat some of the parameters as constants and others as model parameters. If we know N testing points: $\left[\sigma_{\max}, R_i, a_i, \left(\frac{\mathrm{d}a}{\mathrm{d}N} \right)_i \right]$, then the error function is

$$Error = \sum_{i=1}^{N} \left[\left(\frac{\mathrm{d}a}{\mathrm{d}N} \right)_i - \frac{AM_i^m}{1 - \left(K_{\max i} / K_{\mathrm{C}} \right)^n} \right]^2 \tag{5.57}$$

Based on these equations one can determine an optimal set of the parameters. Nonlinear curve fitting theory can be used to solve this problem (Zhou and Cui, 2003; Wang and Cui, 2009b).

5.4.3 Estimation Method from a-N Curve

From the crack growth rate, one can determine the a-N curve

$$N = \int_{a_0}^{a} \frac{1 - (K_{\max} / K_{\mathrm{C}})^n}{AM^m} \mathrm{d}a \tag{5.58}$$

If we know N testing points $[a_i, N_i]$, namely the a-N curve is available, we can estimate the parameters in the crack growth rate model by the error function as follows,

$$Error = \sum_{i=1}^{N}\left[N_i - \int_{a_0}^{a_i}\frac{1-\left(K_{\max}/K_{\mathrm{C}}\right)^n}{AM^m}\mathrm{d}a\right]^2 \tag{5.59}$$

In these equations, one needs to calculate a lot of integrations which is a difficult task. However, we can estimate some parameters by tensile test data of the material and leave only A, m, n to be determined by the method introduced here.

In order to estimate the model parameters (A, m, n), the nonlinear least-square fitting method is usually applied, which includes the establishment and the optimization calculation of the error function. The particular explanation of the mathematical manipulation of the nonlinear least-square fitting method will be introduced below by the flow diagram of the calculation shown in Fig. 5.13. To minimize the value of the error function, the Jacobian matrix must be calculated based on the explicit expression of the integral function inside the right side of Eq. (5.59), which could be obtained by executing the operation of numerical integration depending on the variable a first. Setting the $\boldsymbol{Jacobian}^{\mathrm{T}}$ equal to zero, we can obtain the optimal value of the parameters (A, m, n) by solving the nonlinear equations group.

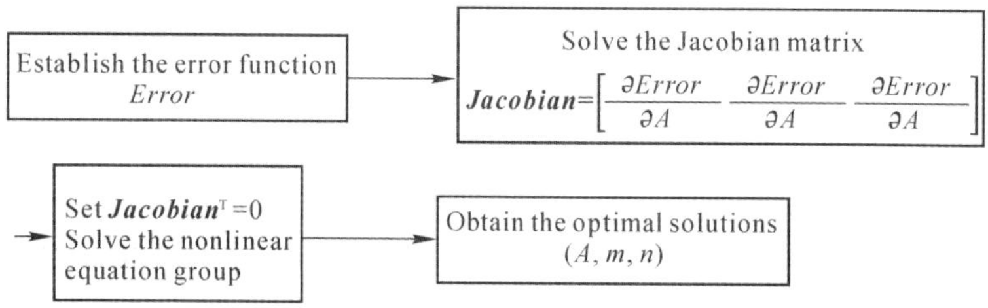

Fig. 5.13 Flow diagram of the Nonlinear Least-Square Fitting Method (Cui, *et al.*, 2011) (with the permission of Elsevier)

However, the operation of the numerical integration calculation and the nonlinear equation group solution mentioned above will become very difficult when the error function is complex. In such cases, a genetic algorithm may be an effective way to avoid the difficulty of these mathematical manipulations. The schematic diagram of a genetic algorithm is illustrated in Fig. 5.14. The procedure will start with coding in some way and then the initial population (A_0, m_0, n_0) can be obtained. By substituting (A_0, m_0, n_0) into Eq. (5.59), we can calculate the fitness of every individual, based on which some individuals are selected as the next generation population and others are eliminated. Subsequently, crossover and mutation are executed according to the genetic operators and the next generation population (A_1, m_1, n_1) is produced. From the heredity of enough generations, the optimal individual (A_n, m_n, n_n) can be obtained when the optimum criterion is satisfied.

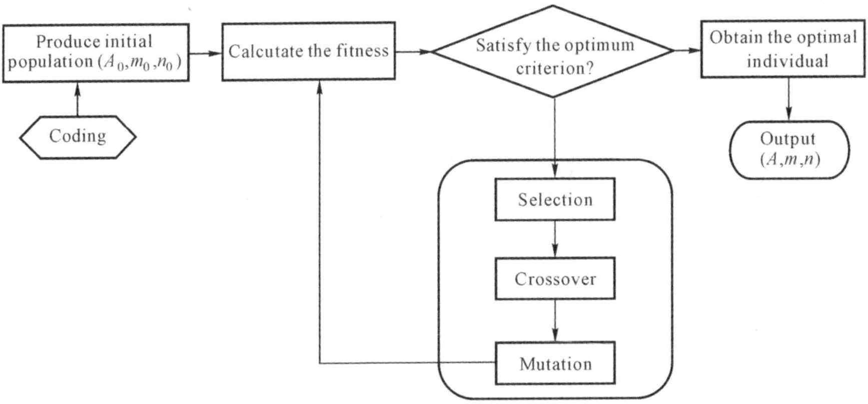

Fig. 5.14 Schematic diagram of genetic algorithm (Cui, *et al.*, 2011) (with the permission of Elsevier)

To obtain crack growth rates from crack length versus data from the cycles more quickly and conveniently, a simple and generally suitable approach is used to calculate straight-line slopes between the data points, as shown in Fig. 5.15. If the data points are numbered by 1, 2, 3, $\cdots$, j, then the crack growth rate for the segment ending at point number j is

$$\left(\frac{\mathrm{d}a}{\mathrm{d}N}\right)_j \approx \left(\frac{\Delta a}{\Delta N}\right)_j \approx \frac{a_j - a_{j-1}}{N_j - N_{j-1}} \tag{5.60}$$

The corresponding ΔK is calculated from the average crack length during the interval as follows,

$$\Delta K_j = Y(a)\Delta\sigma\sqrt{\pi a_{\text{avg}}} \tag{5.61}$$

The engineering estimation approach is valid only if the crack length is measured at fairly short intervals. Also, curve-fitting methods of evaluating $\frac{\mathrm{d}a}{\mathrm{d}N}$, which are more sophisticated than simple point-to-point slopes, are sometimes used to smooth the scatter in the a versus N data. Details are given in the ASTM Standard.

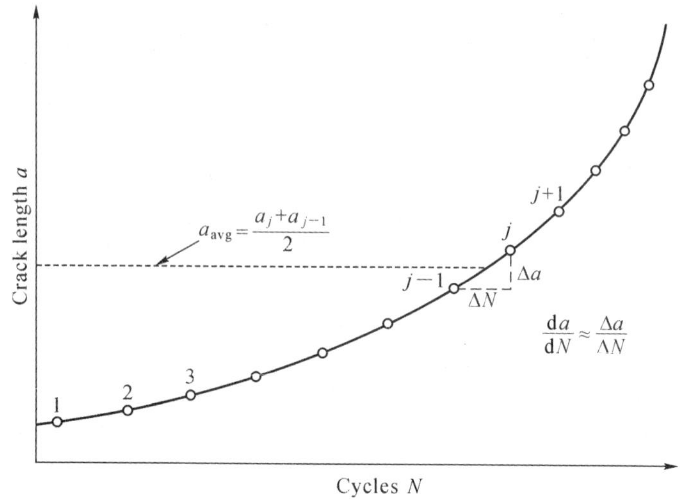

Fig. 5.15 Crack growth rates obtained from adjacent pairs of a versus N data points

5.4.4 Estimation Method from S-N Curve

From the crack growth rate, one can determine the *S-N* curve

$$N_f = \int_{a_0}^{a_f} \frac{1-(K_{\max}/K_C)^n}{AM^m} da \tag{5.62}$$

If we know N testing points [$\sigma_{\max i}$, R_i, N_i], from $\sigma_{\max i}$, K_C, σ_u, r_e, one can calculate a_{fi} and the error function is

$$Error = \sum_{i=1}^{N} \left[N_{fi} - \int_{a_0}^{a_{fi}} \frac{1-(K_{\max}/K_C)^n}{AM^m} da \right]^2 \tag{5.63}$$

Using the same method as explained in Fig. 5.13, one can determine a best set of model parameters. Cui (2002b) comprehensively discussed the relation between the crack growth rate curve and stress versus the life curve.

5.4.5 Estimation Method from ε-N Curve

The strain-based approach differs significantly from the stress-based approach. Remember that the stress-based approach emphasizes nominal (average) stress, rather than local stresses and strains, and it employs elastic stress concentration factors and empirical modifications thereof. Employment of the cyclic stress-strain curve is a unique feature of the strain-based approach, as is the use of a strain versus life curve, instead of a nominal stress versus life curve (Dowling, 2007). In the following, the subscript "*l*" stands for the local value of the parameter.

The cyclic stress-strain curve can be expressed as

$$\begin{cases} \varepsilon_a = \dfrac{\sigma_a}{E} + \left(\dfrac{\sigma_a}{H'}\right)^{1/m'} \\ \varepsilon_{\max} = \dfrac{\sigma_{\max}}{E} + \left(\dfrac{\sigma_{\max}}{H'}\right)^{1/m'} \end{cases} \tag{5.64}$$

The Manson-Coffin equation considering the mean stress effect is adopted here to express the ε-N curve, namely,

$$\varepsilon_a = \frac{\sigma_f' - \sigma_m}{E}(2N)^b + \varepsilon_f' (2N)^c \tag{5.65}$$

If the cyclic stress-strain curve and the strain versus life curve by tests for a material have been obtained (the parameters $E, H', m'\sigma_f', \varepsilon_f', b, c$ are given), namely we know N testing points $[\varepsilon_{lai}, N_i]$, then the value of σ_{lai} and σ_{lmi} can be calculated respectively, according to Eqs. (5.64) and (5.65). Using the relation $\sigma_{lmi} = \sigma_{l\max i} - \sigma_{lai}$ and Eq. (5.64), the values of $\sigma_{l\max i}$ and $\varepsilon_{l\max i}$ will be determined.

The concept of a "crack tip damage zone" proposed by Ma and Lu (1999) will be used here for reference in deducing the relation between ε-N data and S-N data. The basic assumptions include the following:

(1) A small crack tip damage zone with the length of x^* exists inside the plastic zone, in which the stress/strain gratitude is very small and can be regarded as constant. The stress/strain at the middle point ($0.5x^*$) can be taken as its mean value.

(2) The material in the crack tip damage zone will experience fatigue failure and the crack will propagate by an increment x^* after fatigue stress cycles N.

(3) The material inside the damage zone is extremely damaged but the stress/strain gratitude outside it decreases quickly and the material damage can be neglected.

For a center-cracked specimen under cyclic stress, the released energy when the crack propagates can be expressed as $K^2_{\max i}/(E\cdot 2x^*_i)$ while the work inside the crack tip damage zone is $2\int_{x_i^*}\frac{1}{2}\sigma_{l\max i}\cdot 2v(x)\mathrm{d}x$, where $v(x)$ is the displacement normal to the crack face in the damage zone. The two values should be equal to each other, namely,

$$K_{\max i}^2/(E\cdot 2x_i^*)=2\int_{x_i^*}\frac{1}{2}\sigma_{l\max i}\cdot 2v(x)\mathrm{d}x \tag{5.66}$$

Ma and Lu (1999) deduced the expressions of stress and strain in the crack tip damage zone for power-law material, which can be written as

$$\begin{cases}\sigma_l(0.5x_i^*)=\sigma_Y\left[\dfrac{K_i^2}{\pi(1+n')\sigma_Y^2\cdot 0.5x_i^*}\right]^{n'/(n'+1)}\\ \varepsilon_l(0.5x_i^*)=\dfrac{\sigma_Y}{E}\left[\dfrac{K_i^2}{\pi(1+n')\sigma_Y^2\cdot 0.5x_i^*}\right]^{1/(n'+1)}\end{cases} \tag{5.67}$$

Then the relation between the stress and strain in the crack tip damage zone can be expressed as

$$\sigma_{l\max i}\varepsilon_{l\max i}=\frac{K_{\max i}^2}{\pi(1+n')E\cdot 0.5x_i^*} \tag{5.68}$$

Substituting the calculated values of $\sigma_{l\max i}$ and $\varepsilon_{l\max i}$ into Eq.(5.68), the values of $K_{\max i}$ and x_i^* can be obtained by combining Eqs.(5.66) and (5.68). Furthermore, according to the following equation,

$$K_{\max i}=\sqrt{\pi\, r_{\mathrm{e}}\left(\sec\frac{\pi}{2}\frac{\sigma_{\max i}}{\sigma_V}+1\right)}\left(1+Y(a_i)\sqrt{\frac{a_i}{2r_{\mathrm{e}}}}\right)\sigma_{\max i} \tag{5.69}$$

where $a_i=a_0+\sum_i x_i^*$, the value of $\sigma_{\max i}$ can be obtained. By repeatedly employing the procedures introduced above, the ε-N curve can be transformed to the S-N curve. And the crack growth rate model parameters for this material can be determined using the method introduced in the above two parts.

At present, the FCP rates are still mainly measured by experiments, especially the threshold of the SIF range, which are expensive and time consuming. Therefore, some researchers attempt to find methods for predicting the FCP rates from the tensile properties of the material, *e.g.* Farahmand and Nikbin (2008); Kang, *et al.* (2006); Tiryakioglu and Hudak (2008); Firrao, *et al.* (2007). And the threshold effective SIF range can be accordingly estimated.

5.4.6 Estimation Methods from Available Static Test Properties

5.4.6.1 Estimation method of threshold value of SIF range

Minakawa and McEvily (1981) put forward an empirical formula to predict the ΔK_{th} of aluminum alloys and adopted by Ling and Zheng (1990), namely,

$$\Delta K_{\text{th}} = 3.4 - 1.5 \times 10^{-3} \cdot \sigma_{0.2} \quad \text{MPa} \cdot \text{m}^{\frac{1}{2}} \quad (R \approx 0) \tag{5.70}$$

Vosikovski (1979) proposed an empirical formula for steels,

$$\Delta K_{\text{th}} = 11.4 - 4.6 \times 10^{-3} \cdot \sigma_{0.2} \quad \text{MPa} \cdot \text{m}^{\frac{1}{2}} \quad (R \approx 0) \tag{5.71}$$

Farahmand and Nikbin (2008) pointed out that the value of ΔK_{th} for a wide range of alloys was found to be proportional to the final elongation, ε_f, and established an empirical relationship describing ΔK_{th} as a function of ε_f as follows,

$$\Delta K_{\text{th}\,0} = -0.6337 + 31.43 \cdot \varepsilon_f \quad \text{MPa} \cdot \text{m}^{\frac{1}{2}} \tag{5.72}$$

The empirical formulas for ΔK_{th} introduced above are put forward under a certain load ratio. But the value of ΔK_{th} is well known to vary with the load ratio. The following equations (Wang, *et al.*, 2008) are largely adopted to represent the effect of the load ratio on ΔK_{th},

$$\Delta K_{\text{th}} = \Delta K_{\text{th}\,0} - B_1 R \tag{5.73}$$

$$\Delta K_{\text{th}} = \Delta K_{\text{th}\,0} (1 - R)^{\gamma} \tag{5.74}$$

where $\Delta K_{\text{th}\,0}$ is the threshold SIF range value corresponding to R=0 and B_1 and γ are material constants.

Later, a simple engineering approach was put forward by Zhang, *et al.* (2003) from theoretical analysis, for prediction of the effect of the load ratio on the fatigue threshold ΔK_{th} based on the observation of the crack tip plasticity during a fatigue load cycle. Furthermore, an improved equation was proposed by Huang and Moan (2007) and Huang, *et al.* (2008) from a comprehensive analysis of different kinds of materials. The equation is written in a segmented function for three regions of the load ratio, written as Eqs. (5.45) and (5.46).

5.4.6.2 Estimation method of fracture toughness

As K_{IC} test is costly and time consuming and requires sophisticated skills, many researchers had tried to develop the simplified method to estimate the value of K_{IC} by tensile test results (Kang, *et al.*, 2006; Farahmand and Nikbin, 2008; Oh, 1995; 1996; Tiryakioglu and Hudak, 2008; Firrao, *et al.*, 2007). Among the methods for developing the relation between the fracture toughness and tensile test results, it is worth noticing that artificial neural networks have shown a remarkable performance for modeling the complex nonlinear correlation compared to conventional fitting models. Kang, *et al.* (2006) made a comparison between a neural network model and conventional fitting models, as listed in Table 5.1.

Furthermore, there are many formulas to estimate the fracture toughness from Charpy energy or from mechanical properties. Here two formulas were selected.

Table 5.1 Comparison between neural network models and conventional fitting models (Kang, *et al.*, 2006) (with the permission of John Wiley and Sons, Inc.)

Fitting model	Input variable	Regression equation	*R*-squared
Multi-Linear Regression	σ_Y, σ_u, EL, CPO, T	$K_{IC} = 57.835 + 0.231 \cdot \sigma_Y - 0.204 \cdot \sigma_u + 4.418 \cdot EL + 0.436 \cdot T - 13.141 \cdot CPO$	0.437
3D Paraboloid Regression	σ_Y, T	$K_{IC} = 257.0170 - 0.4870 \cdot \sigma_Y + 0.4012 \cdot T + 0.0004 \cdot \sigma_Y^2 - 0.0003 \cdot T^2$	0.414
	σ_u, T	$K_{IC} = 400.5300 - 0.7763 \cdot \sigma_u + 0.4065 \cdot T + 0.0005 \cdot \sigma_u^2 - 0.0003 \cdot T^2$	0.409
	EL, *T*	$K_{IC} = 4.5916 + 8.3046 \cdot EL + 0.4245 \cdot T - 0.1420 \cdot EL^2 - 0.0003 \cdot T^2$	0.395
3D Gaussian Regression	σ_Y, T	$K_{IC} = 484.195 \cdot \exp\left\{-0.5 \cdot \left[\left(\frac{\sigma_Y + 3371.847}{3120.605}\right)^2 + \left(\frac{T - 327.552}{280.971}\right)^2\right]\right\}$	0.366
	σ_u, T	$K_{IC} = 561.967 \cdot \exp\left\{-0.5 \cdot \left[\left(\frac{\sigma_u + 3103.888}{2788.695}\right)^2 + \left(\frac{T - 328.224}{279.590}\right)^2\right]\right\}$	0.367
	EL, *T*	$K_{IC} = 240.514 \cdot \exp\left\{-0.5 \cdot \left[\left(\frac{EL - 27.696}{17.306}\right)^2 + \left(\frac{T - 305.834}{264.465}\right)^2\right]\right\}$	0.403
Neural network	σ_Y, σ_u, *EL*, *CPO*, *T*	—	0.812

Girenko, *et al.* (1985) proposed a formula to estimate the fracture toughness form Charpy energy data. The yield strength range of covered materials is 200–1,700 MPa, the covered Charpy energy is 2–150 J.

$$K_{\mathrm{IC}} = 2.5\sqrt{C_V} \quad (\mathrm{MPa} \cdot \mathrm{m}^{\frac{1}{2}}) \tag{5.75}$$

Barsom, *et al.* (1970) proposed a formula to estimate the fracture toughness from Charpy energy data. The yield strength range of covered materials is 270–1,700 MPa, the covered Charpy energy is 31–121 J.

$$K_{\mathrm{IC}} = \sqrt{0.64(C_V / \sigma_y - 0.01)} \;\; \sigma_y \quad (\mathrm{MPa} \cdot \mathrm{m}^{\frac{1}{2}}) \tag{5.76}$$

5.4.7 Estimation of *A* and *m*

When all the parameters explained above have been determined by experimental or empirical methods, the values of A and m can be determined through the curve fitting method based on the experimental crack growth data under a certain load ratio, *e.g.* R=0 through the following method.

Change the form of the model to

$$\frac{\mathrm{d}a}{\mathrm{d}N} \cdot [(1 - \left(\frac{K_{\max}}{K_{\mathrm{C}}}\right)^n] = A \cdot M^m \tag{5.77}$$

In the crack growth rate curve under a certain R, the relationship of $\frac{\mathrm{d}a}{\mathrm{d}N} \cdot \left[1 - \left(\frac{K_{\max}}{K_{\mathrm{C}}}\right)^n\right]$ and M can be obtained. Consequently, a curve between $\log\left[\frac{\mathrm{d}a}{\mathrm{d}N} \cdot \left(1 - \left(\frac{K_{\max}}{K_{\mathrm{C}}}\right)^n\right)\right]$ and $\log M$ can be drawn. As $\log\left[\frac{\mathrm{d}a}{\mathrm{d}N} \cdot \left(1 - \left(\frac{K_{\max}}{K_{\mathrm{C}}}\right)^n\right)\right] = \log A + m \cdot \log M$, the values of parameters A and m can then be determined by linear curve fitting.

5.5 Capabilities of the UFLP Method

5.5.1 *The Quantitative Analysis of the Improved Crack Growth Rate Model*

In order to investigate the performance of the improved constitutive model, a quantitative analysis is carried out in this section. Fig. 5.16 and Fig. 5.17 respectively depict the effects of exponent γ and OverLoad Ratio (OLR) on predicted results when a single overload occurs. Fig. 5.16 indicates that the level of fatigue crack growth retardation will be changed with exponent γ and the actual phenomenon can be adequately simulated by fixing a relevant exponent γ. And as is well known, the crack growth retardation will be more significant with the increase in the OLR, which can be clearly observed in Fig. 5.17. Furthermore, the results of the cycle-by-cycle integration procedure can be markedly different under various loading sequences. In order to verify the simulation capability of the model on this point, Fig. 5.18 describes six basic types of loading mode and Fig. 5.19 depicts the clusters of simulation curves of crack length (a) versus load cycles (N). A quite different tendency of a-N curves can be found for the six basic types of loading mode, as has been explained in the foregoing sections.

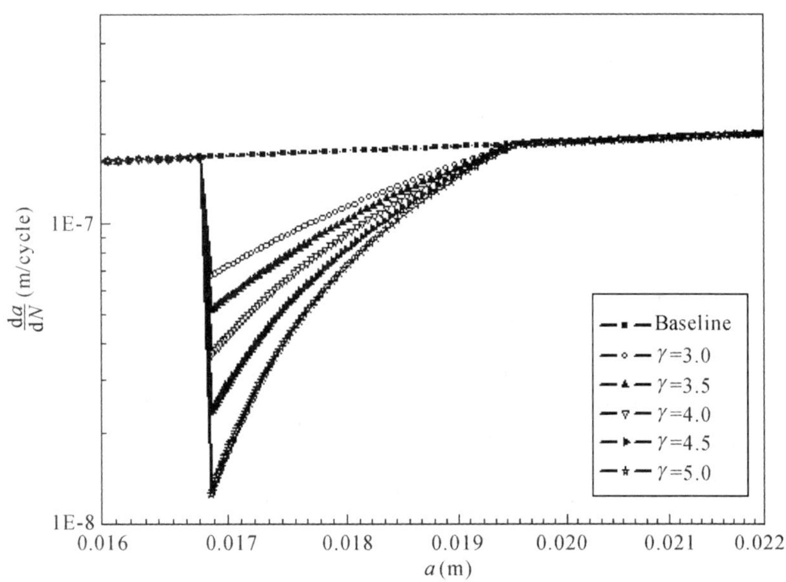

Fig. 5.16 The effect of γ on predicted results (Chen, *et al.*, 2012) (with the permission of John Wiley and Sons, Inc.)

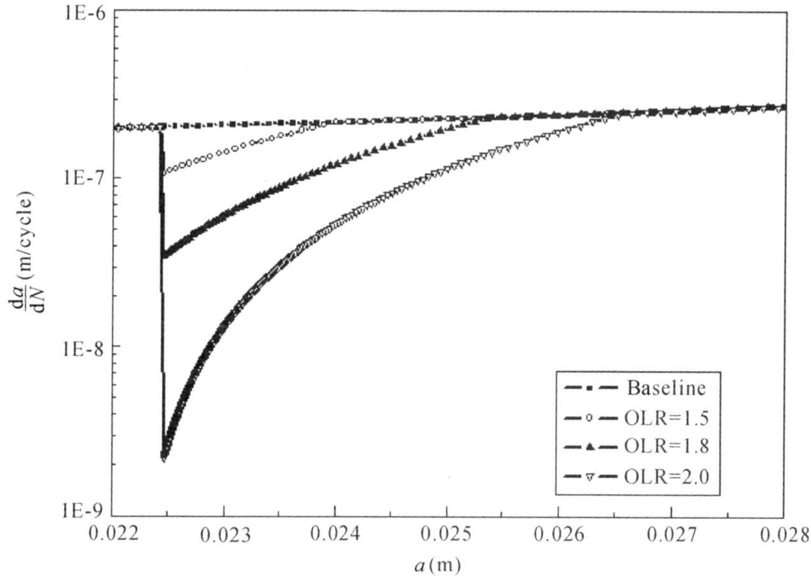

Fig. 5.17 The effect of OLR on predicted results (Chen, *et al.*, 2012) (with the permission of John Wiley and Sons, Inc.)

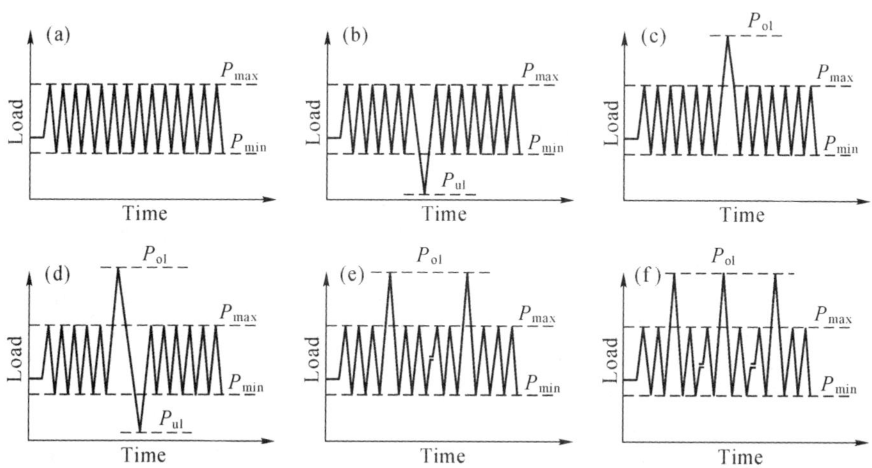

Fig. 5.18 Six basic types of loading mode (Chen, *et al.*, 2012) (with the permission of John Wiley and Sons, Inc.)

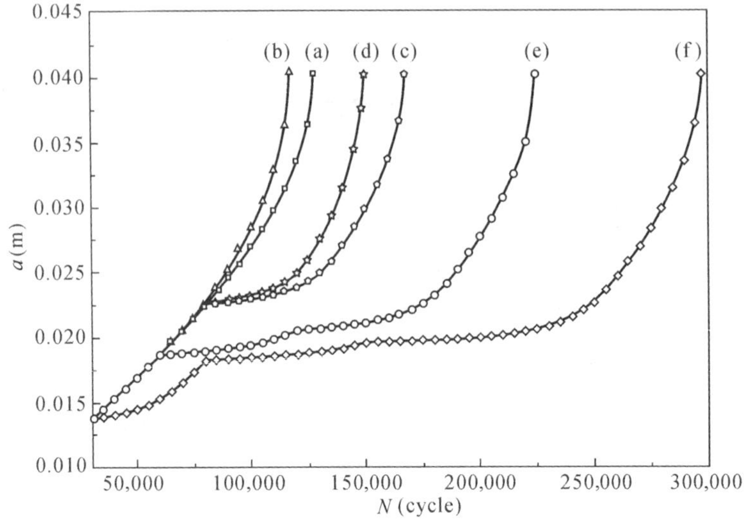

Fig. 5.19 Predicted *a*-*N* curves under six basic types of loading model (Chen, *et al.*, 2012) (with the permission of John Wiley and Sons, Inc.)

5.5.2 Model Validation by Test

5.5.2.1 Constant amplitude loading

The approximate estimation approach of model parameters introduced above will be validated in this part by a wide range of alloys, which include aluminum alloy (Paris, *et al.*, 1999; Newman, *et al.*, 1999; Zhao, *et al.*, 2008), titanium alloy (Jha and Ravichandran, 2000), and steel (Wang and Cui, 2010; Dinda and Kujawski, 2004; Taheri, *et al.*, 2003). For detailed information on test conditions you can refer to these references, but engineering estimation methods (Farahmand and Nikbin, 2008; Kang, *et al.*, 2006; Oh, 1995; Bulloch, 1995) are used for those test parameters not described in references. In all examples, the model parameters are derived from experimental data under stress ratio R=0.1 and then used to estimate the crack growth rate curves under other stress ratios. The comparisons between test data and prediction curves are illustrated in Figs. 5.20–5.29, which show that the prediction curves agree well with test data and the approximate estimation approach is accordingly validated.

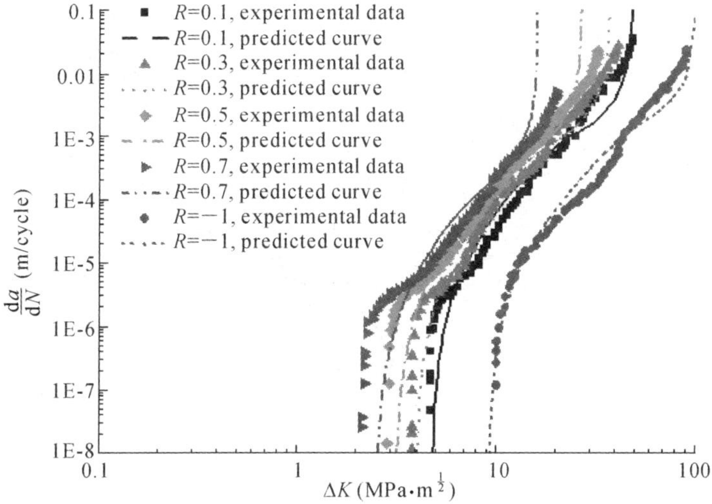

Fig. 5.20 Comparison between predicted crack growth rate curves and test data of 2324-T39 aluminum alloy (Paris, *et al.*, 1999; Cui, *et al.*, 2011) (with the permission of Elsevier)

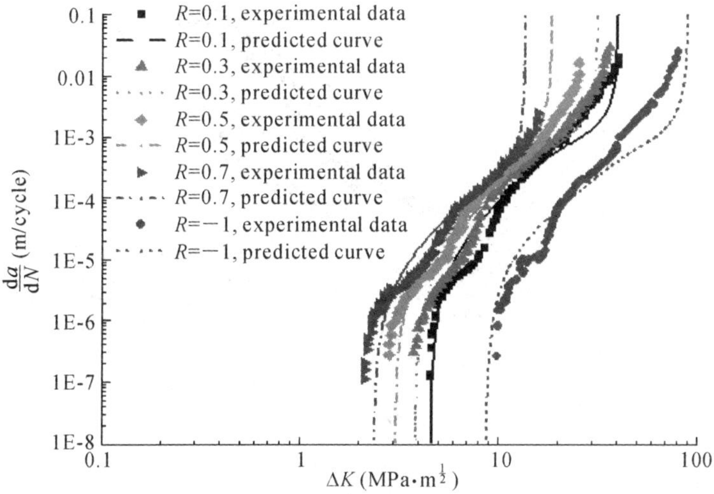

Fig. 5.21 Comparison between predicted crack growth rate curves and test data of 6013-T651 aluminum alloy (Paris, et al., 1999; Cui, *et al.*, 2011) (with the permission of Elsevier)

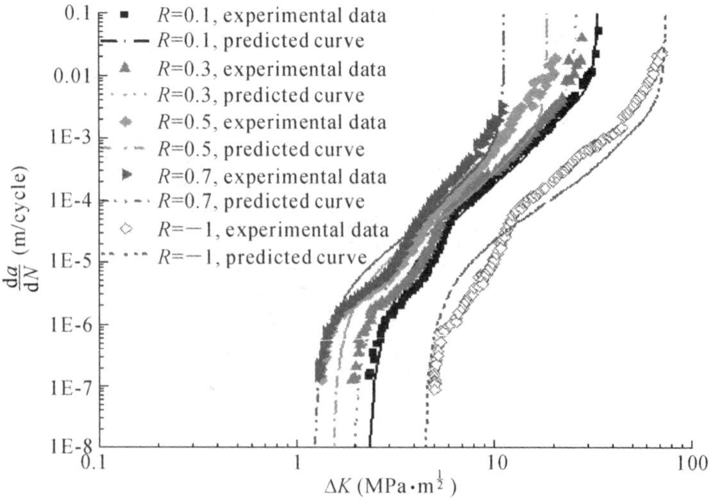

Fig. 5.22 Comparison between predicted crack growth rate curves and test data of 7055-T7511 aluminum alloy (Paris, *et al.*, 1999; Cui, *et al.*, 2011) (with the permission of Elsevier)

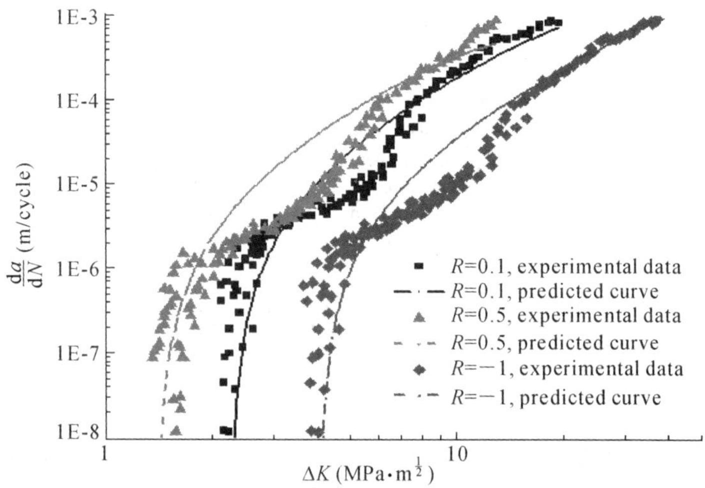

Fig. 5.23 Comparison between predicted crack growth rate curves and test data of 7075-T6 aluminum alloy (Newman, *et al.*, 1999; Cui, *et al.*, 2011) (with the permission of Elsevier)

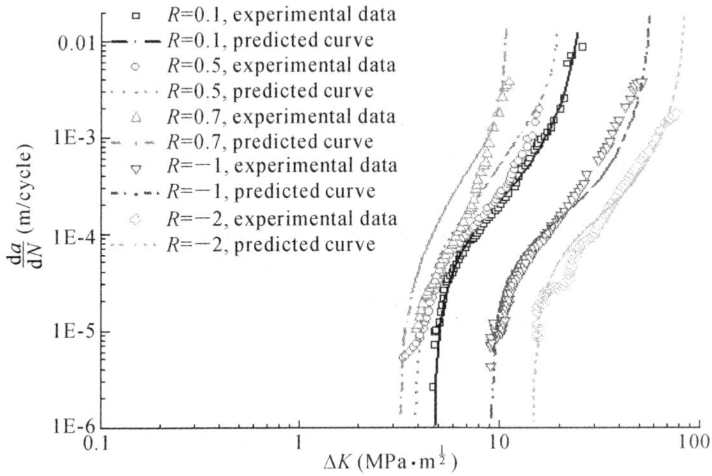

Fig. 5.24 Comparison between predicted crack growth rate curves and test data of 7075-T651 (Zhao, *et al.*, 2008; Cui, *et al.*, 2011) (with the permission of Elsevier)

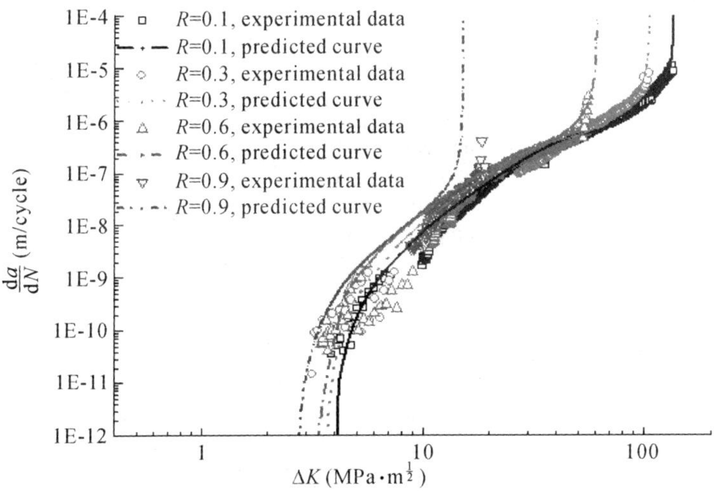

Fig. 5.25 Comparison between predicted crack growth rate curves and test data of HTS-A steel (Wang and Cui, 2010b) (with the permission of Elsevier)

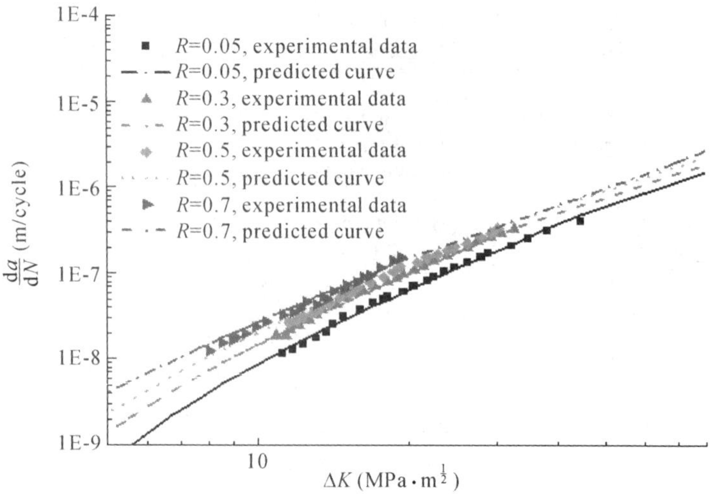

Fig. 5.26 Comparison between predicted crack growth rate curves and test data of 300M steel (Dinda and Kujawski, 2004; Cui, *et al.*, 2011) (with the permission of Elsevier)

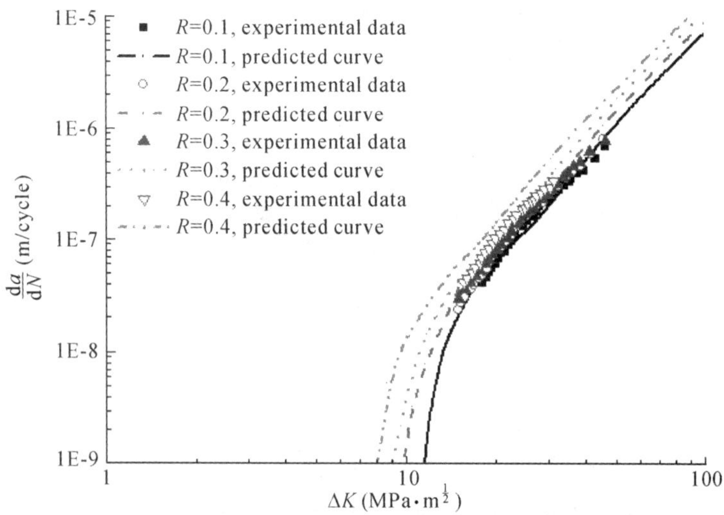

Fig. 5.27 Comparison between predicted crack growth rate curves and test data of 350WT steel (Taheri, 2003; Cui, *et al.*, 2011) (with the permission of Elsevier)

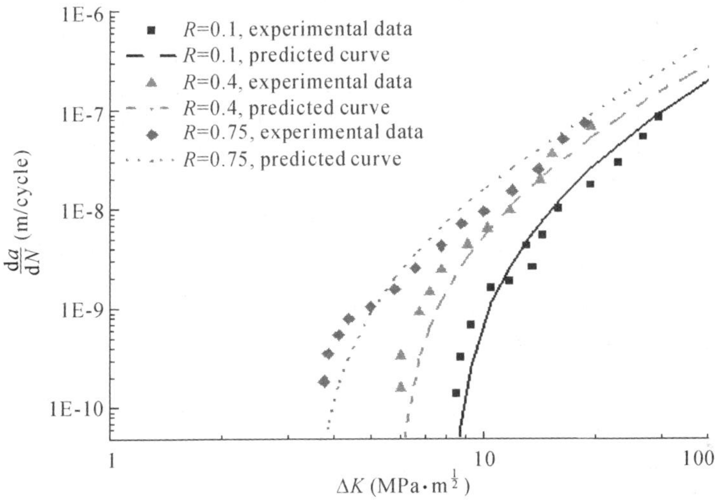

Fig. 5.28 Comparison between predicted crack growth rate curves and test data of CrMoV (Bulloch, 1995; Cui, *et al.*, 2011) (with the permission of Elsevier)

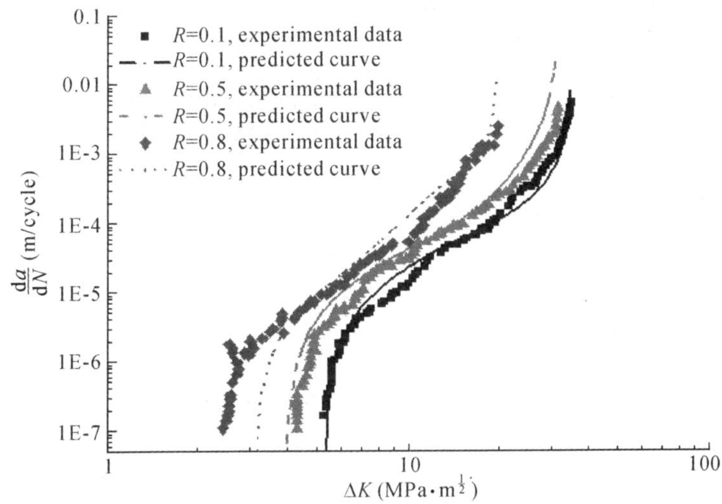

Fig. 5.29 Comparison between predicted crack growth rate curves and test data of Ti 10V Fe 3Al (Jha and Ravichandran, 2000; Cui, *et al.*, 2011) (with the permission of Elsevier)

Another example in the following section is made to validate the improved model for welded structures. Fracture mechanics based computations are extremely sensitive to the initial crack shape and size (Nykänen, *et al.*, 2009). A realistic semi-elliptical initial crack shape is usually assumed for calculation, which has been widely accepted. And the surface crack under fatigue loading always changes its crack front as it grows. The changes in the defect aspect ratio depend principally upon initial configuration, relative crack depth and loading condition, but also depend weakly upon the stress ratio, loading frequency, growth rate exponent, mean stress and the crack tip stress conditions, which alter with changes in the crack shape (Lin and Smith, 1999). Many investigators have carried out the study for predicting crack shape changes (Newman, 1979; Mahmoud, 1988; 1989; Wu, 1985). It is necessary to know in advance that the crack shape changes before estimating the remaining fatigue life using fracture mechanics theory, as the SIF along the crack front are normally expressed by the function of the crack aspect ratio. Generally, the surface crack is assumed to keep its semi-elliptical profile when growing, for simplicity, so that the crack profile changes combined with the semi-elliptical assumption were always investigated in the past and proved to be reasonable. We will also adopt this assumption for the current study.

Then the basic assumptions of the methodology for welded joints include:

(1) The semi-elliptical configuration is used to estimate the fatigue life of welded joints with surface damage problems. It is assumed that the crack will propagate keeping the 2D semi-elliptical profile and every point along the crack front will advance following the same crack growth model with a different growth rate. Then the growth characteristics of two points at the bottom and at the end of the crack are enough to describe the growth of the crack front.

The change in the crack aspect ratio, or the growth rate relationship of the bottom and end of the crack must be known before estimation, because the SIF of the bottom and the end of the crack are dependent on the crack aspect ratio. Many studies have been carried out on this problem and the empirical or numerical results literature can be adopted currently for further analysis. In the analysis of growing a single macro-crack, the empirical equation proposed by Lin and Smith (1999), as follows, is recommended.

$$\frac{a}{c}=\left\{g\left(\frac{a}{t}\right)-\left(\frac{a_0}{a}\right)^{2.5}\left[g\left(\frac{a_0}{t}\right)-\left(\frac{a_0}{c_0}\right)^{-2.5}\right]\right\} \tag{5.78}$$

where

$$g(u)=1.331+0.706u^2+0.155u^4+0.013u^6 \tag{5.79}$$

Eqs. (5.78)–(5.79) are used for growing a single semi-elliptical surface crack, while the random coalescence of cracks at the weld toe has not been taken into account. Typically, the experimental crack aspect curve proposed by Engesvik and Moan (1983) is usually adopted to consider the coalescence effect and, in practice, a continuous edge crack with zero crack aspect ratio is always applied for simplicity.

(2) The plane strain condition can be assumed approximately when estimating the growth rate of the bottom point. And the corresponding crack opening level can be calculated under the plane strain condition, while the actual 3D stress/strain constraint will be considered when calculating the growth rate of the end point, which is related to the plate thickness, yield stress and SIF during propagation.

(3) As the welded structures usually have crack-like defects, the total fatigue life of the weld joints is the summation of the surface crack propagation life and

through-thickness crack propagation life when the deepest point of the crack reaches the plate bottom.

The stress concentration magnification factor of semi-elliptical cracks at the toe of fillet welded joints can be expressed as follows:

$$K = M_K K_{\text{plate}} \tag{5.80}$$

where K_{plate} is the SIF in the vicinity of crack tips in a plate and M_K is a magnification factor for welded joints. Bowness and Lee (2000) recommended the empirical expression of M_K of a semi-elliptical crack and the calculation methods for SIFs of surface cracks can be referred to in Chapter 7.

As an example, the surface crack in a T-joint is illustrated in Fig. 5.30. And the symbols for joint and crack geometry parameters are labeled in the figure.

The test data given in NIMS (1979) for some cruciform joints will be used for comparison study here. The coalescence effect is considered. Table 5.2 gives the parameters of cruciform joints. The comparison and correlation between predicted results and test data (NIMS, 1979) are shown in Figs. 5.31(a) and (b) respectively, which can prove the rationality of the model used for welded structures.

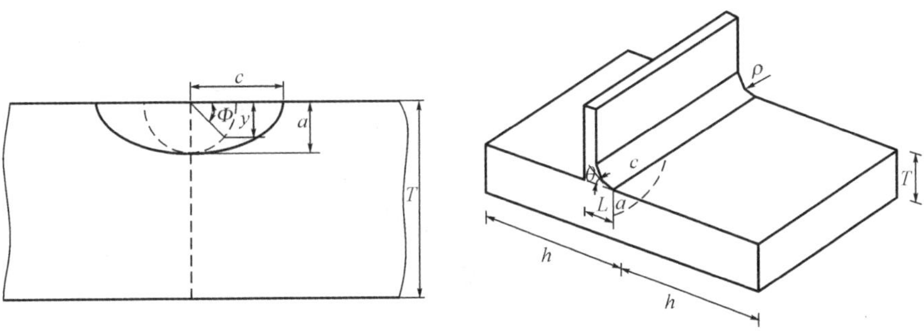

(a) (b)

Fig. 5.30 Illustration of surface crack in T-joint. (a) Surface crack sketch; (b) 3D T-joint model (Wang and Cui, 2010b) (with the permission of Woodhead)

Table 5.2 Parameters of cruciform joints used for comparison study between predicted results and test data (NIMS, 1979) (with the permission of NIMS)

σ_Y (MPa)	σ_b (MPa)	T (mm)	θ (°)	l (mm)
485	570	9	39.2	11
485	570	20	50.9	20
485	570	40	37.4	30.4

*Note: During the prediction, R=0; a_0=0.2 mm and l is calculated by summing the leg length and half of stiffener thickness in the original data of NIMS (1979)

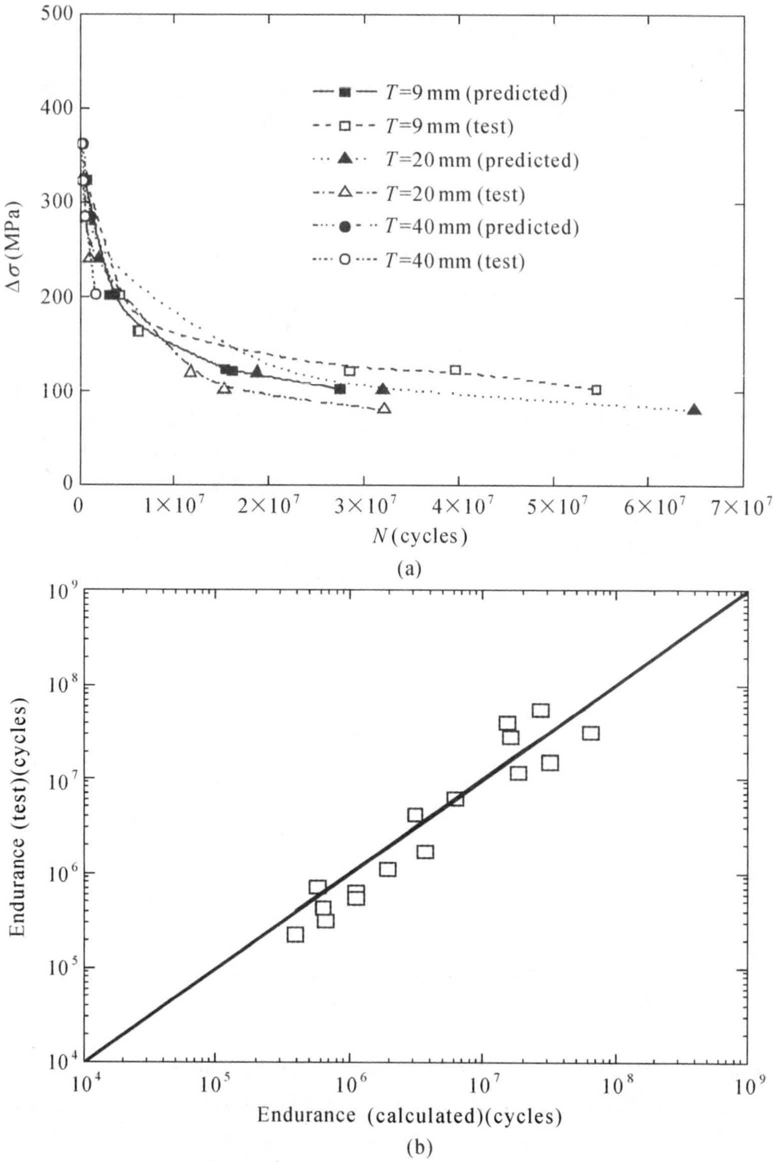

Fig. 5.31 Comparison (a) and correlation (b) between predicted results and test data (NIMS, 1979; Wang and Cui, 2010b) (with the permission of Woodhead)

5.5.2.2 Variable amplitude loading

The capacity of explaining the load interaction effects has been primarily investigated by the qualitative analysis above. This section is to further validate the applicability of this model under VA loading by comparing the prediction results with the fatigue crack growth test data. Several common loading types are involved including single and multiple overloads, underload, combination of overload and underload, and block loading.

(1) Single and multiple overloads

Crack growth retardation due to overload is the most commonly observed load interaction effect and has the most notable influence on fatigue life. The accurate FLP requires accounting for the crack growth retardation due to encountering overloading. The performance of the improved crack growth rate model can be validated by comparing the prediction with corresponding test data including D12Cz aluminum alloy (Schijve, *et al.*, 2004), 7075-T651 aluminum alloy (Zhao, *et al.*, 2008), 18G3A steel (McEvily, *et al.*, 2004), 350WT steel (Huang, *et al.*, 2008). The comparisons are illustrated in Figs. 5.32–5.36, which show that the prediction curves agree well with test data both in the case of single overload and multiple overloads.

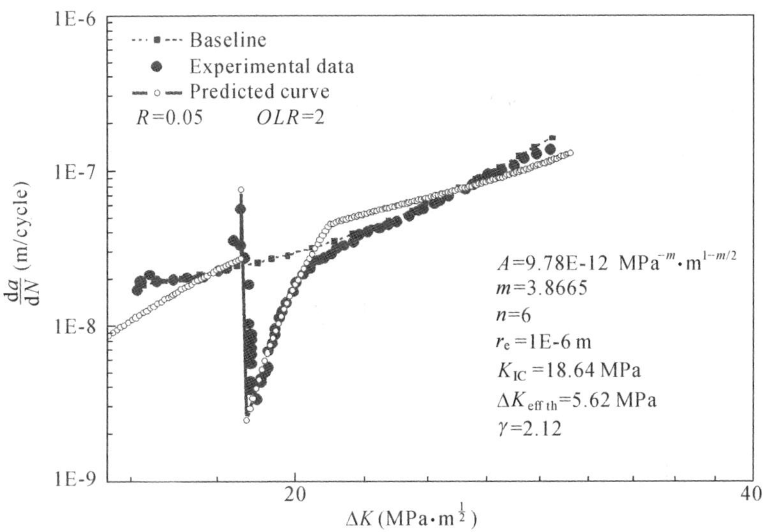

Fig. 5.32 Comparison of prediction with test data of 18G3A steel under single overload (Schijve, *et al.*, 2004; Chen, *et al.*, 2012) (with the permission of John Wiley and Sons, Inc.)

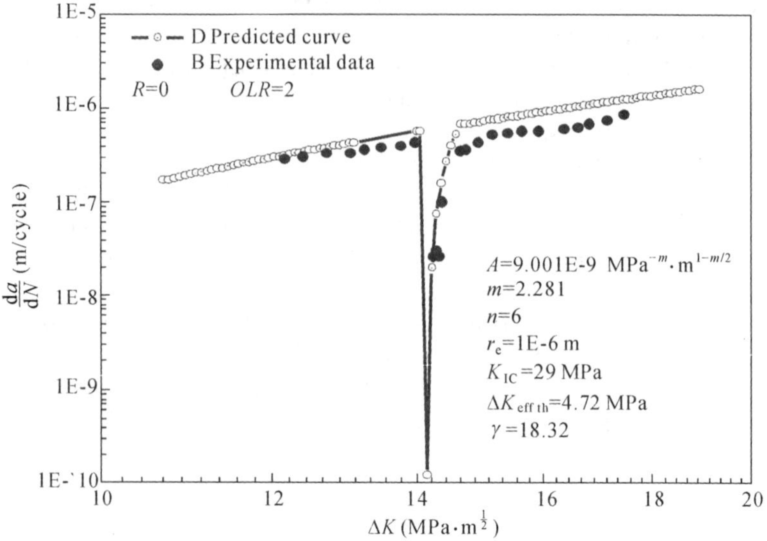

Fig. 5.33 Comparison of prediction with test data of 7075-T651Al alloy under single overload (Zhao, *et al.*, 2008; Chen, *et al.*, 2012) (with the permission of John Wiley and Sons, Inc.)

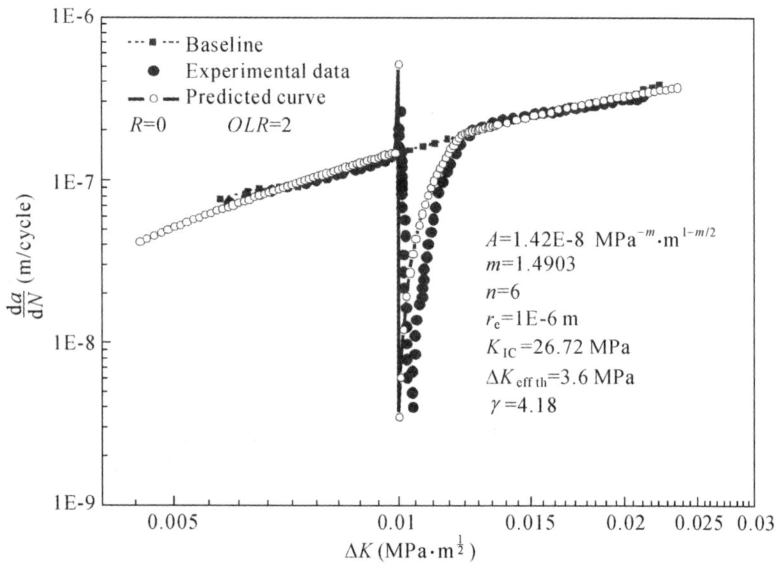

Fig. 5.34 Comparison of prediction with test data of D16Cz Al alloy under single overload (McEvily, *et al.*, 2004; Chen, *et al.*, 2012) (with the permission of John Wiley and Sons, Inc.)

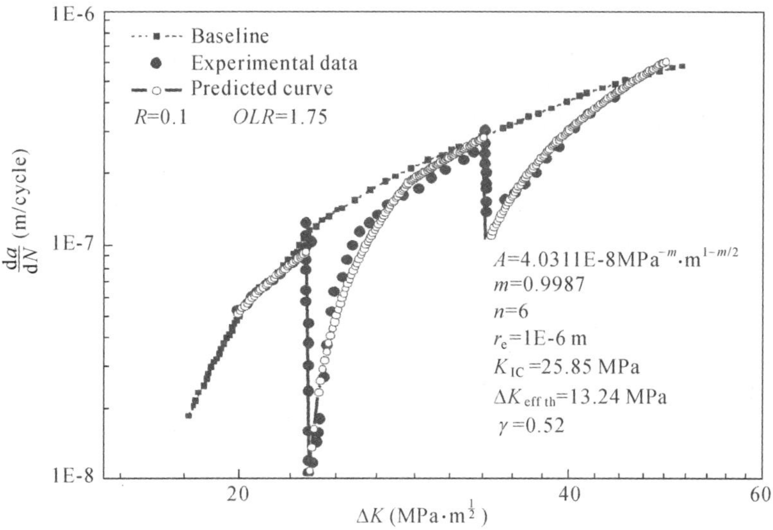

Fig. 5.35 Comparison of prediction with test data of 350WT steel under multiple overloads (Huang, *et al.*, 2008; Chen, *et al.*, 2012) (with the permission of John Wiley and Sons, Inc.)

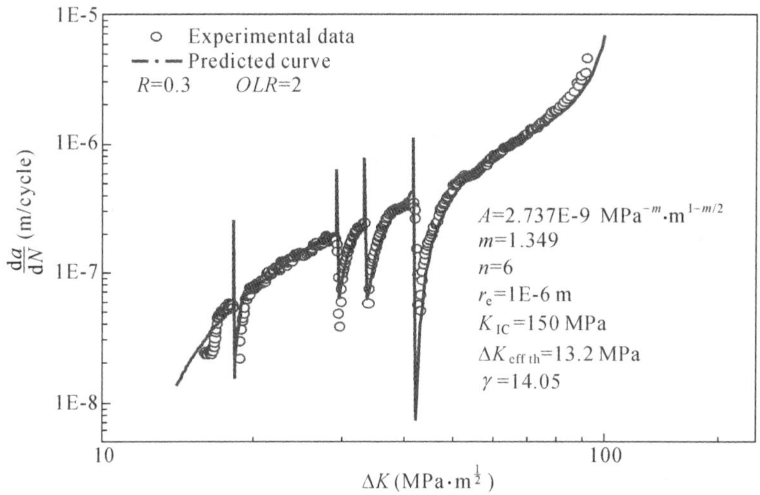

Fig. 5.36 Comparison of prediction with test data of HTS-A steel under multiple overloads

(2) Underload

A compressive underload causes a lesser degree of fatigue crack growth acceleration in the subsequent fatigue cycles by introducing a lesser extent of tensile compressive residual stress at the crack tip, which decreases the level of the crack opening load (Yuen and Taheri, 2004). Many studies are dedicated to the overload retardation effect and its benefit for lifetime structures but the acceleration effect of the underload cannot be ignored (Silva, 2005). The Wheeler model has been successful in simulating crack growth retardation while it cannot take account of fatigue growth acceleration due to underload. Herein, a preliminary study has been carried out by introducing the modification factor, Φ, as written in Eq. (5.26) to decrease the conservative prediction results. The comparison is made between the results of analytical predictions and the experimental data in Figs. 5.37–5.38, and good agreements are found.

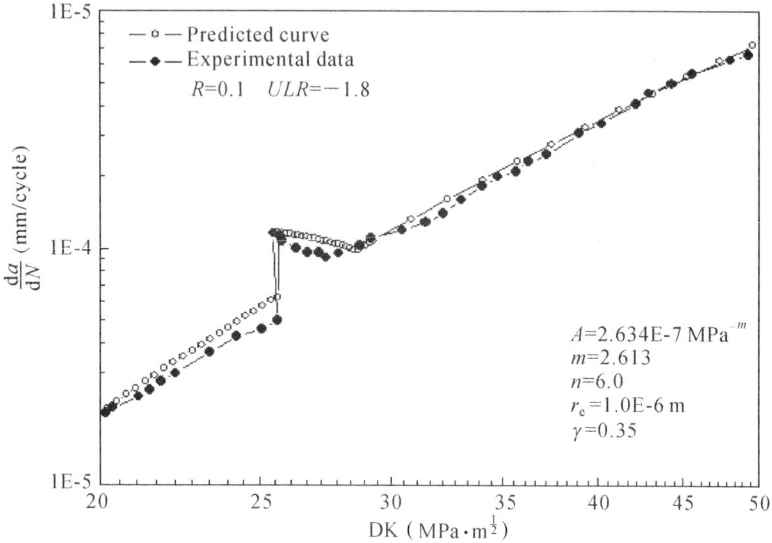

Fig. 5.37 Comparison of prediction with test data of 304L stainless steel under underload (Kalnaus, *et al.*, 2009; Chen, *et al.*, 2012) (with the permission of John Wiley and Sons, Inc.)

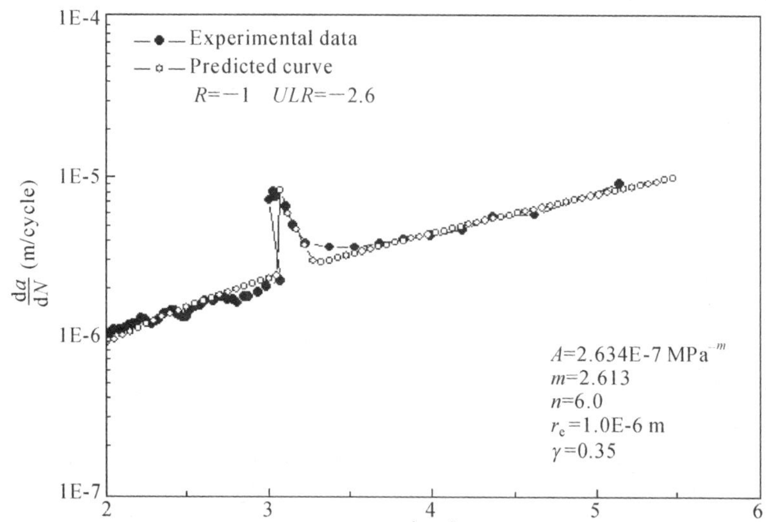

Fig. 5.38 Comparison of prediction with test data of low carbon steel under single underload (Makabe, *et al.*, 2004; Chen, *et al.*, 2012) (with the permission of John Wiley and Sons, Inc.)

(3) Combination of overload and underload

For simple load histories containing combinations of overload and underload cycles, most available test results suggest that an underload cycle applied prior to overloading results in little change in the amount of retardation, but if an underload cycle occurs immediately after an overload the amount of retardation is reduced significantly (Bacila, *et al.*, 2007). The fatigue tests were performed on a compact-tension specimen of HASTELLOY C-2000 alloy (Lee, *et al.*, 2009). Experimental results are illustrated in Fig. 5.39. It is evidently shown that FCP is dependent on the preceding loading sequence and the crack growth rate can be quite different under various loading cases. In the cases 1–3, significant crack growth retardation due to a tensile overload and tiny crack growth acceleration due to a compressive underload occur compared with the crack growth rate under constant amplitude loading. In the cases 4–5, the results indicate that an underload applied immediately after an overload reduces the post-overload retardation more significantly than an underload which immediately precedes an overload. The prediction results are depicted in Fig. 5.39 for cases 1–5, together with test data. Then the capability of the improved crack growth rate model in simulating the load interaction effects can be well validated through these comparisons.

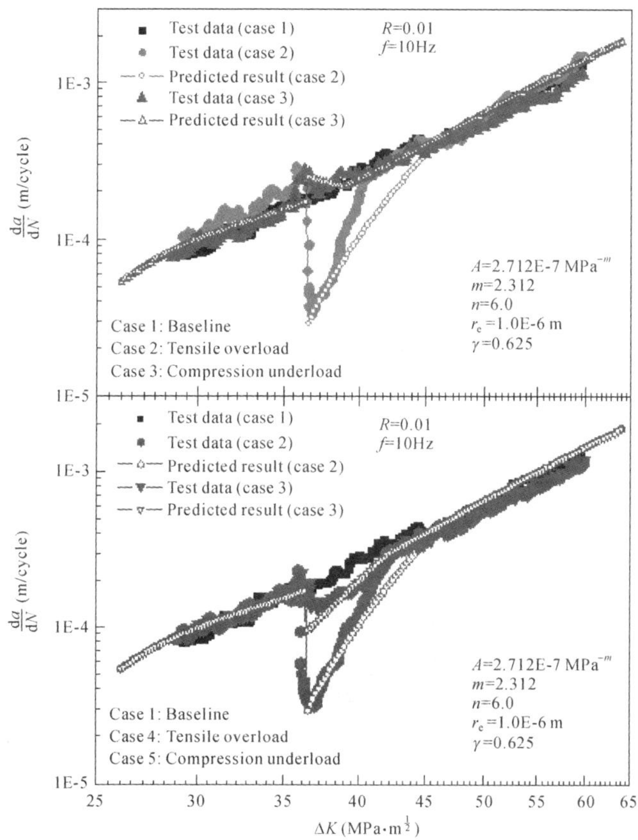

Fig. 5.39 Comparison of prediction results with test data of C-2000 alloy for five basic loading cases (Lee, *et al.*, 2009; Chen, *et al.*, 2012) (with the permission of John Wiley and Sons, Inc.)

(4) Block loading

A common feature of more realistic load spectra is the presence of overloads, underload or mean stress variation. Generally, a block of overloads gives more severe crack growth retardation following CA baseline cycles than a single overload (Khan, *et al.*, 2009) and mean stress has a significant effect on the fatigue life. To account for the influence mentioned above, the experiments have been carried out under five basic types of loading spectra with different loading cycles N_1 and N_2 (Borregoa, *et al.*, 2005). The schematic representation of block loading is listed in Table 5.3. It is indicated that the fatigue crack growth rate can obviously differ with the variation in mean stress by comparison between loading sequences (a) and (b). Crack acceleration is more pronounced for case (c) than case (e), suggesting an increase in crack acceleration with the number of overloads in each block. In addition, the results of case (c) and case (d) can show that the fatigue life can be diverse under the same mean stress due to the load interaction effects. The good agreement between the predicted results and the experimental data proves that the improved crack growth rate model is able to deal with block loading, as shown in Fig. 5.40.

Table 5.3 The schematic of block loading (Chen, *et al.*, 2012) (with the permission of John Wiley and Sons, Inc.)

Loading spectrum	No.	N_1 (cycle)	N_2 (cycle)	N_2/N_1
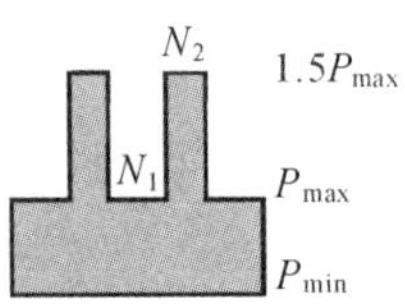	(a)	∞	0	0
	(b)	0	∞	∞
	(c)	1,000	500	0.5
	(d)	200	100	0.5
	(e)	1,000	250	0.25

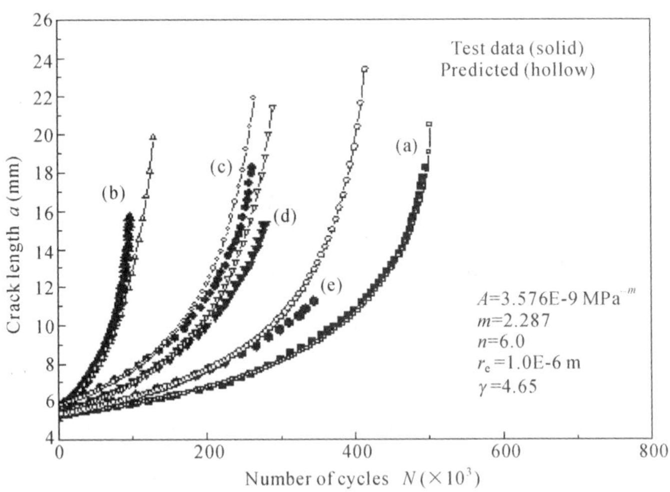

Fig. 5.40 Comparison of prediction with test data of AlMgSil-T6 aluminium alloy under block loading (Borregoa, *et al.*, 2005; Chen, *et al.*, 2012) (with the permission of John Wiley and Sons, Inc.)

5.6 Summary

The first step of the procedure for the UFLP method based on FCP theory is the establishment of a relatively "correct" fatigue crack growth rate model. From the description of the fatigue mechanism, the model should reflect the influencing factors related to loading history, material, structure and also environment. By comparing the existing models, the modified constitutive relationship developed by McEvily and his co-workers has been found to be able to explain many phenomena observed. Consequently, this constitutive relationship is further generalized in several ways: 1) by introducing an unstable fracture condition; 2) by defining a virtual strength to replace the yield stress; 3) the slope of the fatigue crack growth rate curve is regarded as a variable rather than a fixed value of 2 for different materials; 4) the SIF range at the threshold level is written as the function of the load ratio; 5) by defining an overload and underload parameter to explain the load sequence effect; 6) some critical fatigue phenomena such as specimen thickness effect, the unstable fracture condition, *etc*., are discussed.

Studies on the engineering approaches to determine the parameters in the improved model are carried out. The ability of the model and its parameter determination method are validated by comparing the predicted crack growth rate curves with the experimental results on a wide range of alloys. After the definition of the crack growth rate model, the accumulated crack length can be calculated on a cycle-by-cycle basis for a given loading history. This could accurately account for many loading effects. In particular when the sequence effect becomes significant, the loading history must be defined in such a way that the real order of the loading cycles can be reproduced. The discussions on the standardized loading history will be conducted in Chapter 6.

References

Bacila, A., Decoopman, X., Mesmacque, G., Voda, M. & Serban, V. A. (2007). "Study of underload effects on the delay induced by an overload in fatigue crack propagation", International Journal of Fatigue, 29: 1781–1787.

Bao, H. & McEvily, A. J. (1998). "On plane stress-plane strain interactions in fatigue crack growth", International Journal of Fatigue, 20(6): 441–448.

Barsom, J. M. & Rolfe, S. T. (1970). "Correlations between K_{IC} and Charpy V-Notch test results in the transition temperature range", ASTM-STP-466, 281–302.

Borrego, L. P., Ferreira, J. M. & Costa, J. M. (2008). "Partial crack closure under block loading", International Journal of Fatigue, 30: 1787–1796.

Borregoa, L. P., Costa, J. M. & Ferreira, J. M. (2005). "Fatigue crack growth in thin aluminium alloy sheets under loading sequences with periodic overloads", Thin-Walled Structures, 43: 772–788.

Bowness, D. & Lee, M. M. K. (1996). "Prediction of weld toe magnification factors for semi-elliptical cracks in T-butt joint", International Journal of Fatigue, 22: 369–387.

Bulloch, J. H. (1995). "Near threshold fatigue crack propagation behavior of CrMoV turbine steel", Theoretical and Applied Fracture Mechanics, 23: 89–101.

Chen, F. L., Wang, F. & Cui, W. C. (2012). "Fatigue life prediction of engineering structures subjected to variable amplitude loading using the improved crack growth rate model", Fatigue and Fracture of Engineering Materials and Structures, 35(3): 278–290.

Cui, W. C. (2002a). "A state-of-the-art review on fatigue life prediction methods for metal structures", Journal of Marine Science and Technology, 7: 43–56.

Cui, W. C. (2002b). "Relation between crack growth rate curve and *S-N* curve for metal fatigue", Journal of Ship Mechanics, 6(6): 93–106.

Cui, W. C. & Huang, X. P. (2003). "A general constitutive relation for fatigue crack growth analysis of metal structures", Acta Metall Sinica, 16: 342–354.

Cui, W. C., Wang, F. & Huang, X. P. (2011). "A unified fatigue life prediction method for marine structures", Marine Structures, 24: 1–29.

Dinda, S. & Kujawski, D. (2004). "Correlation and prediction of fatigue crack growth for different *R*-ratios using K_{max} and ΔK^+ parameters", Engineering Fracture Mechanics, 71: 1779–1790.

Dominguez, J., Zapatero, J. & Moreno, B. (1999). "A statistical model for fatigue crack growth under random loads including retardation effects", Engineering Fracture Mechanics, 62: 351–369.

Dong, D. S., Mei, X. & Shi, L. D. (2000). "The relationship between the fracture mechanics parameters K_C and K_{IC} basing on the energy conservation principle", Journal of Shanghai Maritime University, 21(4): 50–55 (in Chinese).

Dowling, N. E. (1972). "Fatigue failure predictions for complicated stress-strain histories", J Mater, 7: 71–87.

Dowling, N. E. (2007). Mechanical Behavior of Materials-Engineering Methods for Deformation, Fracture, and Fatigue (3rd ed.). Upper Saddle River, New Jersey: Pearson Prentice Hall, 586.

Engesvik, K. M. & Moan, T. (1983). "Probabilistic analysis of the uncertainty in the fatigue capacity of welded joints", Engineering Fracture Mechanics, 18(4): 743–762.

Farahmand, B. & Nikbin, K. (2008). "Predicting fracture and fatigue crack growth properties using tensile properties", Engineering Fracture Mechanics, 75: 2144–2155.

Firrao, D., Matteis, P., Scavino, G., Ubertalli, G., Ienco, M. G., Pinasco, M. R., Stagno, E., Gerosa, R., Rivolta, B., Silvestri, A., Silva, G. & Ghidini, A. (2007). "Relationships between tensile and fracture mechanics properties and fatigue properties of large plastic mould steel blocks", Materials Science and Engineering, A468–470: 193–200.

Girenko, V. S. & Dyadin, V. P. (1985). "Relationship between impact strength and fracture mechanics criteria delta sub (Ic), K sub (Ic) of structural steels and their welded joints", AUTOM WELDING. 9: 13–20.

Guo, W. (1999). "Three-dimensional analyses of plastic constraint for throughthickness cracked bodies", Engineering Fracture Mechanics, 62: 383–407.

Huang, X. P. & Moan, T. (2007). "Improved modeling of the effect of *R*-ratio on crack growth rate", International Journal of Fatigue, 29: 591–602.

Huang, X. P., Torgeir, M. & Cui, W. C. (2008). "An engineering model of fatigue crack growth under variable amplitude loading", International Journal of Fatigue, 30: 2–10.

Irwin, G. R. (1960). "Plastic zone near a crack and fracture toughness", in: Proceedings of the 7th Sagamore Research Conference on Mechanics & Metals Behavior of Sheet Material, New York, 4: 463–478.

Ishihara, S. & McEvily, A. J. (2002). "Analysis of short fatigue crack growth in cast aluminum alloys", International Journal of Fatigue, 24: 1169–1174.

Jha, S. K. & Ravichandran, K. S. (2000). "Effect of mean stress (stress ratio) and aging on fatigue-crack growth in a metastable beta titanium alloy, Ti-10V-2Fe-3Al", Metallurgical and Materials Transactions, 3: 703–714.

Kalnaus, S., Fan, F., Jiang, Y. & Vasudevan, A. K. (2009). "An experimental investigation of fatigue crack growth of stainless steel 304L", International Journal of Fatigue, 31: 840–849.

Kang, J. Y., Choi, B. I. & Lee, H. J. (2006). "Application of artificial neural network for predicting plain strain fracture toughness using tensile test results", Fatigue and Fracture of Engineering Materials and Structures, 29: 321–329.

Khan, S. U., Alderliesten, R. C. & Benedictus, R. (2009). "Crack growth in fiber metal laminates under variable amplitude loading", ICAF Symposium-Rotterdam, May 27–29.

Kim, J. H. & Song, J. H. (1992). "Crack growth and closure behaviour of surface cracks under axial loading", Fatigue and Fracture of Engineering Materials and Structurest, 15: 477–489.

Lee, S. Y., Choo, H., Liaw, P. K., Oliver, E. C. & Paradowska, A. M. (2009). "In situ neutron diffraction study of internal strain evolution around a crack tip under variable-amplitude fatigue-loading conditions", Scripta Materialia, 60: 866–869.

Li, X. Y., Cui, W. C. & Zhang, W. M. (2006). "A modified constitutive relation fatigue crack growth", Journal of Ship Mechanics, 10: 54–61 (in Chinese).

Lin, X. B. & Smith, R. A. (1999). "Finite element modelling of fatigue crack growth of surface cracked plates (Part II: Crack shape change)", Engineering Fracture Mechanics, 63: 523–540.

Ling, C. & Zheng, X. L. (1990). "Prediction of fatigue crack propagation rates of aluminum alloys from tensile properties", Journal of Northwestern Polytechnical University, 8(1): 115–120 (in Chinese).

Liu, Q., Wang, F., Huang, X. P. & Cui, W. C. (2006). "Three dimensional FE analysis of the plastic zone size near the crack tip", Journal of Ship Mechanics,

10(5): 90–99 (in Chinese).

Ma, J. F. & Lu, G. Z. (1999). "An analytical model for determining ΔK_{eff}", Mechanical Science and Technology, 18(3): 383–389.

Mahmoud, M. A. (1988). "Quantitative prediction of growth patterns of surface fatigue cracks in tension analysis", Engineering Fracture Mechanics, 30: 735–746.

Mahmoud, M. A. (1989). "Growth patterns of surface fatigue cracks under cyclic bending: a quantitative analysis", Engineering Fracture Mechanics, 31: 357–369.

Makabe, C., Purnowidodo, A. & McEvily, A. J. (2004). "Effects of surface deformation and crack closure on fatigue crack propagation after overloading and underloading", International Journal of Fatigue, 26: 1341–1348.

Matos, P. F. P., McEvily, A. J., Moreira, P. M. G. P. & Castro, P. M. S. T. (2007). "Analysis of the effect of cold-working of rivet holes on the fatigue life of an aluminum alloy", International Journal of Fatigue, 29: 575–586.

McEvily, A. J., Ishihara, S. & Mutoh, Y. (2004). "On the number of overload induced delay cycles as a function of thickness", International Journal of Fatigue, 26: 1311–1319.

McEvily, A. J., Bao, H. & Ishihara, S. (1999). "A modified constitutive relation for fatigue crack growth", in: Wu, X. R. & Wang, Z. G. (ed.) Proceedings of the 7th International Fatigue Congress (Fatigue'99). Beijing: Higher Education Press, 329–336.

Minakawa, K. & McEvily, A. J. (1981). "On near-threshold fatigue crack growth in steels and aluminum alloys", in: "Fatigue Threshold" EMAS, U.K., 633–636.

Newman, J. C., Crews, J. H., Bigelow, C. A. & Dawicke, D. S. (1995). "Variations of a global constraint factor in cracked bodies under tension and bending loads", ASTM STP, 1224: 21–42.

Newman, J. C. A. (1984). "Crack opening stress equation for fatigue crack growth", International Journal of Fatigue, 24(3): R131–135.

Newman, Jr. J. C., Phillips, E. P. & Everett, R. A. (1999). "Fatigue analyses under constant and variable amplitude loading using small-crack theory", NASA/TM-1999-209329, ARL-TR-2001.

Newman, Jr. J. C. & Raju, I. S. (1979). "Analyses of surface cracks in finite plates under tension or bending loads", NASA Technical Paper, 1578.

NIMS (1979). "Data sheets on fatigue properties of non-load-carrying cruciform welded joints of SM50B rolled steel for welded structure", Effect of Specimen Size, NIMS Fatigue Datasheet No. 13, http://smds.nims.go.jp/MSDS/pdf/sheet/FBJ.pdf.

Nykänen, T., Marquis, G. & Björk, T. (2009). "A simplified fatigue assessment method for high quality welded cruciform joints", International Journal of Fatigue, 31(1): 79–87.

Oh, H. K. (1995). "Determination of fracture toughness by means of the uniaxial tensile test", Journal of Materials Processing Technology, 54: 372–374.

Oh, H. K. (1996). "Determination of fracture toughness by uniaxial tensile test",

Engineering Fracture Metchanics, 55(5): 865–868.
Paris, P. C., Tada, H. & Donald, J. K. (1999). "Service load fatigue damage—a historical perspective", International Journal of Fatigue, 21: S35–46.
Prawoto, Y. (2009). "Designing steel microstructure based on fracture mechanics approach", Materials Science and Engineering, 507: 74–86.
Putra, I. S. & Schijve, J. (1992). "Crack opening stress measurements of surface cracks in 70752T 6 aluminium alloy plate specimen through electron fractography", Fatigue and Fracture of Engineering Materials and Structures, 15: 323–338.
Rama, C. M. A., Palani, G. S. & Nagesh, R. (2004). "State-of-art review on fatigue crack growth analysis under variable amplitude loading", Institution of Engineers (Civil Engg. Division), 85: 118–129.
Rama, C. M. A., Palani, G. S. & Nagesh, R. (2007). "Remaining life prediction of cracked stiffened panels under constant and variable amplitude loading", International Journal of Fatigue, 29: 1125–1139.
Schijve, J. (2003). "Fatigue of structures and materials in the 20th century and the state of the art", Int J Fatigue, 25: 679–702.
Schijve, J., Skorupa, M., Skorupa, A., Machniewicz, T. & Gruszczynski, P. (2004). "Fatigue crack growth in the aluminium alloy D16 under constant and variable amplitude loading", International Journal of Fatigue, 26: 1–15.
Silva, F. S. (2005). "The importance of compressive stresses on fatigue crack propagation rate", International Journal of Fatigue, 27: 1441–1452.
Skorupa, M. (1998). "Load interaction effect during fatigue crack growth under variable amplitude loading—a literature review (Part I: Empirical trends)", Fatigue and Fracture of Engineering Materials and Structures, 21: 987–1006.
Suresh, S. (1998). Fatigue of Materials (2nd ed.). Cambridge, UK: Cambridge University Press.
Taheri, F., Trask, D. & Pegg, N. (2003). "Experimental and analytical investigation of fatigue characteristics of 350WT steel under constant and variable amplitude loadings", Marine Structures, 16: 69–91.
Tiryakioglu, M. & Hudak, D. (2008). "On estimating the fracture toughness of Al-7Si-Mg alloys by sharp-notch tensile test results", Materials Science and Engineering, A498: 501–503.
Voorwald, H. J. C. & Torres, M. A. S. (1991). "Modeling of fatigue crack growth following overloads", International Journal of Fatigue, 13(5): 423–427.
Vosikovski, O. (1979), "The effect of stress ratio on fatigue crack growth rates in steels", Engineering Fracture Mechanics, 11: 595–602.
Wang, F. & Cui, W. C. (2009). "Approximate method to determine the model parameters in a new crack growth rate model", Marine Structures, 22(4): 744–757.
Wang, F. & Cui, W. C. (2010a). "Effect of three dimensional stress state on unstable fracture condition and crack opening level in a new crack growth model", Acta Metallurgica Sinica (English Letters), 23(1): 41–49.
Wang, F. & Cui, W. C. (2010b). "On the engineering approach to estimate the

parameters in an improved crack growth rate model for fatigue life prediction", Ship and Offshore Structures, 3(8): 227–241.
Wang, Y. F., Cui, W. C., Wu, X. Y., Wang, F. & Huang, X. P. (2008). "The extended McEvily model for fatigue crack growth analysis of metal structures", International Journal of Fatigue, 30: 1851–1860.
Wu, S. X. (1985). "Shape change of surface during fatigue growth", Engineering Fracture Mechanics, 22: 897–913.
Yuen, B. K. C. & Taheri, F. (2004). "The effects of loading frequency, tensile overload and compressive underload on the fatigue crack propagation behavior of polymethyl methacrylate", Polymer Testing, 23: 491–500.
Yuen, B. K. C. & Taheri, F. (2006). "Proposed modifications to the Wheeler retardation model for multiple overloading fatigue life prediction", International Journal of Fatigue, 28: 1803–1819.
Zhang, B. & Guo, W. L. (2005). "Numerical simulation of surface crack propagation considering the crack closure effects and the three-dimensional stress constraints", Chinese Journal Computational Mechanics, 22(6): 716–721 (in Chinese).
Zhang, J., He, X. D. & Du, S. Y. (2003). "A simple engineering approach in the prediction of the effect of stress ratio on fatigue threshold", International Journal of Fatigue, 25: 935–938.
Zhao, W., Zhang, J. X. & Jiang, Y. Y. (2008). "A study of fatigue crack growth of 7075-T651 aluminum alloy", International Journal of Fatigue, 30: 1169–1180.
Zhou, C. & Cui, W. C. (2003). "Determination of fatigue crack growth rate using existing data", Shipbuilding of China, 44: 74–79 (in Chinese).

6

Description of Fatigue Loading

6.1 The Nature of Fatigue Loading

Fatigue is defined as a process of cycle by cycle accumulation of damage in a material undergoing fluctuating stresses and strains (Almar-Naess, 1985). A significant feature of fatigue is that the load is not large enough to cause immediate failure. Instead, failure occurs after a certain number of load fluctuations have been experienced, *i.e.* after the accumulated damage has reached a critical level, see Fig. 6.1 for an illustration.

The service life of a component or a structure depends on both the load conditions and the fatigue strength. Hence, in order to obtain a proper fatigue design, it is important to consider the real structural loading history, $P(t)$. From the definition of fatigue, one should know that different from monotonic loading which caused static failure, fatigue loading is basically of a cyclic nature. Furthermore, since it is a future event, the fatigue loading history for the designed structure can never be known *a priori*. Many engineering methods are based on finding the worst case scenario, where "worst" should often be interpreted as a certain severe load condition.

From Chapter 1 we know that generally speaking there are two types of FLP methods. One is CFD analysis where fatigue loading is expressed as a spectrum. The other is FCP analysis where fatigue loading is expressed as a time history. In this chapter, technical problems related to fatigue loading description and determination are discussed.

In this book we distinguish the fatigue load history from the fatigue load spectrum while many other authors do not make this distinction, *e.g.* Schijve (2009). In his book, the fatigue load on a structure in service is also referred to as the load spectrum. When we say a fatigue load history, it means a time-domain description, *e.g.* $P(t)$. When we say a fatigue spectrum, it means a frequency-domain description, *e.g.* a Weibull distribution. The fatigue load history is used in the FCP process while the fatigue load spectrum is used in CFD calculation, see Fig. 6.2 for the comparison.

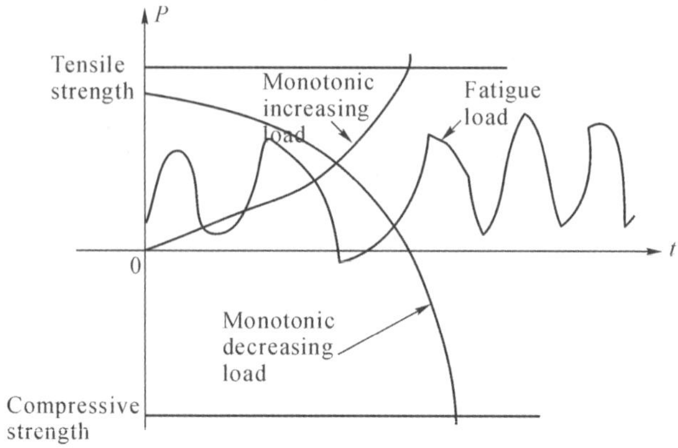

Fig. 6.1 Illustration of the nature of fatigue loading

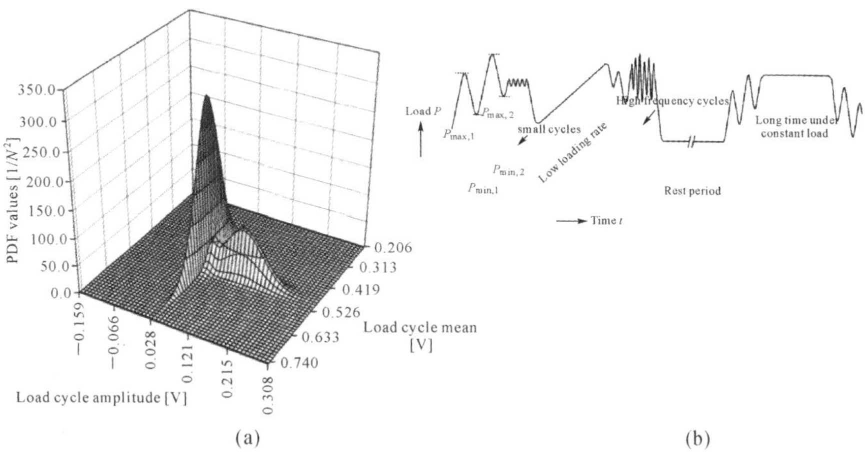

Fig. 6.2 A comparison of fatigue load history and fatigue load spectrum. (a) 2D load spectrum; (b) A typical fatigue load history (Schijve, 2009)

6.2 Load Spectra for CFD Analysis

Current FLP of marine structures at the design stage is generally performed by calculating the damage on the basis of the structure's dynamic load spectrum, the structure's shape and material endurance using *S-N* curves. The structure's shape and material endurance are usually known, so only the structure's dynamic load spectrum is needed to calculate the damage. The current fatigue strength assessment method for ship structures can be summarized as follows (adopted from *Common Structural Rules for Tankers*, 2006).

The following assumptions are made in the fatigue strength assessment:

(1) A linear cumulative damage model, *i.e.* Palmgren-Miner's Rule, has been

used in connection with the *S-N* data.

(2) Both nominal, hot spot or notch stresses can be used by choosing the corresponding appropriate *S-N* curves.

(3) The long term stress ranges of a structural detail can be characterized using a modified Weibull probability distribution parameter, ξ.

If we assume that only the stress amplitude S_a is affecting the fatigue damage, the probability density function of the stress amplitude is $f(S_a)$, and the *S-N* curve can be described by $N=N(S_a)$, then according to Palmgren-Miner's Rule, the fatigue damage can be calculated by the following equation,

$$D = N_T \cdot \int_0^{\infty} \frac{f(s)}{N(s)} \mathrm{d}s \tag{6.1}$$

For example, if the *S-N* curve is expressed by Eq. (6.2),

$$\log(N) + m\log\left(\frac{S}{S_R}\right) = 7.0 - 0.69897 \cdot \frac{m}{m_0} \tag{6.2}$$

where m is the inverse of the slope of the *S-N* curve, $m = \begin{cases} 3 & N \leq 10^7 \\ 5 & N > 10^7 \end{cases}$, m_0 is the inverse of the slope of the *S-N* curve for $N \leqslant 10^7$, $m_0 = 3$; S_R is the reference fatigue strength which corresponds to $N=2\times10^6$, MPa.

If the long term stress ranges of a structural detail can be represented by Eq. (6.3),

$$f(s) = \frac{\zeta}{W}\left(\frac{s}{W}\right)^{\zeta-1} \exp\left[-\left(\frac{s}{W}\right)^{\zeta}\right] \quad (s \geq 0) \tag{6.3}$$

where ζ is the shape parameter of the Weibull distribution, W is the scale parameter of the Weibull distribution. Then, by substituting Eqs. (6.2) and (6.3) into Eq. (6.1), one can derive

$$D = \frac{N_T}{A} \sum_{n=1}^{N_{\text{load}}} P_n \cdot W_n^3 \cdot \mu_n \cdot \Gamma\left(1 + \frac{3}{\zeta_n}\right) \tag{6.4}$$

where N_T is the total number of cycles of a ship encountered in a design life of usually 25 years, N_{load} is the number of working conditions, P_n is the time fraction of the working condition n, and

$$\mu_n = 1 - \left[\Gamma_0\left(1 + \frac{3}{\zeta_n}, Z_n\right) - Z_n^{-2/\zeta_n} \Gamma_0\left(1 + \frac{5}{\zeta}, Z_n\right)\right] \Big/ \Gamma\left(1 + \frac{3}{\zeta_n}\right) \tag{6.5}$$

$$A = S_R^3 \times 2 \times 10^6 \tag{6.6}$$

$$Z_n = \left(\frac{0.5848 S_R}{S_0} \right)^{\zeta} \cdot \ln N_0 \tag{6.7}$$

$$\Gamma(x) = \text{complete Gamma function} = \int_0^{\infty} u^{x-1} \mathrm{e}^{-u} \mathrm{d}u \tag{6.8}$$

$$\Gamma_0(x, z) = \text{incomplete Gamma function} = \int_0^{Z} u^{x-1} \mathrm{e}^{-u} \mathrm{d}u \tag{6.9}$$

It is generally agreed that the structural load variations with respect to fatigue should be characterized in the time domain since, in most cases, the range (or amplitude) of a load, stress or strain cycle plus its respective max or mean values can be considered as dominating parameters. Furthermore, the sequence or arrangement of load cycles of different ranges and means must not be neglected. However, due to its complexity, the fatigue life of structures is rarely estimated directly on the basis of measured or simulated load time series. These are usually transformed into loading spectra before the fatigue-life analysis is performed. A loading spectrum represents a distribution of the load-cycle parameters according to their frequency of occurrence in a summarized manner. Analyses in the frequency domain give an insight into the frequency content of a load signal which is particularly instructive for a flexible structure, but do not deliver the above-mentioned values (Berger, *et al.*, 2002).

The main problem in fatigue-life analysis is thus to determine the loading spectra that are realized when the structure operates under real operating conditions. In order to determine the loading spectrum the load-cycles must first be extracted from the simulated or measured load time series using a suitable counting method which will be introduced in Section 6.4.

Later, researchers found that not only the amplitude but also the mean value will affect the fatigue life. Two methods can be used for handling the mean stress effect. One is the mean stress correction method which is used in current practice (*e.g.* CSR for tankers and bulk carriers (2006)). The other method is to explicitly use two parameters to describe the *S-N* surface instead of the *S-N* curve (Sadandana, *et al.*, 2009).

In the second method, the rainflow counting method is regarded as the best method for estimating the loading spectrum because, in this method, the load-cycles are described by two parameters: the amplitude S_a and the mean value S_m of the corresponding closed hysteresis loop.

The loading spectrum then represents a distribution of the load-cycles in a 2D space S=(S_a, S_m). When using the *S-N* surface, the fatigue damage can be calculated by the following formula,

$$D = \sum_{ij} N_{ij} / N_{f,ij}(S_{ai}, S_{mj}) = \iint_{S=(S_a, S_m)} N_{\text{tot}} \cdot f(S_a, S_m) / N_f(S_a, S_m) \mathrm{d}S_a \mathrm{d}S_m \tag{6.10}$$

Under the presumption that the material characteristics are known, the multivariate probability density function (p.d.f.) $f(S_a, S_m)$ of load cycles must be estimated on the basis of the extracted rainflow load cycles for the purpose of calculating the fatigue damage. The multivariate p.d.f. of load cycles must adequately represent the probability distribution of load cycles and must enable extrapolation to the region where there are no measured data. For this purpose, Klemenc and Fajdiga (2000) have developed a method to estimate the statistical dependency of two parameters of random load states (the amplitude and the mean of load cycles) by means of the multivariate p.d.f. of rainflow load cycles that are represented as vectors $\boldsymbol{S}(S_a, S_m)$. The multivariate p.d.f. of load cycles is modeled with a mixture of multivariate normal functions. A comparison of different methods for an estimation of the unknown parameters of the normal mixture is carried out. Agreement of numerically modeled and experimentally obtained data is checked by comparison of corresponding marginal probability distributions of load cycle amplitudes and means. The maximum likelihood method was recognized to be the most appropriate for the determination of the multivariate p.d.f. of load cycles.

6.3 Generating the Whole Life Loading History from Short Time Measurement

FLP based on FCP theory requires information about the fatigue loading history over the entire design life. However, in many cases the measured or computed load history only covers a small portion of the design life. Therefore, a method to extrapolate the measured loads to longer periods of time, typically to a design life, is required. The simplest method is to repeat this load block until we reach the design life. This has the drawback that only the cycles in the measured signal will appear in the extrapolation, even though other cycles are also possible. In particular, this can be critical for the most damaging large amplitude cycles.

In order to allow more extreme loads than the observed ones, statistical theories have to be employed. In this section, the method based on statistical extreme value theory proposed by Johannesson (2006) is recommended for extrapolation of a measured or computed time signal to a longer time period. The method can be applied to any signal and be used for any purpose, *e.g.* the extrapolated time signal could be the input to a fatigue test, or the load input to a fatigue analysis.

6.3.1 Method for Extrapolation of a Load History

The methodology here will be to repeat the measured load block, but modify the highest maxima and lowest minima in each block. The random regeneration of each block is based on statistical extreme value theory. An example showing the principle of the method is shown in Fig. 6.3, where three repetitions of a measured block are compared to three extrapolated blocks.

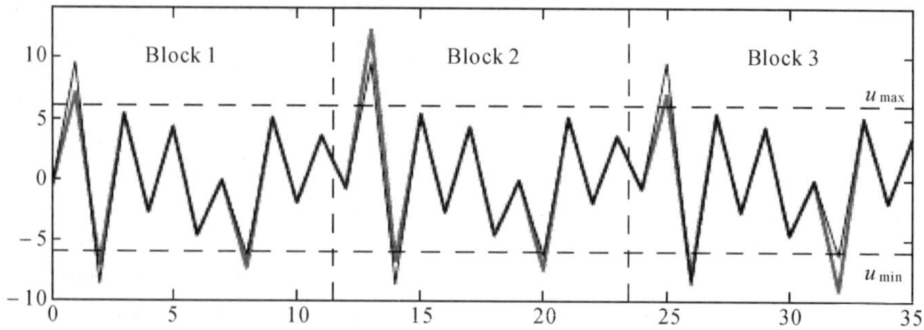

Fig. 6.3 Three repetitions of a measured block (black), compared to three extrapolated blocks (gray). The horizontal dashed lines represent the threshold levels, u_{min}=–6 and u_{max}=6, where the extrapolation starts (Johannesson, 2006) (with the permission of Blackwell Publishing)

The theoretical basis for the method is the so-called Peak Over Threshold (POT) technique in statistical extreme value theory, see *e.g.* Castillo (1988). Only the extreme excesses over a high level u are modeled, *i.e.* the height of the excursions above u, see Fig. 6.4. If we fix a high load level and study the excesses over this level, then under certain conditions these excesses approximately follow an exponential distribution for high enough threshold levels u. We then have the approximation for the excess $Z = \text{Max} - u \in \exp(m)$, with cumulative distribution function

$$F(z) = 1 - \exp\left(-\frac{z}{m}\right), \quad m=\text{“mean excesses over } u\text{”} \tag{6.11}$$

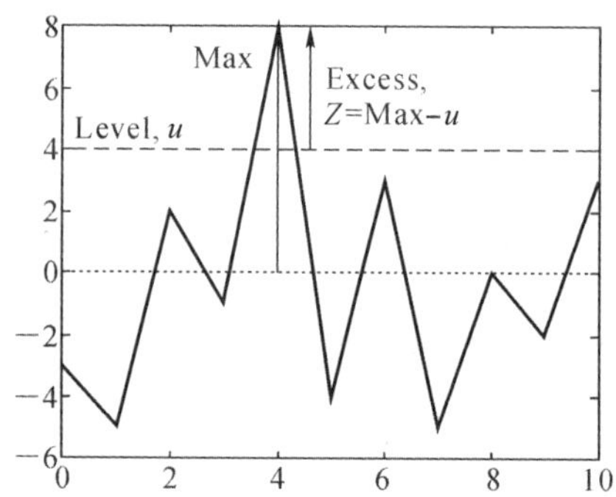

Fig. 6.4 Definition of excess over a threshold u (Johannesson, 2006) (with the permission of Blackwell Publishing)

The estimation of the parameter in the exponential distribution is the sample mean of the excesses $z_1, \cdots, z_N$

$$m = \frac{1}{N}\sum_{i=1}^{N} z_i \tag{6.12}$$

The family of possible distributions for excesses in extreme value theory is the Generalized Pareto Distribution (GPD),

$$F(z)=1-\left(1+\frac{\xi z}{\sigma}\right)^{-1/\xi} \tag{6.13}$$

where the shape parameter $-\infty<\xi<\infty$ and the scale parameter $\sigma>0$. When $\xi<0$, the GPD has an upper endpoint, $0<z<-\sigma/\xi$, while for $\xi\geqslant 0$, $z>0$. The special case of $\xi=0$ is the exponential distribution. The estimation of the parameters in the GPD is known to be problematic, often giving large systematic errors, especially for small sample sizes. Care should be taken in the selection of the moment method or maximum likelihood.

Mathematically, the extreme value approximations are formulated as asymptotic results. For a random process $\{X(t):t\in[0,T]\}$ satisfying certain mixing conditions, it can be shown that the excesses over a level $u(T)$ converge in distribution to a GPD as $T\to\infty$ and $n(T)/T\to 0$, where $n(T)$ is the number of excesses over $u(T)$. The parameters of the limiting GPD depend on the distributional properties of the process.

As a matter of fact, the exponential distribution of excesses is equivalent to modeling the global maximum as a Gumbel distribution, which works well in many applications. In this section, the model for excesses over a high level will be the exponential distribution, but it is straightforward to replace it by a GPD.

The generation of a *k*-fold extrapolated signal (sequence of turning points) can be performed in the following steps.

(1) Start with a time signal.

(2) Extract the turning points of the time signal, where the small cycles should be removed by using a rainflow filter.

(3) Choose threshold levels $u_{\min}$ and $u_{\max}$ for the POT extrapolation. The choice of levels is discussed in Section 6.3.3. Extract the excesses under $u_{\min}$ and the excesses over $u_{\max}$.

(4) Estimate the mean excesses $m_{\min}$ and $m_{\max}$ under and over the thresholds $u_{\min}$ and $u_{\max}$, respectively.

(5) Generate an extrapolated load block by simulating independent excesses as exponential random numbers. Replace each observed excess under $u_{\min}$ by a simulated exponential number with mean $m_{\min}$ and each observed excess over $u_{\max}$ by a simulated exponential number with mean $m_{\max}$.

(6) Repeat step (5) until you have generated *k* extrapolated load blocks.

(7) The *k*-fold extrapolated signal is obtained by putting the generated load blocks after each other.

From the *k*-fold extrapolated load time signal, we can also obtain the *k*-fold extrapolated load spectrum by applying the rainflow cycle counting algorithm which will be introduced in Section 6.4.2. In order to see the method for generating a *k*-fold extrapolated signal more clearly, a flow chart is given in Fig. 6.5 (Li, *et al.*, 2011).

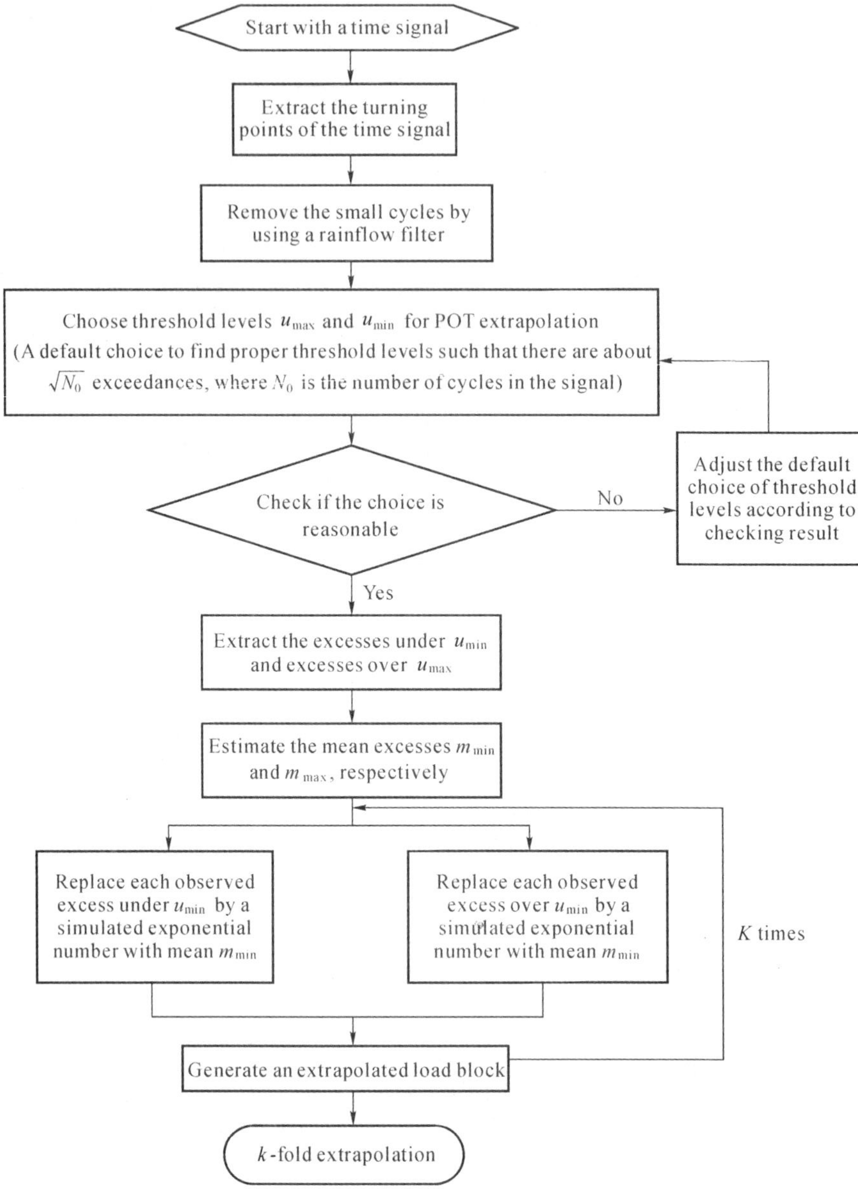

Fig. 6.5 Flow chart of the generation of k-fold extrapolated signal (Li, *et al.*, 2011)

6.3.2 Choice of Threshold Levels

For the extrapolation, we need to choose the threshold levels where the extrapolation shall start (see Fig. 6.4). The choice of proper threshold levels is a difficult problem that demands some judgments. Note that the levels need to be chosen high enough for the extreme value theory to be a reasonable approximation, but also low enough to obtain a sufficient number of exceedances. Hence, it is important to verify that the exponential assumptions are fulfilled for the threshold levels chosen.

A useful diagnostic tool for assessing the threshold is the mean excess plot. The estimate of the mean exceedance over the threshold is plotted as a function of the threshold level (see Fig. 6.6). For threshold levels that are too low, the extreme value approximation is not good enough, giving a systematic error, and for very high levels there is little information giving large scatter, which is seen through the confidence intervals. Hence, a proper level should be a compromise and be chosen somewhere in between, in a region where the estimate is stable, *i.e.* the mean excess plot is approximately horizontal. Unfortunately, the mean excess plots may be hard to interpret, but they are still a valuable tool together with the exponential probability plots.

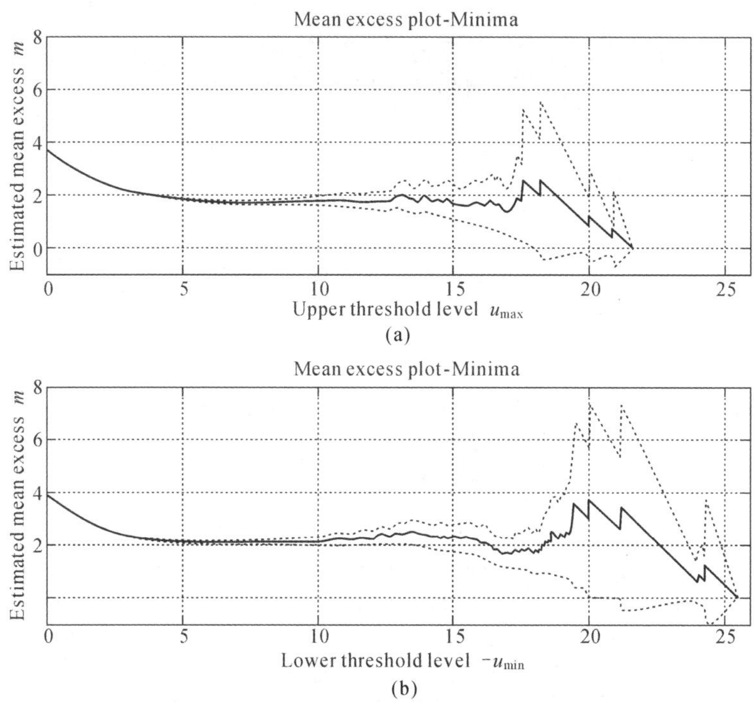

Fig. 6.6 Mean excess plots, *i.e.* the estimate of mean excesses as a function of the threshold level, for Norway measurements. The dashed lines represent 95% confidence intervals for the estimate. The threshold level should be chosen in a region where the estimate is stable, *i.e.* where the plot is approximately horizontal (Johannesson, 2006) (with the permission of Blackwell)

The choice of proper threshold levels is the most tricky part of the extrapolation. Here one needs to use experience together with the diagnostic tools to ensure that the exponential distribution is a good approximation. A default choice that seems to work well in many cases is to find threshold levels such that there are about $\sqrt{N_0}$ exceedances, where N_0 is the number of cycles in the signal.

6.3.3 Examples of Extrapolation of Load Histories

In order to demonstrate the extrapolation, a full scale test of wave bending moments on a surface ship is used to validate the method (Hu, 2007). Fig. 6.7 shows the measurements of the longitudinal strain in a deck position of a surface ship for 15 min.

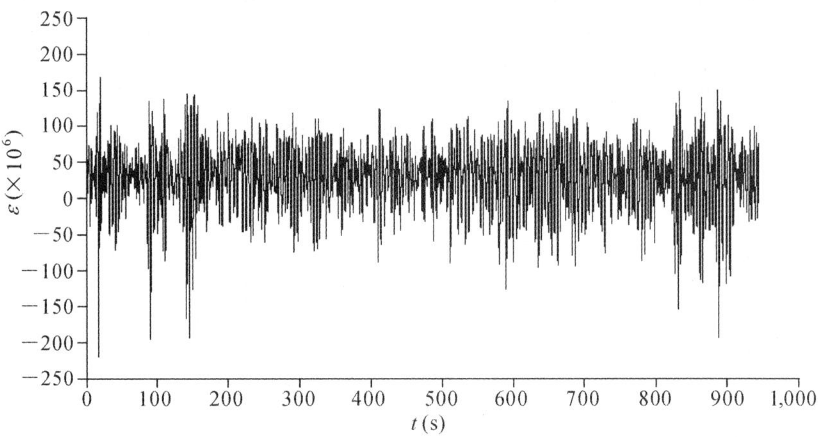

Fig. 6.7 The measured load signals used for demonstration of extrapolation (Hu, 2007)

For this example, we first choose the first three minutes to be extrapolated to a 5 times longer time period and compare this with the original measurement. First, the turning points of the time signal are extracted. Furthermore, the signal should be rainflow filtered in order to remove small oscillations that do not contribute to the fatigue damage. It is also advantageous for the POT analysis, since it removes small oscillations during excursions. A default choice for the rainflow filter is to set the threshold range to 5% of the total range of the signal, see Fig. 6.8. Then the threshold levels u_{min} and u_{max} for the POT extrapolation are chosen. We choose 121×10^6 and -111×10^6 as u_{max} and u_{min}, respectively, see Fig. 6.9. Here we use a default choice to find threshold levels such that there are about $\sqrt{N_0}$ exceedances, where N_0 is the number of cycles in the signal. For assessing whether the thresholds are proper, we use the mean excess plot method to check it and the result is shown in Fig. 6.10, where the chosen threshold levels are in the regions where the estimates are almost stable. With the measured time period increasing, *e.g.* 15 min were chosen to be extrapolated in this example, the horizontal region in Fig. 6.10 is much clearer. The comparison of the measured time signal and extrapolated time signal for 15 min is shown in Fig. 6.11.

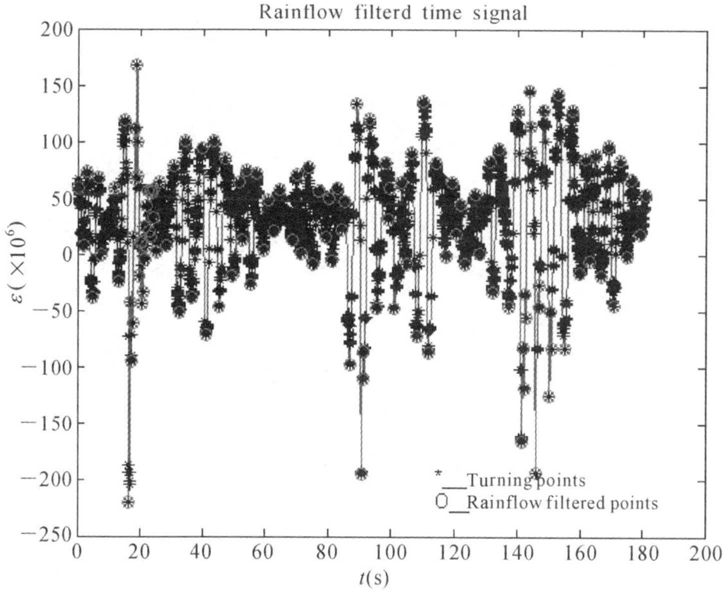

Fig. 6.8 The rainflow filtered load signal for first three minutes

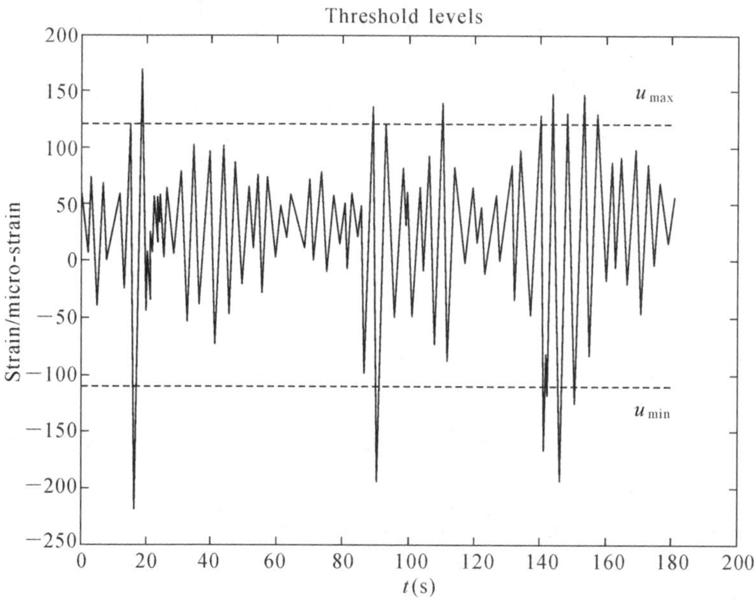

Fig. 6.9 Threshold levels

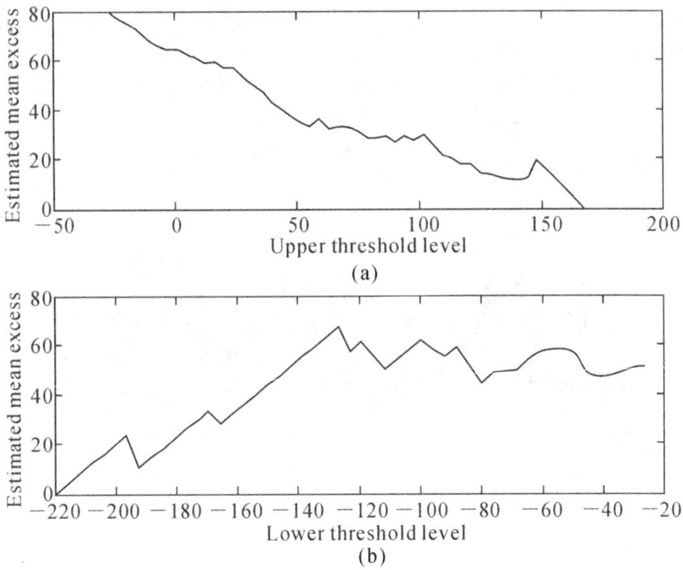

Fig. 6.10 Mean excess plot

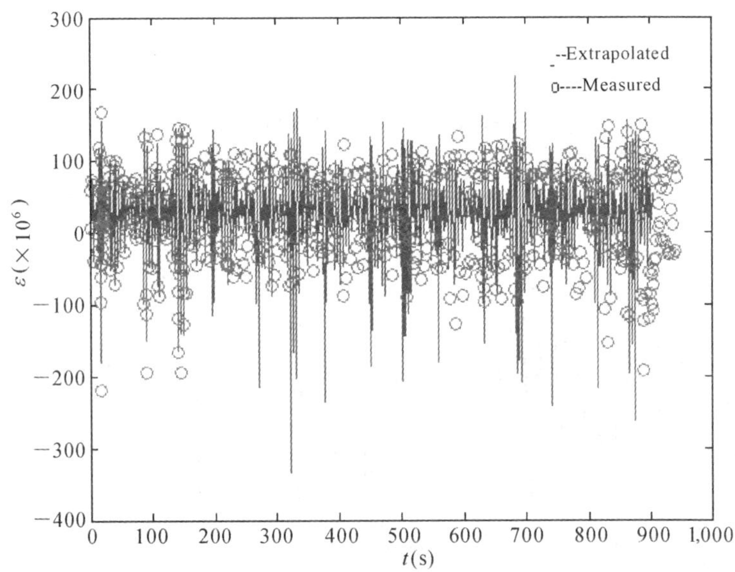

Fig. 6.11 A comparison of measured time signal and extrapolated time signal for 15 min

We observe that the extrapolation looks reasonable, since it agrees well with the observed load time history. Furthermore, the load time history is extrapolated to higher amplitudes and is smoother compared to the observed ones. These are the properties of the recommended extrapolation method.

The extrapolation method is based on a random simulation of the high maxima and low minima. Hence, the resulting extrapolated turning points, and hence also the load time histories, will be different for each new simulation. As the number of simulations or the number of folds increases, the load spectra will converge. Johannesson (2006) has compared ten load spectra for 10-fold extrapolation and the load spectrum for 100-fold extrapolation and good agreement has been found.

6.4 Cycle Count Methods

6.4.1 Definitions

Cycle counting is the process of reducing a complex variable amplitude load history into a number of constant amplitude stress excursions that can be related to available constant amplitude test data. This is a necessary step that needs to be carried out in order to predict fatigue crack growth in components subjected to variable amplitude loading. The method of cycle counting often used depends on the occurrences in the particular sequence that are considered to be significant in terms of fatigue damage (Etube, 2001; Schijve, 2009).

Cycle counting is used to summarize lengthy, irregular load-versus-time histories by providing the number of cycles of various single quantities, such as amplitude, or a pair of quantities such as (maximum, minimum) or (maximum, mean) or (maximum, range). For a given load-versus-time history $S(t)$, as shown in Fig. 6.12, the following six quantities are frequently used to describe the cycle:

S_{max} — the maximum value of the cycle;

S_{min} — the minimum value of the cycle;

$S_a=(S_{max}-S_{min})/2$ — the amplitude of the cycle;

$R=S_{min}/S_{max}$ — the ratio of the cycle;

$\Delta S=S_{max}-S_{min}$ — the range of the cycle;

$S_m=(S_{max}+S_{min})/2$ — the mean value of the cycle.

But only two of them are independent. At the beginning of the study of fatigue problems, it was thought that only the amplitude or range affects the fatigue life, for example Wöhler for the *S-N* curve (Wöhler, 1860) and Paris for the crack growth rate equation (Paris, *et al.*, 1961). However, nowadays more and more people have realized that both parameters will affect the fatigue life and therefore, in order to describe the cycle, a pair of two independent quantities must be used. In that case, the pair of (maximum, range) is mostly recommended by Sadananda, Vasudevan and co-workers (Vasudevan, *et al.*, 1994; Sadananda and Vasudevan, 1997; 2003; 2004; 2005; Vasudevan and Sadananda, 1999; 2001; 2009) and many others (Noroozi, 2005; Holtz, 2003; Stoychev and Kujawski, 2005; Maymon, 2005; Zhang, *et al.*, 2005).

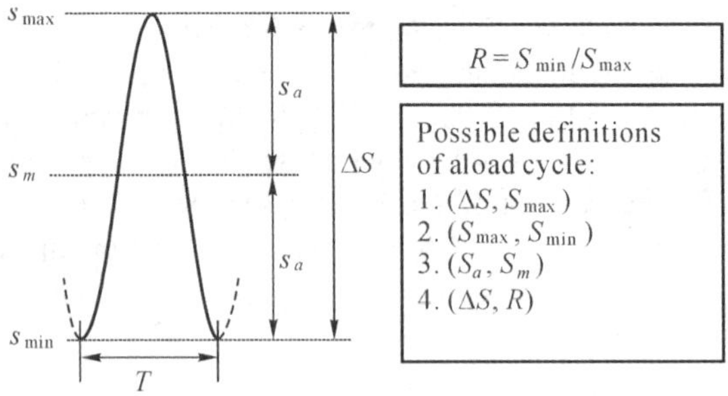

Fig. 6.12 Definition of a load cycle

The definition of a cycle varies with the method of cycle counting. In fatigue analysis, a cycle is the load variation from the minimum to the maximum and then to the minimum load. At the moment, there is no frequency information contained in any cycle counting method but this can easily be added if three independent quantities are used. Furthermore, some of the counting methods can also give information of the order of the cycles. Cycle counts can be made for all time histories such as force, moment, torque, stress, strain, acceleration, deflection or other loading parameters of interest.

Cycle counting yields the marginal probability density distribution of the single quantity, like amplitude or the joint probability density distribution of the pair such as (maximum, minimum) or (amplitude, mean) or (maximum, range).

Up to now, many different cycle counting methods have been developed such as level crossing counting, range counting, rainflow counting, see Refs. (Schijve, 2009; ESDU 95006, 1995). One of the most important considerations in cycle counting is that the basis of the counting method needs to be compatible with the understanding of the relevance of stress fluctuations to the fatigue process. Using this criterion, it is now almost the common consensus that the rainflow cycle counting method is the best cycle counting method (ASTM Standard, 1995; Amzallag, *et al*., 1994).

6.4.2 Rainflow Cycle Counting

The rainflow counting algorithm was first developed by Matsuiski and Endo (1968). Downing and Socie (1982) created one of the more widely referenced and utilized rainflow cycle counting algorithms in 1982, which was included as one of many cycle counting algorithms in ASTM E 1049-85 (1995). Rychlik (1987) gave a mathematical definition for the rainflow counting method, thus enabling closed-form computations from the statistical properties of the load signal.

In principle, range counting includes counting of all successive load ranges, also small load variations occurring between adjacent larger ranges. It might be thought that small load variations can be disregarded in view of a negligible contribution to fatigue damage. A fundamental counting problem arises if a small load variation occurs between larger peak values. This situation is illustrated in Fig. 6.13. A two-parameter range counting procedure will count the ranges AB, BC and CD, and store this information in a matrix. Now, consider the situation where the intermediate range BC would not occur. Then the large range AD would only be counted. Traditionally, it was thought that fatigue damage is related to load ranges but now it is found that the maximum value is also important (*e.g.* Vasudevan, *et al.*, 2001; Sadananda, *et al.*, 2009). It should be expected that the fatigue damage of the large range AD alone is larger than that for the three separate ranges AB, BC and CD. This has led to the so-called rainflow counting method of Matsuiski and Endo (1968). The intermediate small load reversal BC is counted as a separate cycle and then removed from the major load range AD. This larger range can then be counted as a separate load range, see Fig. 6.13(b). If four successive peak values are indicated by P_i, P_{i+1}, P_{i+2} and P_{i+3}, the rainflow count requirement for counting and removing a small range from a larger range is

$$P_{i+1} < P_{i+3} \text{ and } P_{i+2} > P_i \tag{6.14a}$$

If the intermediate small load reversal occurs in a descending load range, see Fig. 6.13(c), the requirement is

$$P_{i+1} > P_{i+3} \text{ and } P_{i+2} < P_i \tag{6.14b}$$

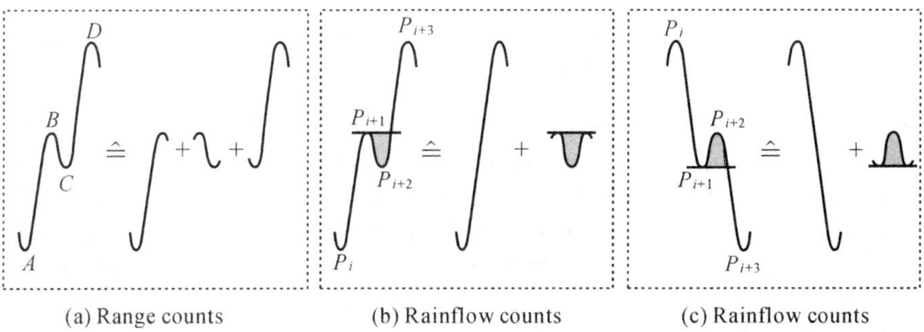

Fig. 6.13 Intermediate load reversal as part of a larger range (Schijve, 2009)

In other words, the peak values of the intermediate small load reversal should be inside the range of the two peak values of the larger range. Successive rainflow counts are indicated in Fig. 6.14. In Fig. 6.14(a), five rainflow counts can be made. After counting and removing these small cycles, Fig. 6.14(b) is obtained. In this figure three rainflow counts can again be made, but now of larger ranges. Removing these cycles leads to Fig. 6.14(c) in which again two still larger load reversals can be counted and removed. In the final residue, Fig. 6.14(d), no further counts are possible. The ranges of the residue must be counted separately at the end of the counting procedure. The rainflow count results can be stored in a similar two-parameter matrix.

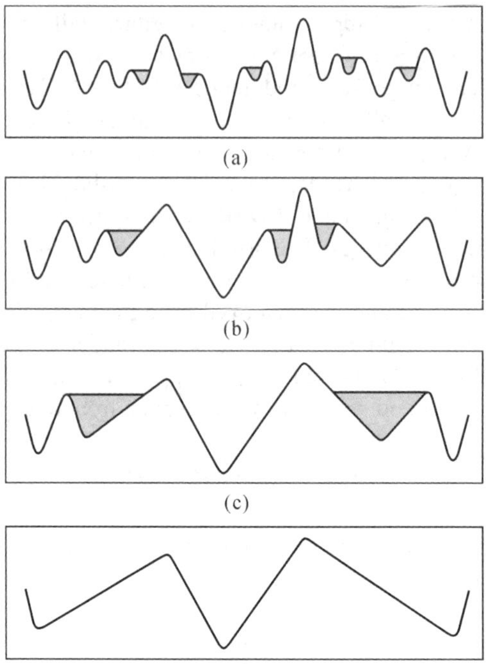

(a)

(b)

(c)

(d)

Fig. 6.14 Successive rainflow counts (Schijve, 2009). (a) 5 small cycles counted; (b) 3 cycles counted; (c) 2 cycles counted; (d) Residue

The rainflow count procedure has found some support by considering cyclic plasticity. A short load sequence is given in Fig. 6.15(a), which leads to counting two intermediate load reversals by the rainflow count method, as indicated in this figure. The corresponding plastic behavior is schematically indicated in Fig. 6.15(b), which could apply to local plasticity at the material surface during the initiation period, or to crack tip plasticity during crack growth. The intermediate load reversals C_1 and C_2 are causing hysteresis loops inside the major hysteresis of the major cycle between A and B. It is thus assumed that the intermediate plasticity loops do not affect the major loop. This reasoning gives somewhat speculative support to the rainflow counting method.

The rainflow counting method can also be illustrated in Fig. 6.16. The original method was so-called because, if the stress or strain history is presented vertically, the algorithm can be considered analogous to the behavior of rain flowing from a pagoda style roof.

Rules for this method are as follows (ESDU95006, 1995).

Let X denote the range under consideration; Y, the previous range adjacent to X; and S the starting point in the history.

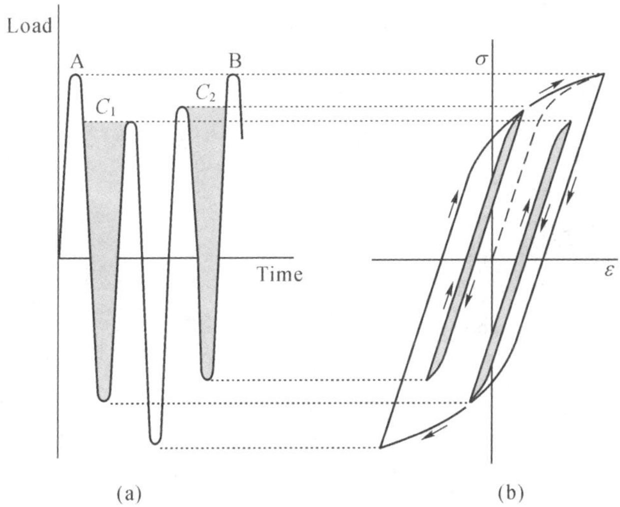

Fig. 6.15 Hysteresis loops associated with rainflow counts (Schijve, 2009). (a) Load history sample with two intermediate load reversals; (b) Corresponding plastic deformation loops

(1) Read the next peak or valley. If out of data go to Step (6).

(2) If there are fewer than three points go to Step (1). Form ranges *X* and *Y* using the three most recent peaks and troughs that have not been discarded.

(3) Compare the absolute values of ranges *X* and *Y*.

1) If $X<Y$, go to Step (1).

2) If $X \geq Y$, go to Step (4).

(4) If range *Y* contains the starting point *S*, go to Step (5); otherwise, count range *Y* as one cycle; discard the peak and trough of *Y*; and go to Step (2).

(5) Count range *Y* as one-half cycle; discard the first point (peak or trough) in range *Y*; move the starting point to the second point in range *Y*; and go to Step (2).

(6) Count each range that has not previously been counted as one half cycle.

These rules may be converted to an algorithm suitable for computation. A commercial program for rainflow cycle counting is also available now (*e.g.* Durability, 2002). Fig. 6.17 is the result of applying "RAINFLOW" code developed by Nieslony (2009) to count the number of cycles for the actual measurement given in Fig. 6.7. For the left half cycles, every two half cycles are merged as an equivalent complete cycle and they are put at the end of the series of cycles. For each cycle, we use both maximum value and minimum to represent it and there are two curves in Fig. 6.17.

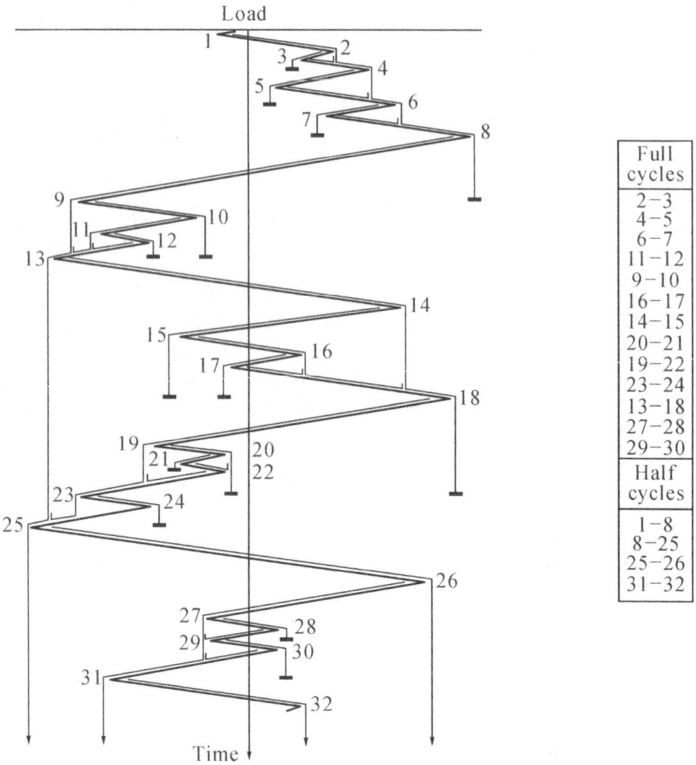

Fig. 6.16 The rainflow cycle counting method

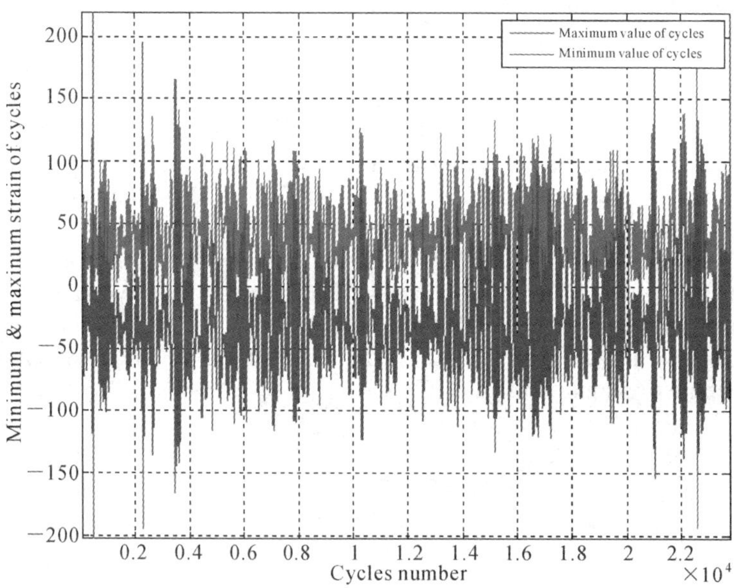

Fig. 6.17 Rainflow cycle counted results for the practical measurement given in Fig. 6.7

However, one significant aspect is not yet mentioned in the above algorithm. Information about the sequence in which the counts were made is not emphasized. Some indirect information about sequences is retained in the rainflow count method. Each range counted by the rainflow procedure and stored in the matrix combines two peak values which may have been separated by intermediate load reversals in the original load-time history. However, these smaller ranges should have occurred between those two peak values in order to satisfy the rainflow count Eq.(6.14). Intermediate larger ranges did not occur because of the counting condition in the same equation. Anyway, the question must be considered whether the sequence is important for fatigue predictions.

The rainflow count method is generally suggested to give a better statistical reduction of a load time history defined by successive numbers of peaks and valleys if compared to the level crossing counts and the range counts. Two main reasons for preferring the rainflow count method are: 1) an improved handling of small intermediate ranges, and 2) an improved coupling of larger maxima and lower minima to range counts as will be illustrated below. In spite of these advantages, it cannot be said that the rainflow count method is fully based on rational arguments. It is easily recognized from Fig. 6.14 that larger ranges counted by the rainflow method are separated by more intermediate smaller load ranges. The counting residue mentioned before consists of an increasing/decreasing series of positive peaks, and a decreasing/increasing series of negative peaks. These peaks may be separated by many intermediate peaks. It then becomes questionable whether counts of the residue are still meaningful for the evaluation of fatigue damage accumulation. It depends on the memory of the material during fatigue damage accumulation. The problem may be less serious for a flat spectrum with a large number of severe peak loads. The significance of the basic idea of the rainflow counting method is illustrated by two simple examples. Fig. 6.18 shows the load variation in a workshop machine due to switching on and off. The basic load cycle is a static block type load, but at the moment of switching some vibrational loads are introduced. The rainflow procedure will count several small load cycles, but instead of the deterministic static load cycle it also counts an increased basic load cycle between P_{max} and P_{min}, see Fig. 6.18. Although the small vibrational load cycles could be practically harmless by themselves, they increase the severity of the basic load cycle. From a fatigue point of view, the material feels the increased basic cycle, and this increased load cycle is recognized by the rainflow counting method. A similar case was already discussed in relation to the superposition of random loads on a deterministic load cycle in Fig. 6.6. The rainflow count method recognizes the largest cycle of the flight between the maximum peak in flight and the most severe downward load on the ground.

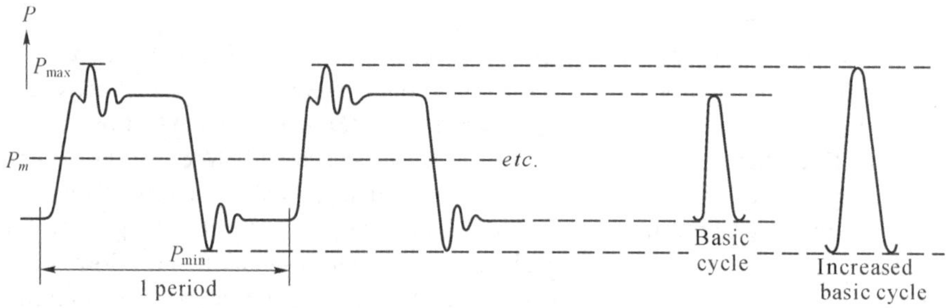

Fig. 6.18 Transient load fluctuations during on/off switching of machinery

A second illustrative example is shown in Fig. 6.19. A high-frequency load cycle (amplitude B) is superimposed on a low-frequency base line load cycle (amplitude A, frequency ω_1). According to the rainflow counting method, one cycle with an amplitude of $A+B$ will be counted in each period of the base line period, and that makes this cycle more damaging. This can be important depending on the damage done by the small superimposed high-frequency cycles. If ω_2 is much larger than ω_1, the number of the high-frequency cycles will contribute the major part to the fatigue damage, and the base line cycle is no more than a varying mean load, probably with a limited effect only.

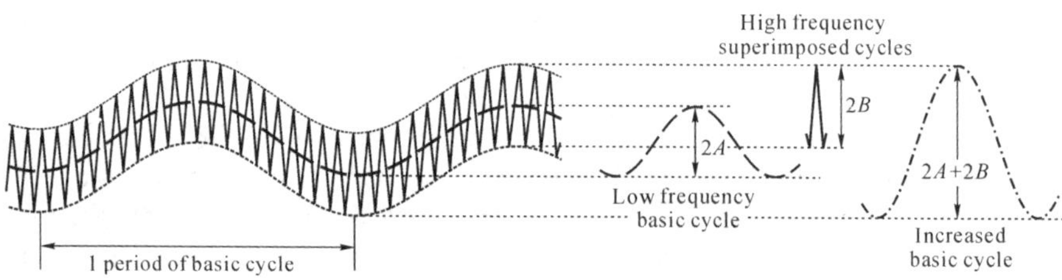

Fig. 6.19 High-frequency cyclic load superimposed on a low-frequency base line cycle

6.4.3 A Practical Example

Let us apply the rainflow cycle counting method to the two load-time series shown in Fig. 6.11. For each time series, we will obtain a certain number of complete cycles and some half cycles. For these half cycles, every two half cycles are merged as an equivalent complete cycle and they are put at the end of the series of cycles. For each cycle, we use both maximum value and minimum to represent it. The counted results are shown in Fig. 6.20.

For the counted maximum values and the ranges, they can be plotted as a spectrum. Fig. 6.21(a) shows a comparison of the load spectra of the 5-fold extrapolations to the 15 minutes of the measured ones for the maximum value of each cycle while Fig. 6.21(b) shows the range. As is customary, the load spectra are plotted as the cumulative number of cycles larger than a given amplitude as a function of the amplitude. We observe that the extrapolation looks reasonable since it agrees well with the observed load spectrum. Furthermore, the load spectra are extrapolated to higher amplitudes, and are smoother compared to the observed ones. These are the properties of the recommended extrapolation method.

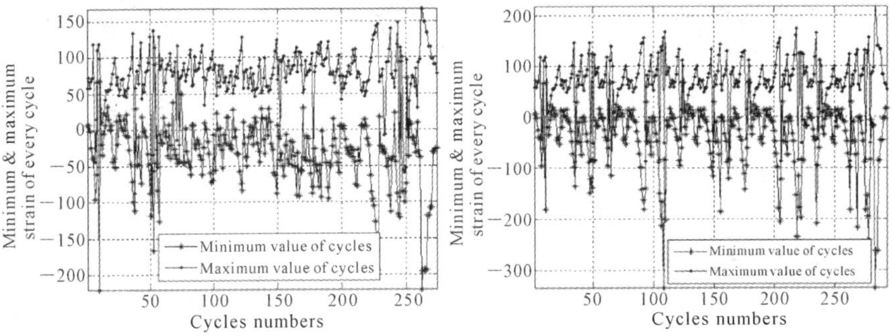

(a) Original measurement 274 cycles (b) Extrapolated time signal 294 cycles

Fig. 6.20 Rainflow cycle counted results for two load time signals given in Fig. 6.11 (threshold range to 5% of the total range)

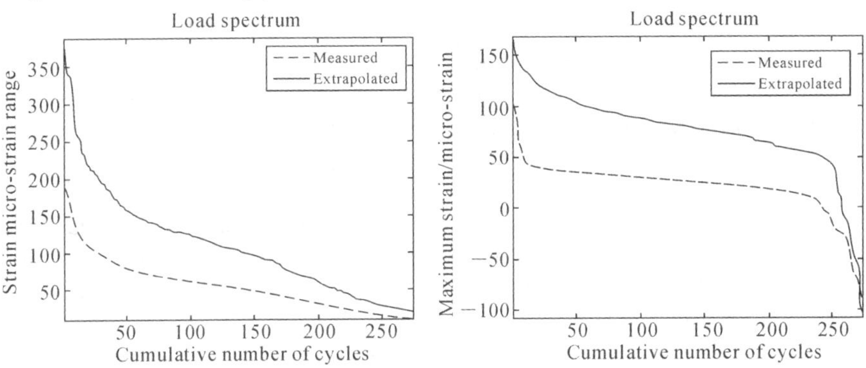

(a) The maximum values of all cycles (b) The ranges of all cycles

Fig. 6.21 Five-fold extrapolated load spectra compared to the measured load spectra

Now let us take the original measurement of 15 min as the source data for extrapolation and to check the effect of folds on the distribution properties. The extrapolation method is based on a random simulation of the high maxima and low minima. Hence, the resulting extrapolated turning points, and hence also the load spectra, will be different for each new simulation. As the number of simulations or the number of folds increases, the load spectra will converge. Fig. 6.22 shows the load spectra for the range of cycles for five different fold numbers: 1-fold (original measurement of 15 min), 4-fold (1 h), 96-fold (1 day). Due to the limit of the hard disk, longer extrapolations are not carried out here.

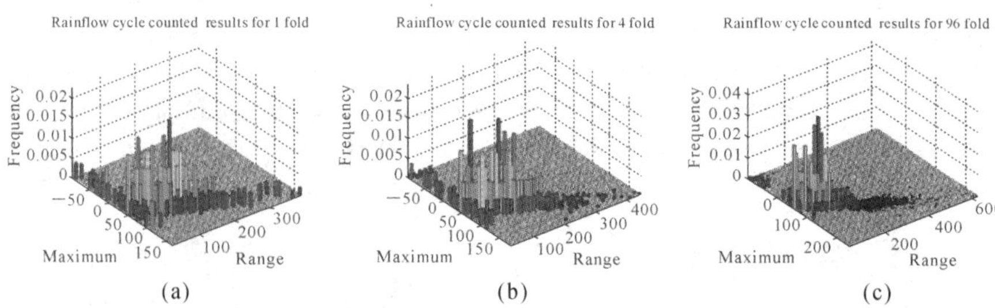

Fig. 6.22 Effect of fold number on the load spectra for the range of cycles

6.5 SLHs for FCP Analysis

6.5.1 Definition

In this book, the acronym SLH is specifically used for "standardized load-time histories" but not including "standardized load sequences" and "load spectra" by Heuler and Klätschke (2005). More specifically, it distinguishes between "load-time history" for processes with actual time information (*e.g.* multiaxial, real-time processes) and "load sequence" as a succession of loading events without explicit time information (*e.g.* load turning points, typically uniaxial applications) (Heuler and Klätschke, 2005). An SLH may be defined as a load-time history that does not cover the total expected in-service load-time history, but only a representative fraction (return period) (Heuler, *et al.*, 2005). Thus this load-time history is thought to be repeated in a test or a numerical simulation until final failure or the anticipated usage would be safely covered. If the respective design lives are taken into account, for example the number of, say, 50 years for offshore installations, the number of repetitions of the standardized load-time history can be estimated for a test at realistic stress levels. The simplest repetitions of the standardized load-time history may also be replaced by the extrapolation method introduced in Section 6.2.

Furthermore, the product life is also determined by the interaction between the strength of the component and its mission profile, *i.e.* load-time history. The importance of load-time history, *i.e.* variable amplitude loading, on durability and light-weight design was recognized by Ernst Gaßner, who in 1939 formulated a procedure for simulating variable amplitude loading, the historical blocked program sequence with a Gaussian-like distribution of loads. This sequence was frequently used as a standard until the 1970s and was then substituted by random load sequences applied with modern servohydraulic actuators. In the meantime different standardized load histories for different application areas were developed on the basis of extensive service load measurements, mainly for material testing and comparisons (Heuler and Schütz,1998).

The respective SLH has to reflect these characteristics in an appropriate manner. It is this mixture, sequence, relative frequency and magnitude of modes, events and scaled load cycles plus time and phase information for multiaxial problems which is "standardized"—generally not the actual load, stress or strain level which has to be assigned to a specific problem. It is preferable—but not always possible—to base an SLH on quantities which are representative of the external loading acting on a structure or vehicle (Berger, *et al.*, 2002).

6.5.2 History

SLHs have been developed and applied to fatigue studies for many years because of various advantages (Berger, *et al.*, 2002). The first SLH proposed and extensively applied was Gaßner's 8 level blocked program test which can be dated back to the 1930s (Heuler, *et al.*, 2005). However, the significant progress was made in the 1970s. With the need for optimum light-weight design, the aircraft industry was originally the main driver for these efforts. Two of the most well-known SLHs are the TWIST and FALSTAFF sequences for transport and fighter aircraft, respectively, which have been and are still being applied for numerous studies on materials, joints and other structural elements. HELIX/FELIX and the two TURBISTAN sequences are further examples from the aerospace field. Because of the relatively high number of cycles of the TWIST and HELIX/FELIX sequences, shortened versions of these SLHs have been also devised, as was done later with the WISPER and other sequences.

Starting in the mid 1980s, SLHs for non-aerospace applications have been developed covering areas such as wind turbines, offshore platforms or steel mill drives. For automotive applications, the CARLOS series of SLHs have been presented including the very recent load sequence for car trailer couplings, CARLOS-TC.

In the US, activities were mainly centered on the derivation of test load sequences to be used for evaluation and development of FLP methodology (Heuler, *et al.*, 2005). Bodies like the SAE Fatigue and Evaluation Committee took a pragmatic approach by selecting test load sequences from existing strain measurements, which were felt to be typical for the ground vehicle industry. The early quite short sequences published in 1977 were followed later by biaxial sequences that again were based on a set of measurements made available from different sources, but were more realistic with regard to spectrum length and distribution of small and large cycles. Similarly, in the US aircraft community a variety of load sequences have been generated and distributed without the formal framework of European SLHs, for example sequences typical for fighters (F-16) or transports (C-5) with different mission-dependent load cycle distributions. These load sequences are frequently used for validation of numerical FLP concepts and software packages.

6.5.3 Basis of Generation of SLHs

It is evident that "load" measurements are most essential, but on top of that some further input is required for the derivation of a meaningful SLH. These include load measurements, usage statistics, general definitions and non-technical items (Berger, *et al*., 2002).

(1) Load measurements

Measurements from several similar structures, vehicles, *etc*. should be available, because otherwise it would be impossible to identify the characteristic features of the load time histories, the sequence and mixture of events and load variations. For the existing SLHs, the respective data bases that have been collected and carefully assessed by the groups of experts are considered sufficient with respect to their statistical significance.

(2) Usage statistics

At best, the measurements available cover all relevant modes of operation and environmental conditions, but very probably not in the correct percentages. Therefore, additional information has to be collected and introduced with regard to usage statistics such as the anticipated distribution of weather statistics (sea states, maximum wave heights) for certain geographical locations for offshore applications, *etc*.

(3) General definitions

A number of SLH parameters have to be defined such as the extreme values of loads which may have to be extrapolated from the data pool, truncated for certain reasons or determined by physical limits. Similarly, the block size or return period of the SLH has to be agreed which must not be too short in order to avoid unrealistically truncated spectrum shapes (close to a constant amplitude test). This is more or less equivalent to the requirement of creating a spectrum shape where the (calculated) fatigue damage is realistically distributed over large, medium and small cycles of the history. With respect to applicability considerations, it is often necessary (and possible) to omit small "non-damaging" (at least in a relative sense) load variations. "Rules of thumb" have evolved that give some guidance for the selection of omission levels (Heuler and Seeger, 1986). The aspect of a realistic distribution of damage over the load ranges such as the block size (number of cycles per block) and equivalent usage present in the spectrum should be adequately taken into account.

(4) Non-technical items

Experience shows that the development of an SLH should be a cooperative effort of several contributors in order to consider the various aspects and requirements effectively. The principles followed and the range of application should be clearly documented and published.

6.5.4 Generation of Load-Time Histories

This section is mainly adapted from Heuler and Klätschke (2005).

(1) General considerations

It is generally agreed that the structural load variations should be characterized in the time domain since in most cases the range (or amplitude) of a load, stress or strain cycle plus its respective max or mean value can be considered as fatigue-relevant. Furthermore, the sequence or mixture of load cycles of different ranges and means must not be neglected.

Many structural loading environments can be described as sequences of different modes (Tenhave, 1991) which may be a particular flight, driving a car on certain road types, a sea state of a given severity, *etc.* These modes of operation contain load cycles of different, but typical magnitudes and frequencies. Often distinct patterns of grouped load cycles can be distinguished, called a loading event or element, such as braking or cornering of a car, different flight phases or manouevers of an aircraft. The occurrence of modes and loading events has to be defined in terms of frequency, severity and mixture/sequence, which in summary establish the so-called operating profile, Fig. 6.23.

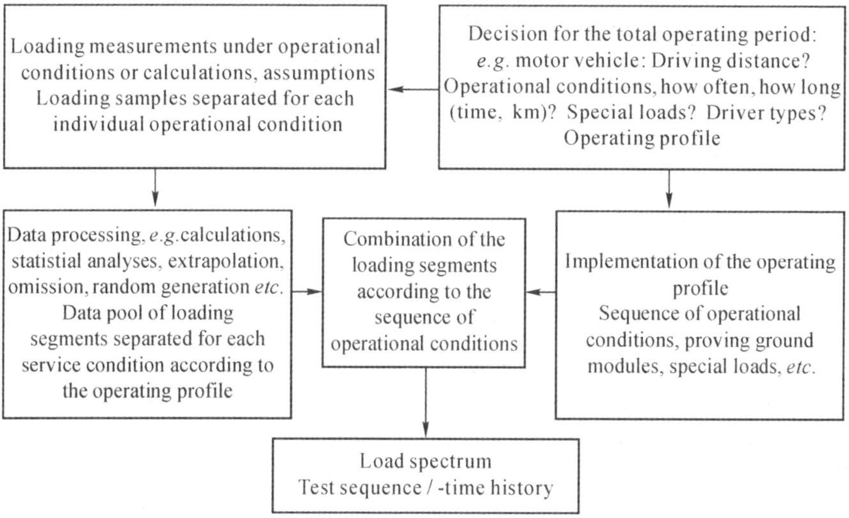

Fig. 6.23 General (simplified) approach for the generation of SLH's (Heuler and Klätschke, 2005) (with the permission of Elsevier)

As a basic input to SLH generation, statistically adequate samples of load measurements under operational conditions have to be available for every loading event. By means of data evaluation and processing, the measured loading samples are prepared for compilation according to the operating profile. Thus, the overall load spectra will be defined, but for definition of the final load sequences and/or load time histories further steps have to be taken that are addressed later in this section.

Experience shows that the development of an SLH should be a cooperative effort of several contributors in order to create a statistically significant data base and to consider the various aspects and requirements effectively. The principles adopted and the range of application should be clearly documented and published.

(2) Operating profiles

The operating profile describes the service conditions for the total or a representative fraction of the operating period. A difficulty is the fact that for different sections of, say, an aircraft, different operating profiles may have to be composed. That becomes quite clear by the comparison of *e.g.* airplane wings and landing gears; the wings are loaded by the take-off and landing load (ground-air-ground) cycles and the superimposed flight gust loads. The nose landing gears see, on the other hand, the push back, towing, taxiing, take off, landing impact, braking and landing roll operations. In the automotive area, a similar situation is found *e.g.* for wheel suspension, drive train and body/trailer-coupling components; the operating profiles of the respective SLHs diverge correspondingly.

The approaches for determination of operating profiles depend on the type of structural application concerned. The use of airplanes, helicopters, offshore structures, wind energy plants, *etc.* can, as a rule, be predicted and verified quite well. Operating profiles can therefore be based on reliable data, *e.g.* mission profiles (aircraft), meteorological long-time wind (wind energy plants) or sea state observations (offshore platforms).

(3) Analysis of operational loads

Statistically adequate samples of operational loads have to be available for every load case. These samples are usually acquired by in-service measurements, but they may also be supported by calculations or simple assumptions. Data processing aims at the preparation of the measured samples to be used as a basis for the generation of test load sequences compiled according to the operating profile, comp. Fig. 6.23. Table 6.1 summarizes the procedures that may be applied for analysis purposes.

Today, the two-parametric rainflow counting has the greatest significance for time domain fatigue analyses. The result is the rainflow matrix (with residuum) that contains the number, amplitudes and mean values of load cycles that can be traced back to closed stress-strain hysteresis loops of a representative material element. These load cycles are considered to be the basic damaging elements of a load sequence. Rainflow matrices serve for damage calculations and/or for the reconstruction of uniaxial random load sequences partly under consideration of omission of small, non-damaging load cycles (see Table 6.1).

Table 6.1 Compilation of relevant procedures for load analysis (Heuler and Klätschke, 2005) (with the permission of Elsevier)

Procedures to analyze operational loads	Stiff component	Dynamic response
Uniaxial loading	Rainflow counting	Additional Fourier Analysis (*e.g.* power spectral density)
Rotating loading	Rainflow counting Dwell time and/or revolution-at-level counting	Additional Fourier Analysis (*e.g.* power spectral density)
Loading under environmental impact	Rainflow counting Consideration of dwell time (environmental impact)	Additional Fourier Analysis (*e.g.* power spectral density)
Multiaxal loading	Rainflow counting Correlation analyses, *e.g.* joint density (multiaxial dwell time) counting	Additional Fourier Analysis (*e.g.* auto-, cross spectra, coherence functions, transfer matrices)

(4) Determination of load-time history by extrapolation

The load cycle distributions derived from sample measurements have to be extrapolated according to the requirements of the operating profile. The simplest way is plain repetition (the factor for the cumulative frequency of every loading element) without extrapolation of the maximum load levels. This may not be very suitable for the ocean environment and it is recommended that the statistical extrapolation method introduced in Section 6.2 is used for this purpose.

(5) Generation of load sequences and load time-histories for fatigue tests

For the synthesis of test load sequences and load time histories, approaches compiled in Table 6.2 will be applied with similar categories as for the analysis phase.

With respect to the various loading and test conditions, synthesis of test load sequences by combination and repetition of measured loading segments is always possible. The main advantage of this approach is that all characteristics of the loading conditions are maintained with respect to the correct order, frequency contents, correlation functions or phase relations of load cycles. However, the resulting test sequence and time would be unacceptably long without further processing. Therefore, this approach has to be combined with time reduction methods which again depend on the loading and test conditions.

For uniaxial loading with stiff components (without environmental impact), the most effective time reduction is achieved by reduction of the load time histories to turning points and omission of small load cycles. "Allowable" filter levels for omission have to be defined with care, but the reader should refer to the literature with regard to this topic (Heuler and Seeger, 1986; Conle, 1979). The additional inclusion of environmental exposure (*e.g.* temperature) may partly counterbalance the time reduction effect (comp. FALSTAFF and ENSTAFF).

Table 6.2 Compilation of procedures to synthesize test loads (Heuler and Klätschke, 2005) (with the permission of Elsevier)

Procedures to synthesize test loads	Stiff component	Dynamic response
Uniaxial loading	Combination of loading segments (reduced to turning points) Random rainflow reconstruction Simplified (blocked) test programs based on level crossings and/or range pair distributions	Combination of real-time loading segments Generation of random processes with defined power spectral densities (possibly additional effective-value distributions)
Rotating loading	Combination of loading segments Random rainflow reconstruction (or simplified test programs), taking into account the dwell time and/or revolution-at-level distributions Simplification: *e.g.* block-wise constant speed	Combination of real-time loading segments according to the desired load distributions and acceptable test time
Loading under environmental impact	Combination of loading segments Random rainflow reconstruction (or simplified test programs) under consideration of dwell time	Combination of real-time loading segments according to the desired load distributions and acceptable test time
Multiaxal loading	Combination of measured loading segments Guide function procedure, combined with calculation of correlated variables Calculation of optimised loading segments (simultaneous procedure)	Combination of real-time loading segments according to the desired load distributions and acceptable test time

The omission of low-intensity load segments (instead of single load cycles) is a less effective method that, nevertheless, has to be applied when real-time load histories need to be utilized, *e.g.* in case of dynamic response, connected with multiaxial loading.

For stiff components (negligible dynamic response), however, a reduction in testing time for multiaxial and/or rotating loading is also possible via the much more effective omission of individual load cycles if it is combined with an adaptation of the load-time history characteristics to the limits of the test facility. The procedure described in Ref. (Klätschke and Schütz, 1995) computes optimized time-histories, taking into account the phase relations between 12 channels and the test facility limits in terms of frequency, velocity and acceleration.

In most standardization applications, however, the more general (pseudo-) random generation procedures are preferred based on one, or a set of, load spectra that encompass the desired loading environment. For the generation of a random sequence of load cases, the Markov reconstruction is well suited, because it takes into account successive events. It has been applied, *e.g.* for generation of the sequence of wind speed classes (Tenhave, 1991), sea states (Sonsino, *et al.*, 1988).

For synthesis of uniaxial random load sequences on a cycle-by-cycle basis, random drawing of load cycles from cumulative frequency distributions (Schütz, *et al.*, 1973; Aicher, *et al.*, 1976), Markov-reconstruction for stationary processes (Haibach, *et al.*, 1976) up to sophisticated rainflow reconstruction (Schütz, *et al.*, 1999) have been used. For multiaxial loading combined with stiff test systems/components the so-called guide function procedure is an appropriate concept that was applied in Ref. (Schütz, *et al.*, 1994). Guide functions are direct or processed elements (for example, weighted combinations) of load signals that are representative for specific loading causes. In the framework of this method, as a first step, time histories of guide function values are generated under consideration of their respective joint density distributions. The resulting multi-directional and phase-related load vectors are then calculated from the guide functions (loading causes) in the time domain. The relationship between a given guide function level and the associated set of load vectors are provided by vehicle specific transfer functions. This approach was adopted for the development of CARLOS multi. In this application a stationary random process was generated and included as the (non-correlated) stationary part of road roughness loading in the guide function concept.

The Fourier procedures (like power spectral density) are particularly relevant for applications under dynamic test conditions. In this context it has to be emphasized that power spectra and other results of Fourier analyses are not directly fatigue-relevant, because load cycles are not clearly defined. For stationary Gaussian processes, however, a well-defined relation exists between power spectral densities and cumulative frequency distributions (Bendat and Piersol, 1971; Buxbaum, 1992; Papoulis, 1965). This means that the Fourier tools can be applied for stationary Gaussian processes even if they are only a fraction of more complex load-time histories.

(6) Potential applications of SLHs

The use of loading standards is recommended for the following purposes:

Evaluation of the fatigue strength of specimens or actual components made from different materials or by different manufacturing techniques;

- Optimization of design details;
- Determination of allowable stresses for the preliminary fatigue design of components;
- Investigation of methods to increase fatigue life;
- Verification of models for the prediction of fatigue life and crack propagation;

- Investigations on the scatter of fatigue test lives;
- Round-robin programs on general fatigue and crack propagation problems with several participating laboratories;
- Basis for comparisons with current in-service measurements;
- Fatigue tests with prototypes as long as no design specific load assumptions are available;
- New application-updating/amendment of directives of authorities.

Essential advantages of the application of service-like loading standards are:

- Availability at every stage of development;
- Reduction of the number of test parameters, comparability of results;
- Reliable (relative) fatigue estimates for current design and loading based on fatigue life curves determined under comparable loading conditions;
- Possible use as a main element of loading specifications for suppliers.

SLHs are usually applied in the pre-design stage or for studies of a more generic nature. These may include an evaluation of how different materials, detail geometries, surface treatments or manufacturing routes affect the fatigue behavior of specimens and components. SLHs are also frequently used in projects to develop or evaluate numerical life prediction models or in round robin programs with several participating laboratories on fatigue-related experimental or analytical issues. Results generated by use of SLHs are readily compared at least with regard to the important aspect of the loading environment of a given test series. Thus, a database is created which allows the evaluation of individual test results or of scatter bands and may help to reduce the amount of fatigue testing. SLH spectra may also be used as a reference case for comparison with, and assessment of, ad-hoc load measurements which necessarily cover a limited time or usage period in most cases. Also, for the generation of new load sequences background information and/or generation principles of SLHs may be very helpful.

It is a common understanding that SLHs do not refer—in the first instance—to a specific design problem, but comprise the typical features of the loading environment of a certain class of structures, vehicles, *etc*. Nevertheless, test results obtained with SLHs may be used to assess—on a relative basis—particular design problems. Recent developments such as the CARLOS group of SLHs are oriented towards the qualification of real components and structures, even up to the point where a specific SLH represents a core element of fatigue design requirements agreed between OEM's and suppliers. Heuler and Klätschke (2005) provide an extended overview from the literature on examples of the application of SLHs, including:

- round robin programs conducted, for instance, by AGARD or others;
- the generation of tailored new load sequences using the generation principles and usage information of, for example, WISPER, TWIST and HELIX/FELIX;
- the use of SLHs as a reference case for individual load measurements.

6.6 Generating a Pseudo Random Loading History from a Spectra

For many practical structures such as ships, the fatigue loading is random. The actual sequence is unknown. At most, we only know the long term distribution of the fatigue loading. In this situation, according to the UFLP model, the fatigue life is undetermined. However, we can derive the statistical characteristics of the fatigue life. In this section, we will show how this can be done.

If we fix the stress ratio R as 0.1 and assume that the maximum stress σ_{max} follows the Weibull distribution with shape parameter k=1 and scale parameter σ=80 MPa, then through random data generation we can generate a sequence of fatigue loading which follows the given distribution, see Fig. 6.24 as an example. We can calculate the corresponding crack length propagation with the increase in load cycles through cycle-by-cycle integration.

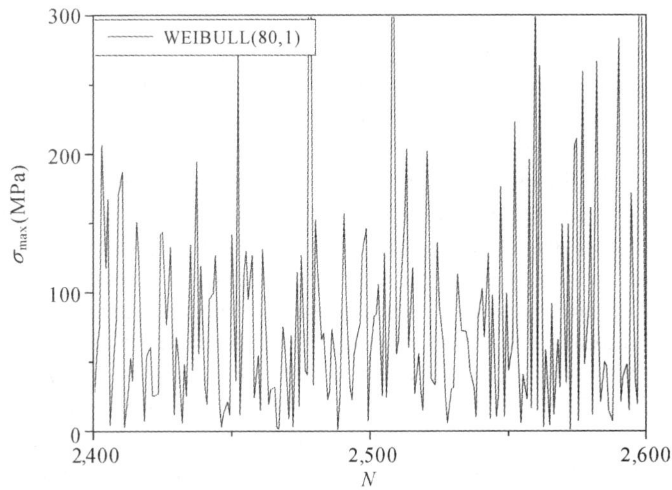

Fig. 6.24 A sample of random fatigue loading generated with a given seed number

Without fixing the stress ratio parameter, we can generate two independent parameters of maximum value and the range at a time according to the given distribution function. This is basically a 2D random number generation.

For the same distribution parameters of the long-term fatigue loading, an infinite number of random load sequences can be generated by varying the seed number. Fig. 6.25 shows the change in the normalized mean and COV (coefficient of variation) of the fatigue life with the number of simulations of load sequences (Cui, 2003). It can be seen that the statistical characteristics of the fatigue life are gradually convergent for a given distribution of the fatigue loading. However, it has a quite high COV around 1, relatively. This indicates that for a given structure, the fatigue life can be quite different if the load sequence is different.

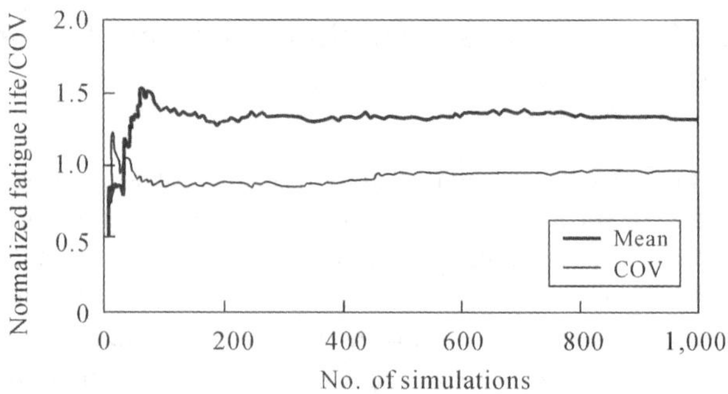

Fig. 6.25 Convergence of mean and COV of fatigue life with the number of simulations of load sequences (Cui, 2003) (with the permission of Sage)

6.7 Summary

FLP requires the information of fatigue loading history or the spectrum over the whole design life. There are basically two types of methods to determine the fatigue loads. One is based on calculation and the other is based on experiment. However, in many cases the measured load history can only include a very short time compared to the entire design life. How to extrapolate the measured loads to a longer time and how to count the information for every cycle are two important problems to be solved in fatigue analysis. In this chapter, a method to extrapolate the measured loads to a longer time has been introduced together with the rainflow cycle counting method. Both methods have been implemented in computer programs. An example calculation is given to illustrate the process. Problems on how to use these two methods to determine the design load are also pointed out.

How to construct the design load from limited practical measurements is a very important issue. However, one may be aware that using different measurement results may result in different design loads if we simply apply the method introduced above. This is certainly not satisfactory from a practical application point of view. In order to overcome this deficiency, we suggest taking into account the corresponding wave data information. In order to determine the design load for a specific ship, we first need to provide the time fraction for each sea state, (S_i, p_i), where i is the sea state number from 0 to 12, 0 can be used as the port time, S_i is the ith sea state number and p_i is the time fraction of the ith sea state. Now, if we know a short time period of sea state k, we can convert this measurement to a different sea sate assuming all the responses are in linear range. Then, for each sea state, we extrapolate the corresponding fraction of the design life and count the cycles. Summarizing all the cycles for different sea states, we can derive the load spectra. This load spectrum will be independent of the short time measurements. If the time fraction for each sea state, (S_i, p_i) has been specified, no matter what sea state is used for the short time measurement, the same design load spectrum will be derived. However, how to construct the time sequence is another challenging problem. This is basically the problem of constructing a so-called SLH (Berger, *et al.*, 2002; Heuler and Klätschke, 2005). This problem needs to be investigated in the future.

References

Aicher, W., Branger, J., van Dijk, G. M., Ertelt, J., Hück, M., de Jonge, J. B., Lowak, H., Rhomberg, H., Schütz, D. & Schütz, W. (1976). "Description of a fighter aircraft loading standard for fatigue evaluation FALSTAFF", Common report of FCW Emmen, LBF, NLR, IABG.

Almar-Naess, A. (1985). Fatigue Handbook. Trondheim: Tapir.

Amzallag, C., Gerey, J. P., Robert, J. L. & Bahuaud, J. (1994). "Standardization of the rainflow counting method for fatigue analysis", International Journal of Fatigue, 16(4): 287–293.

ASTM Standard (1995). "ASTM standard practices for cycle counting in fatigue analysis", American Society for Testing and Materials, Philadelphia, PA, USA. E1049-85 (Reapproved 1990).

Bendat, J. S. & Piersol, A. G. (1971). Random Data Analysis and Measurement Procedures. New York: Wiley.

Berger, C., Eulitz, K. G., Heuler, P., Kotte, K. L., Naundorf, H., Schuetz, W., Sonsino, C. M., Wimmer, A. & Zenner, H. (2002). "Betriebsfestigkeit in Germany—an overview", International Journal of Fatigue, 24: 603–625.

Buxbaum, O. (1992). Betriebsfestigkeit, sichere und wirtschaftliche Bemessung schwingbruchgefährdeter Bauteile (Durability, safe and economic design of fatigue-loaded components) (2 Auflage). Düsseldorf: Verlag Stahleisen mbH.

Castillo, E. (1988). Extreme Value Theory in Engineering. San Diego: Academic Press.

Conle, F. A. (1979). "An examination of variable amplitude histories in fatigue", Ph. D. thesis. University of Waterloo, Canada.

CSR of IACS (2006). "IACS common structural rules for tankers and bulk carriers", the IACS Council for implementation on Apr. 1, 2006.

Cui, W. C. (2003). "A feasible study of fatigue life prediction for marine structures based on crack propagation analysis", Journal of Engineering for the Maritime Environment, 217(5): 11–23.

Downing, S. D. & Socie, D. F. (1982). "Simple rainflow counting algorithms", International Journal of Fatigue, 4(1): 31–40.

Durability (2002). "Rainflow cycle counting software", 2002 Release, Durability, Inc., 1995–2002.

ESDU 95006 (1995). "Fatigue life estimation under variable amplitude loading using cumulative damage calculations", Engineering Sciences Data Unit. ESDU-95006. London, UK.

Etube, L. S. (2001). Fatigue and Fracture Mechanics of Offshore Structures.

London and Bury St. Edmunds, UK: Professional Engineering Publishing Ltd.

Haibach, E., Fischer, R., Schütz, W. & Hück, M. (1976). "A standard random load sequence of Gaussian type recommended for general application in fatigue testing; its mathematical background and digital generation", in: Bathgate, R. G. (ed.) Fatigue Testing and Design. London: The Society of Environmental Engineers, 29.1–29.21.

Heuler, P. & Klätschke, H. (2005). "Generation and use of standardised load spectra and load-time histories", International Journal of Fatigue, 27: 974–990.

Heuler, P., Bruder, T. & Klätschke, H. (2005). "Standardised load-time histories—a contribution to durability issues under spectrum loading", Materialwissenschaft und Werkstofftechnik, 36(11): 669–677.

Heuler, P. & Schütz, W. (1998). "Standardized load-time histories—status and trends", in: Proc. of the 4th Int. Conf. on Low-Cycle Fatigue and Elasto-plastic Behaviour of Materials, Sept. 7–11, 1998, Garmisch-Partenkirchen, 729–734.

Heuler, P. & Seeger, T. (1986). "Criterion for omission of variable amplitude loading histories", International Journal of Fatigue, 8(4): 225–230.

Holtz, R. L. (2003). "Phenomenology of the effective stress intensity related to fatigue crack growth thresholds", International Journal of Fatigue, 25: 891–897.

Hu, J. J. (2007). "The full scale measurement of a surface ship under high sea states", Technical Report No. 07150, China Ship Scientific Research Center, (in Chinese).

Johannesson, P. (2006). "Extrapolation of load histories and spectra", Fatigue and Fracture of Engineering Materials and Structures, 29: 201–207.

Klätschke, H. & Schütz, D. (1995). "Das Simultanverfahren zur Extrapolation und Raffung von mehraxialen Belastungs-Zeitfunktionen für Schwingfestigkeitsversuche (Procedure for extrapolation and squeezing of multiaxial load-time histories)", Materialwiss Werkstofftechnik, 26(8): 404–415.

Klemenc, J. & Fajdiga, M. (2000). "Description of statistical dependencies of parameters of random load states (dependency of random load parameters)", International Journal of Fatigue, 22: 357–367.

Li, S. S., Huang, Z. B. & Cui, W. C. (2011). "On implementing a practical algorithm to generate fatigue loading history or spectrum from short time measurement", Journal of Ship Mechanics, 15(3): 286–300.

Matsuishi, M. & Endo, T. (1968). Fatigue of Metals Subjected to Varying Stress. Fukuoka, Japan: Japan Society Mechanical Engineering.

Maymon, G. (2005). "A 'unified' and a $(\Delta K^{+}\cdot K_{max})^{1/2}$ crack growth models for aluminum 2024-T351", Internal Journal of Fatigue, 27: 629–638.

Nieslony A. (2009). "Determination of fragments of multiaxial service loading strongly influencing the fatigue of machine components", Mechanical Systems and Signal Processing, 23(8): 2712–2721.

Noroozi, A. H., Glinka, G. & Lambert, S. (2005). "A two parameter driving force for fatigue crack growth analysis", International Journal of Fatigue, 27: 1277–1296.

Papoulis, A. (1965). Probability—Random Variables and Stochastic Processes. New York: McGraw-Hill.

Paris, P. C., Gomez, M. P. & Anderson, W. P. (1961). "A rational analytical theory of fatigue", The Trend in Engineering, 13: 9–14.

Rychlik, I. (1987). "A new definition of the rainflow cycle counting method", International Journal of Fatigue, 9: 119–121.

Sadananda, K. & Vasudevan, A. K. (1997). "Short crack growth and internal stresses", International Journal of Fatigue, 19(93): S99–108.

Sadananda, K. & Vasudevan, A. K. (2003). "Fatigue crack growth mechanisms in steels", International Journal of Fatigue, 25: 899–914.

Sadananda, K. & Vasudevan, A. K. (2004). "Crack tip driving forces and crack growth representation under fatigue", International Journal of Fatigue, 26: 39–47.

Sadananda, K. & Vasudevan, A. K. (2005). "Fatigue crack growth behavior of titanium alloys", International Journal of Fatigue, 27: 1255–1266.

Sadananda, K., Sarkar, S., Kujawski, D. & Vasudevan, A. K. (2009). "A two-parameter analysis of *S-N* fatigue life using $\Delta\sigma$ and σ_{max}", International Journal of Fatigue, 31: 1648–1659.

Schijve, J. (2009). Fatigue of Structures and Materials (2nd ed. with CD-Rom). Dordrecht, the Netherlands: Springer Science+Business Media, B.V.

Schütz, D., Lowak, H., de Jonge, J. B. & Schijve, J. (1973). "A standardised load sequence for flight simulation tests on transport aircraft wing structures", LBF-Report FB-106, NLR-Report TR 73.

Schütz, D., Klätschke, H. & Heuler, P. (1994). "Standardized multiaxial load sequences for car wheel suspension components—car loading standard—CARLOS multi", Fraunhofer-Institut für Betriebsfestigkeit LBF, Darmstadt. Report No. FB-201.

Schütz, D., Klätschke, H., Steinhilber, H., Heuler, P. & Schütz, W. (1999). "Standardized load sequences for car wheel suspension components, car loading standard—CARLOS", Fraunhofer-Institut für Betriebsfestigkeit (LBF), Darmstadt, Industrieanlagen-Betriebsgesellschaft mbH (IABG), Ottobrunn, LBF-Report No. FB-191.

Sonsino, C. M., Klätschke, H., Schütz, W. & Hück, M. (1988). "Standardized load sequence for offshore structures-wave action standard history-WASH 1", Fraunhofer-Institut für Betriebsfestigkeit (LBF), Industrieanlagen-Betriebsgesellschaft mbH (IABG). LBF-Report No. FB-181, 1988, IABG-Report No. TF-2347.

Stoychev, S. & Kujawski, D. (2005). "Analysis of crack propagation using ΔK and K_{max}", International Journal of Fatigue, 27: 1425–1431.

Tenhave, A. A. (1991). "WISPER and WISPERX-final definition of two standardised fatigue loading sequences for wind turbine blades", NLR Report CR 91476 L, Amsterdam.

Vasudevan, A. K. & Sadananda, K. (1999). "Application of unified fatigue damage approach to compression-tension region", International Journal of

Fatigue, 21(s1): S263–273.
Vasudevan, A. K. & Sadananda, K. (2001). "Analysis of fatigue crack growth under compression-compression loading", International Journal of Fatigue, 23: S365–374.
Vasudevan, A. K. & Sadananda, K. (2009). "Classification of environmentally assisted fatigue crack growth behavior", International Journal of Fatigue, 31: 1696–1708.
Vasudevan, A. K., Sadananda, K. & Glinka, G. (2001). "Critical parameters for fatigue damage", International Journal of Fatigue, 23: S39–53.
Vasudevan, A. K., Sadananda, K. & Louat, N. (1994). "A review of crack closure, fatigue crack threshold and related phenomena", Materials Science and Engineering, A188: 1–22.
Wöhler, A. (1860). "Versuche über die Festigkeit der Eisenbahnwagen-Achsen", Zeitschrift fur Bauwesen,10(4): 160–161.
Wöhler, A. (1870). "Über die Festigkeitsversuche mit Eisen und Stahl", Zeitschrift für Bauwesen, 20: 73–106.
Zhang, J. Z., He, X. D. & Du, S. Y. (2005). "Analyses of the fatigue crack propagation process and stress ratio effects using the two parameter method", International Journal of Fatigue, 27: 1314–1318.

7

Some Applications and Demonstrations of UFLP

7.1 Introduction

It is now well recognized that small pre-existing defects are an inherent feature of engineering components and structures, especially in welded structures. For an ideal material with no initial flaws or imperfections, the fatigue crack initiation life usually represents the majority of the total fatigue life, but for structures with a pre-existing flaw or cracks, the initiation life is very small and the FCP life represents the majority of the total life. Ships and marine structures are very large welded metal structures operating in a marine environment. The fatigue life of marine structures can be predicted by FCP. Due to the complex nature of the structure and environmental loading, much research on the fatigue life of marine structures predicted by FCP has been done. But no universal or all-encompassing model exists. Here, only the UFLP method is demonstrated.

The UFLP method for marine structures based on FCP has been established and described in prior chapters. For example, Chapter 4 addresses basic fracture mechanics and the SIF calculation. Chapter 5 deals with the crack growth rate model and Chapter 6 introduces the fatigue loading history. Some validations for UFLP have been done in Chapter 5, but these validations are for showing the ability of the UFLP to deal with different fatigue phenomena, such as the mean stress effect, loading sequence, *etc*., merely based on the simple specimen test results and the parameters are determined to give a better prediction of the test data. In this chapter, some more applications of UFLP of specimens and marine structures under different loading conditions are demonstrated emphasizing the consistency of the material constants of the same material.

7.2 The Fatigue Crack Growth Rate of UFLP

One of the key parts of FLP, which is based on fatigue crack growth, is the fatigue crack growth rate model; the basic equation of the UFLP introduced in Chapter 5 can be synthetically expressed as follows,

$$\frac{\mathrm{d}a}{\mathrm{d}N} = \frac{A\left[K_{\max}\cdot\left(1-\Phi f_{\mathrm{op}}\right)-\Delta K_{\mathrm{eff\,th}}\right]^{m}}{1-\left(K_{\max}/K_{\mathrm{C}}\right)^{n}} \tag{7.1}$$

where the expressions of the parameters A, $K_{\max}$, Φ, f_{op}, $\Delta K_{\mathrm{eff\,th}}$, K_{C}, m, n can be referred to Eqs. (5.7), (5.8), (5.13), (5.26)–(5.33), (5.44)–(5.51) respectively.

7.3 FLP of Specimen with Through-Thickness Crack Under Different Fatigue Loading

The accuracy and ability to deal with fatigue phenomena of the UFLP have been validated by some of the experimental data taken from the published literature and the data from our fatigue experiments (Wang F., *et al.*, 2009; 2010; Cui, *et al.*, 2011). The fatigue loading conditions of marine structures can be understood and simulated by the basic constant amplitude cyclic loading plus a number of overloads, underloads, overload-underloads and underload-overloads, occasionally. In this section, more fatigue test cases of specimens with a through-thickness crack for different materials under different loading conditions taken from the literature or from our research reports will be predicted. Furthermore, the fatigue life of a submarine hull with a surface crack at the weld toe under different loading spectra will be simulated.

7.3.1 FLP of D16 Aluminum Alloy Specimens Under Different Spectrum Loading

Schijive, *et al.* (2004) and Skorupa, *et al.* (2007) published the fatigue test results of center cracked plate specimens of clad D16 aluminum alloy under a series of load spectra, and the mechanical properties of D16 are similar to 2024 aluminum alloy. The material constants of UFLP for D16 aluminum alloy are obtained by fitting the experimental crack growth rate data under constant amplitude loading. The mechanical properties and parameters for UFLP of the material of D16 aluminum alloy are listed in Table 7.1 and the crack growth specimen geometry is shown in Fig. 7.1. The specimens are rectangular of approximately (length×width×height) $500\times100\times4$ mm^3 with an initial crack length $2c$=10 mm central through the thickness crack. The thickness of the cladding layer is approximately 0.1 mm. The influence of the clad is neglected when calculating the SIFs of the crack. The fatigue life tests were performed under five types of loading conditions shown in Fig. 7.2 and the loading parameters for each loading spectrum are listed in Table 7.2 respectively. The Constant Amplitude (CA) plus single OverLoad (OL) loading spectrum shows that the overload happens when the half length of the crack is about 10 mm. The load situations of Type I, II, and III, which contain cyclic overload and underload, can simulate and validate the phenomena of crack retardation or acceleration under some typical conditions with overload or underload.

The experimental results of these five types of spectra will be predicted by the UFLP method and analyzed respectively. For simplifying the calculation, the parameter K_C is set as a constant and equals K_{IC} in all of these examples in this chapter.

Fig. 7.3 shows the comparison of predicted crack length via the number of cycles and the test data. It can be seen that the fatigue crack growth of central cracked D16 plate specimens under CA with different stress ratios can be simulated accurately.

Table 7.1 Material constants for UFLP of D16 aluminum alloy

σ_y (MPa)	σ_u (MPa)	A	m	$\Delta K_{th\,0}$ ($\mathrm{MPa \cdot m^{\frac{1}{2}}}$)	K_{IC} ($\mathrm{MPa \cdot m^{\frac{1}{2}}}$)	n	r_e (m)
355	457	7.38E-10	2.068	3.6	51	6	1E-6

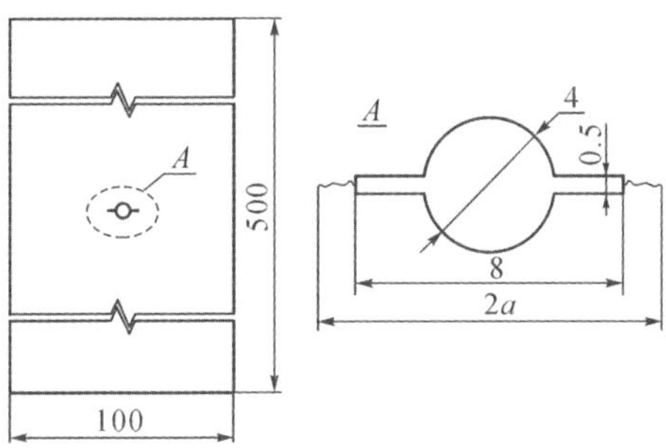

Fig. 7.1 The crack growth specimen geometry (dimension: mm) (Schijve, *et al.*, 2004) (with the permission of Elsevier)

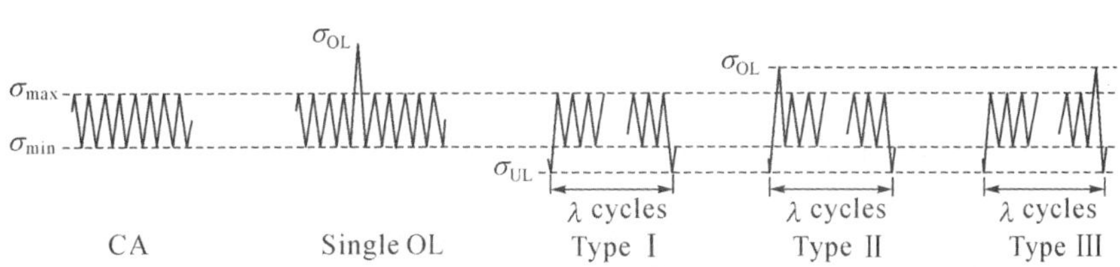

Fig. 7.2 Five fatigue test loading blocks (Schijve, *et al.*, 2004) (with the permission of Elsevier)

Table 7.2 Parameters of five typical loading spectra (MPa) (Schijive, *et al.*, 2004) (with the permission of Elsevier)

Types of loading	σ_{min}	σ_{max}	$\Delta\sigma$	R	λ	σ_{OL}	σ_{UL}
Constant amplitude	0	64	64	0	—	—	—
Constant amplitude	32	96	64	0.33	10	—	—
Single OL	0	64	64	0	—	128	—
Periodic ULs (Type Ⅰ)	32	96	64	0.33	10	—	0
Periodic ULs-OLs (Type Ⅱ)	32	96	64	0.33	10	128	0
Periodic OLs-ULs(Type Ⅲ)	32	96	64	0.33	10	128	0

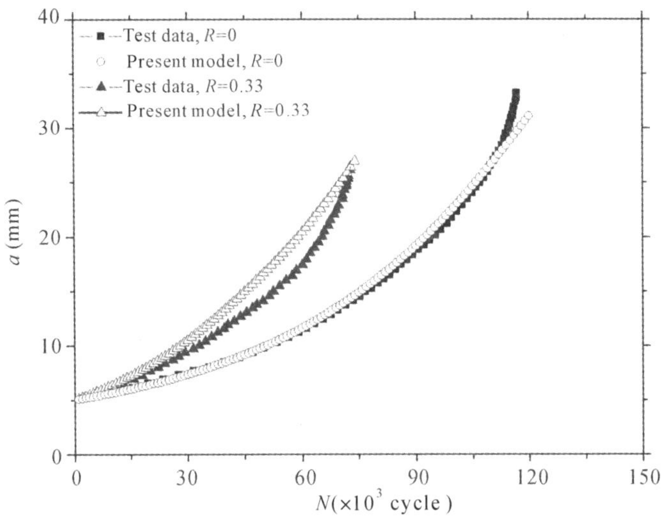

Fig. 7.3 Comparison of predicted results with test data of D16 under CA loading

7.3.1.1 Prediction of crack growth under CA plus single OL

The test of CA+single OL is designed to study the phenomena of crack retardation under single OL. The overload with twice the maximum load was applied when the half crack extends to 10 mm. The test results are shown in Fig. 7.4. The crack growth and the phenomena of retardation are simulated by the UFLP model, and the predicted results are compared with the experimental data and that predicted by the NASGRO model. In the prediction, the parameter γ is set to 17.5 and 17.6 respectively. It can be seen that there are differences between the two crack growth curves corresponding to γ of 17.5 and 17.6 respectively and the curve corresponding to γ= 17.6 agrees well with the experimental data. The phenomena of crack retardation under single OL can be predicted well by the UFLP method.

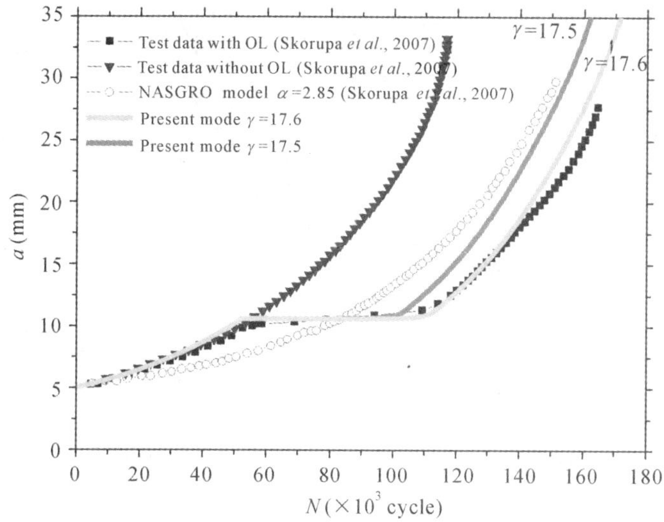

Fig. 7.4 Comparison of predicted results with test data of D16 and results predicted by NASGRO under CA plus single OL

7.3.1.2 Prediction of crack growth under different loading sequences

Type I, Type II, and Type III shown in Fig. 7.2 were used to simulate the phenomena of crack retardation or acceleration under cyclic overload and underload (also the combination of both) loading conditions. Fig. 7.5 shows the fatigue crack growth curves of D16 specimens under the above three types of load conditions. In the calculation, the material constant parameters listed in Table 7.1 are used and γ is set to 8. The experimental data show that the fatigue life of Type III is larger than that of Type II. The crack retardation after underload-overload is more obvious than that of overload-underload with the same minimum and maximum stress. The fatigue life of a central cracked D16 aluminum plate under Type II spectrum is a bit smaller than that of Type III spectrum. At the same time, it can be found that there are some differences between the fatigue crack growth under these three spectra loadings. It shows the effect of overload retardation and underload acceleration, although the effect of underload is not very obvious. The UFLP model solves the problem well by considering the phenomena of crack acceleration due to the decreasing size of the plastic zone under underload conditions. By analyzing the predicted results shown in Fig. 7.5, it can be seen that the predicted results are in good agreement with the test data under these three types of loading sequences.

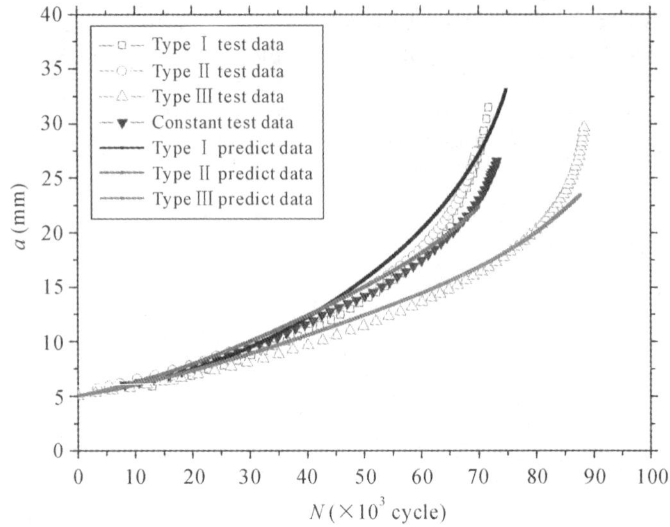

Fig. 7.5 Comparison of predicted results with test data of D16 under different fatigue loading

7.3.2 FLP of Aluminum Alloy Al 7075-T6

Porter (1972) collected fatigue data on center-notched 7075-T6 aluminum alloy specimens made of 305 mm wide, 915 mm long, and 4.1 mm thick plates, for which E=69,600 MPa, yield strength is 520 MPa and ultimate strength is 575 MPa. The initial crack size (2c) was 12.7 mm and the experiments were conducted in laboratory air. Fig. 7.6 shows the profile of a bi-level block loading where the positive integers, p and q, indicate that a block of p constant amplitude cycles is followed by a block of q cycles with different constant amplitude (Ray, 2001). The material constants for UFLP of aluminum alloys Al 7075-T6 are listed in Table 7.3. The comparison between predicted results by UFLP and the test data was shown in Fig. 7.7. Fig. 7.7(a) shows a comparison of the fatigue crack growth under a bi-level block loading with different p, q and Fig. 7.7(b) shows a comparison of the fatigue crack growth under a bi-level block loading with the same p, q but different overload maximum stress. Fig. 7.8 shows a comparison of the fatigue crack growth under different amplitudes of overload with p=29, q=1 for different overload stress ratios σ_2/σ_1, while σ_1 is held at a fixed value of 69 MPa. Similar comparisons are made in Fig. 7.9 for single cycle overload (q=1) with different overload spacing value p, and fixed values of σ_1=69 MPa and σ_2=103.5 MPa. By regulating the value of MPa, it can be found that the results of the UFLP method agree with the test data very well, as shown in Figs. 7.8 and 7.9. Fig. 7.10(b) shows the comparison of predicted results with test data for crack growth under overload-underload block loading shown in Fig. 7.10(a). Fig. 7.11(b) shows the comparison of predicted results with test data for crack growth under underload-overload block loading shown in Fig. 7.11(a). The comparisons show that the crack growth under different bi-level block loadings can be predicted well by UFLP.

The fatigue crack growth curves of center cracked plate specimens made of aluminum alloys under different block loadings have been predicted by the UFLP method. By analyzing the experimental data, it can be found that the larger the overload is, the more obvious the effect of load sequence. The shaping exponent for loading sequence interaction shows great effect on the predicted results. The UFLP model can predict the fatigue crack growth under different loading sequences with different overload, underload, overload-underload and underload-overload, *etc.* by regulating the exponent γ. It provides an effective way to study fatigue crack growth under random loading.

Table 7.3 Material constants for UFLP of aluminum alloys Al 7075-T6

σ_y (MPa)	A	m	$\Delta K_{\text{th}\,0}$ ($\text{MPa}\cdot\text{m}^{\frac{1}{2}}$)	K_{IC} ($\text{MPa}\cdot\text{m}^{\frac{1}{2}}$)	n	r_e (m)
520	7.2E-10	2	2.2	28	6	1E-6

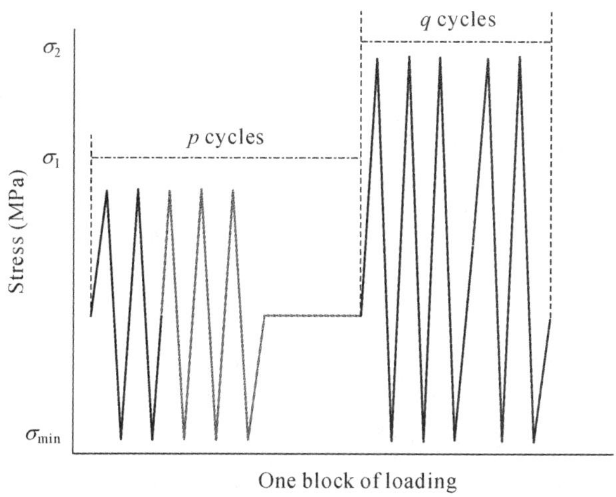

Fig. 7.6 Bi-level fatigue test loading block

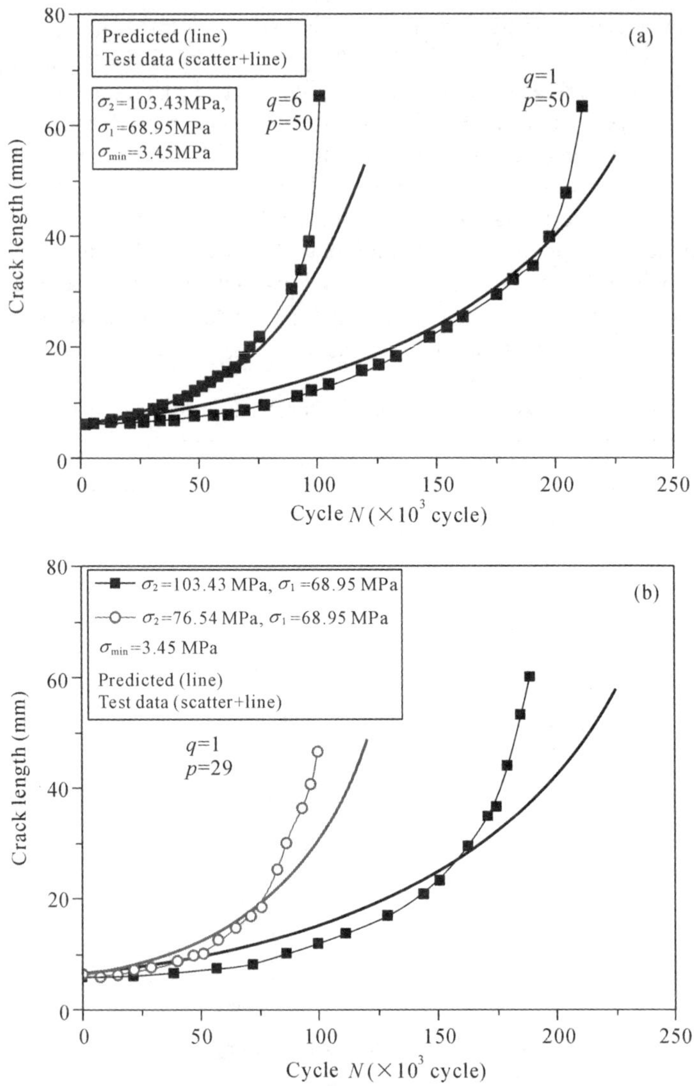

Fig. 7.7 Comparison of predicted values with test data of Al 7075-T6 (low-high). (a) Fatigue crack growth under a bi-level block loading with different *p*, *q*; (b) Fatigue crack growth under a bi-level block loading with the same *p*, *q* but different overload maximum stress

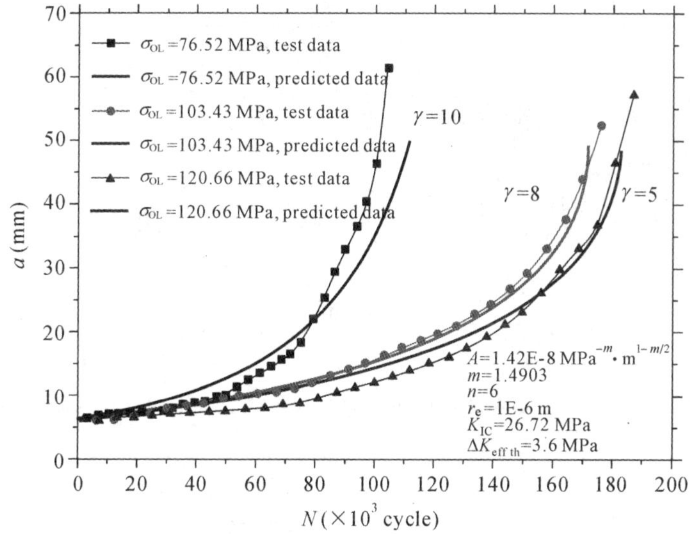

Fig. 7.8 Comparison of predicted results with test data of Al 7075-T6 under different OL loading

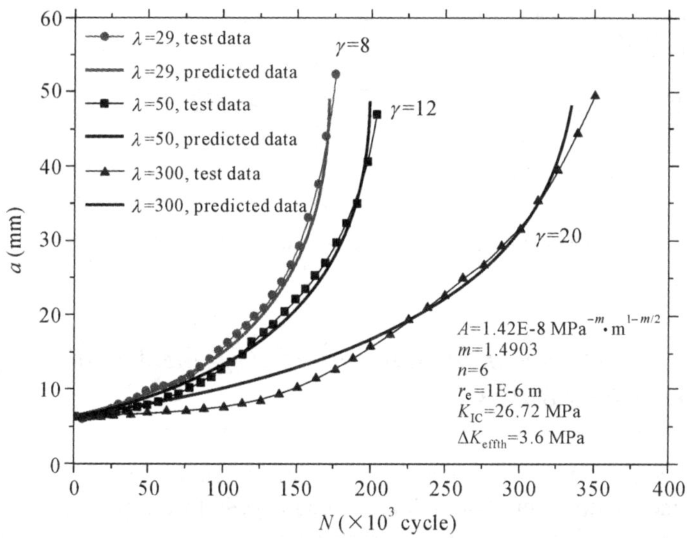

Fig. 7.9 Comparison of predicted results with test data of Al 7075-T6 under different fatigue loading

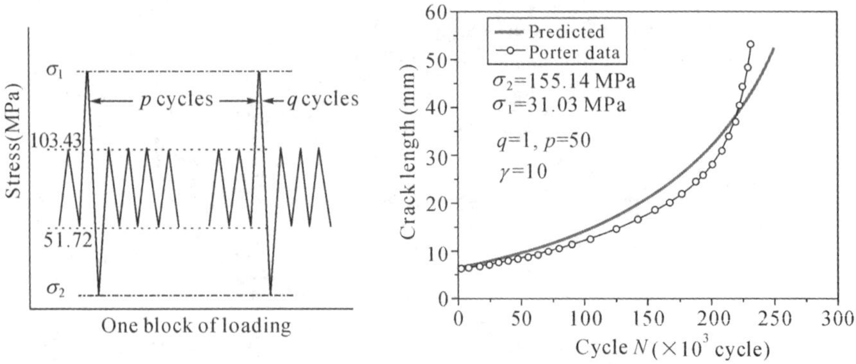

Fig. 7.10 Comparison of predicted values with Al 7075-T6 test data (overload-underload)

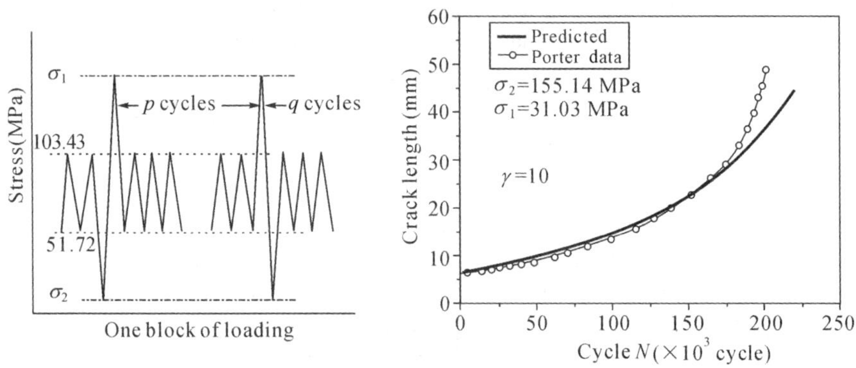

Fig. 7.11 Comparison of predicted values with test data of Al 7075-T6（underload-overload）

7.3.3 FLP of Specimen Made of 350WT Steel Under Different Overload Ratios

Taheri, *et al.* (2003) have conducted an experimental fatigue program consisting of constant and semi-random (variable) amplitude cyclic loadings under constant and constant with different overload loading on serial 350WT steel, which is commonly used in applications such as bridges, pressure vessels and offshore structures, and this provided the experimental fatigue crack growth data of the steels. Some fatigue models were compared to the results obtained from the experiments. The experimental results showed that periodically applied single tensile overloads interspersed in a constant amplitude-loading scenario resulted in crack growth retardation by an amount dependent upon the number and magnitude of the applied overloads.

For predicting the experimental results by the UFLP method, the specimen geometry size and the properties of the material of 350WT steel should be known or determined first.

The specimens are rectangular of approximately (length×width×height) 300×100×5 mm^3 with an initial crack length 2c=20 mm central through the thickness crack. The mechanical properties and parameters for the UFLP of the material of 350WT steel are listed in Table 7.4. The fatigue life tests were performed under constant plus different overloading conditions shown in Table 7.5. The calculated fatigue lives and the test data are listed in Table 7.5. From the comparison, it can be found that the results predicted by the UFLP method agree with the test data very well.

Table 7.4 Material constants for UFLP of 350WT steel

σ_y (MPa)	A	m	$\Delta K_{\text{th }0}$ $(\text{MPa}\cdot\text{m}^{\frac{1}{2}})$	K_{IC} $(\text{MPa}\cdot\text{m}^{\frac{1}{2}})$	n	r_e (m)	γ
350	7.38E-10	2.068	13.24	106	6	1E-6	10

Table 7.5 Loading serials parameters, experimental data and predicted results for steel 350WT

No.	$P_{\max}$ (N)	$P_{\min}$ (N)	P_{OL} (N)	No. of OL	a_0 (mm)	a_f (mm)	No. of cycles tested	No. of cycles predicted
Val 1	57,000	5,700	69,825	2	11.19	35.11	146,000	150,000
Val 2	57,000	5,700	69,825	3	11.35	34.56	165,000	170,000
Val 3	57,000	5,700	82,650	2	11.25	34.43	167,500	170,000
Val 4	57,000	5,700	82,650	3	11.23	34.26	193,000	200,000
Val 5	57,000	5,700	95,475	2	11.25	35.19	223,000	230,000
Val 6	57,000	5,700	95,475	3	11.17	34.48	255,000	260,000

7.3.4 The Fatigue Crack Growth Prediction of Steel HTS-A Under Multi-Level Block Loading

7.3.4.1 The material constants for UFLP of HTS-A

Steel HTS-A is a high strength steel commonly used in pressure hulls of underwater structures. A series of fatigue tests on this steel with CT specimens have been done by our group (Huang, 2007). The experimental fatigue crack growth rate data of the steel under different loading ratios (R=0.1, 0.3, 0.6, and 0.9 respectively) are shown in Fig. 5.25 in Chapter 5. Based on the abundant fatigue crack growth rate data, the material constants for the UFLP model of steel HTS-A are determined and listed in Table 7.6. In the section, fatigue crack growth curves of specimens made of steel HTS-A under constant amplitude cycles with a single overload-underload or a single underload-overload applied at three different crack lengths, and under five-stage multi-level block loadings, are predicted and compared with the test data.

Table 7.6 Material constants for UFLP of steel HTS-A

σ_y (MPa)	A	m	ΔK_{th0} $(\text{MPa}\cdot\text{m}^{\frac{1}{2}})$	K_{IC} $(\text{MPa}\cdot\text{m}^{\frac{1}{2}})$	n	r_e (m)
800	2.18E-10	2.267	13.24	140	6	1E-6

7.3.4.2 Fatigue crack growth prediction under block loading

(1) Overload-Underload

The loading spectrum for testing the retardation of a single overload-underload is shown in Fig. 7.12. The thickness of the standard compact tension specimen is 9.16 mm, and the initial crack length is 13.5 mm. A single overload-underload cycle is applied when crack length, a, grows to a_1=16.1 mm, a_2=21.05 mm and a_3=26.1 mm, respectively. The overload ratios are OL=$\sigma_{\text{OL}}/\sigma_{\text{Bmax}}$=2 and UL=$\sigma_{\text{UL}}/\sigma_{\text{Bmin}}$=–1. The maximum applied loading force is 9 MPa. Some of fatigue test parameters of the overload-underload block loading are listed in Table 7.7.

Fig. 7.13 shows the predicted *a-N* curve and the experimental data under overload-underload spectrum loading. The black line with circle is the test fatigue crack growth curve (*a-N*), and the solid line is the predicted result by the UFLP method. It can be seen that the predicted results agree well with the experimental data.

(2) Underload-Overload

The loading spectrum for testing the retardation of a single underload-overload is shown in Fig. 7.14, The thickness of the standard CT specimen is 9.16 mm, and the initial crack length is 12.0 mm. A single underload-overload cycle is applied when crack length a is extension to a_1=16.1 mm, a_2=21.05 mm and a_3=27.3 mm respectively. The overload ratios are OL=2 and $\text{UL}=-1$. The maximum applied loading force is 9 MPa. Some of the fatigue test parameters of the underload-overload block loading are listed in Table 7.7.

Fig. 7.15 shows the predicted *a-N* curve and the experimental data under underload-overload block loading. The black line with circle is the experimental fatigue crack growth curve (*a-N*), and the solid line is the predicted result by the UFLP method. It can be seen that the predicted results agree well with the experimental data.

Table 7.7 Fatigue test parameters of block loading

Type of load	R	T (mm)	a_0 (mm)	$P_{\max}$ (KN)	F (Hz)
Overload-Underload	0.3	9.16	13.5	9	15
Underload-Overload	0.3	9.16	12.0	9	15

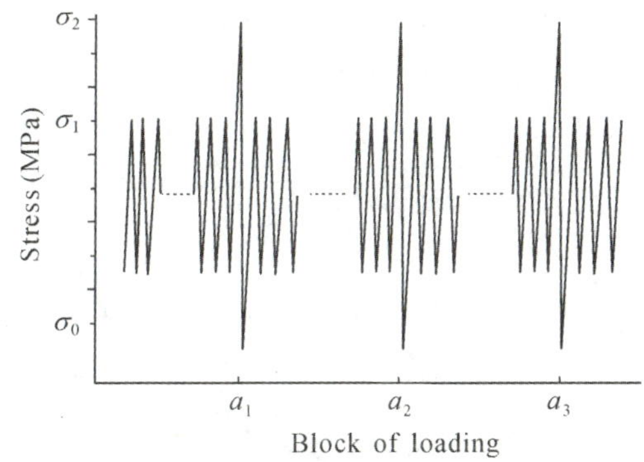

Fig. 7.12 Spectrum of overload-underload

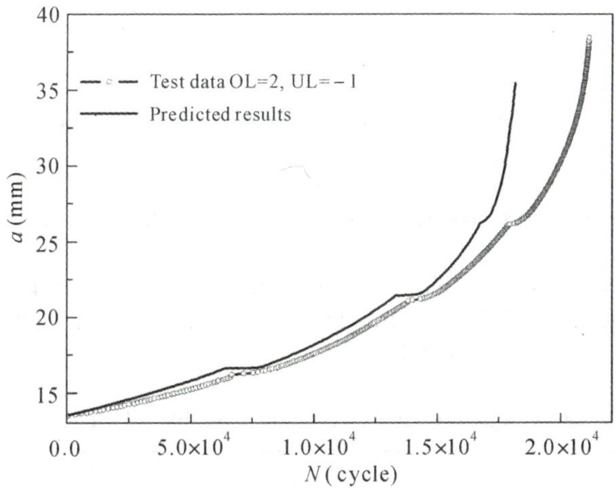

Fig. 7.13 Comparison of predicted *a-N* curves with test data under overload-underload spectrum

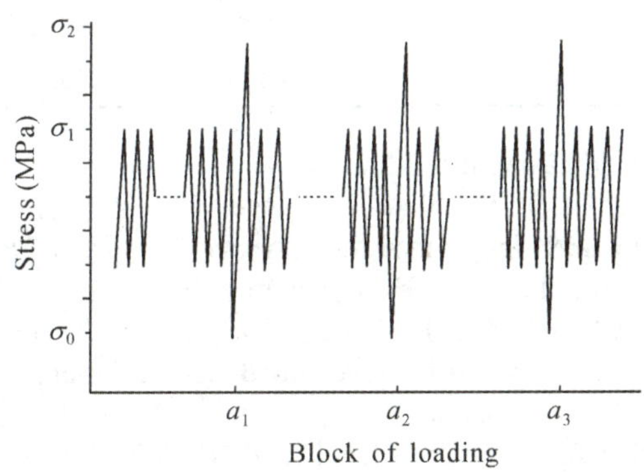

Fig. 7.14 Spectrum of underload-overload

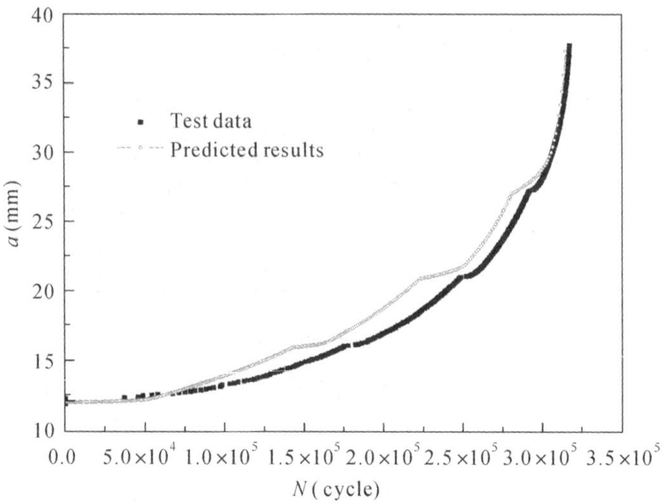

Fig. 7.15 Comparison of predicted *a-N* curves with test data under underload-overload spectrum

7.3.4.3 Fatigue crack growth prediction under multi-level block loading

Four CT specimens were used to examine the fatigue crack growth under five-stage block loading. Fig. 7.16 shows the 5-level block loading, and the test parameters are listed in Table 7.8.

Table 7.8 Fatigue test parameters of multilevel block loading (increase)

	R	T (mm)	a_0 (mm)	P_1 (KN)	f (Hz)
No. 1	0.1	10	13.61	7.42	10
No. 2	0.1	10	13.33	7.72	10
No. 3	0.1	10	13.56	15.12	10
No. 3	0.1	10	13.67	15.08	10

(1) Gradually increasing block loading

The specimen No. 1: the first level load: P_1=7.42 kN, n_1=400 cycles; the second level load is 20% larger than the first level load P_2=1.2P_1=8.09 kN, n_2=600 cycles; the third level load is 20% larger than the second, P_3=1.2P_2=10.68 kN, n_3=600 cycles; the fourth level load is 20% larger than the third, P_3=1.2P_2=12.82 kN, n_4= 300 cycles; the fifth level load is 20% larger than the fourth, P_5=1.2P_4=15.39 kN, n_5=30 cycles.

The specimen No. 2: the first level load: P_1=7.72 kN, n_1=4,000 cycles; the second level load is 20% larger than the first level load P_2=1.2P_1, n_2=6,000 cycles; the third level load is 20% larger than the second, P_3=1.2P_2, n_3=6,000 cycles; the fourth level load is 20% larger than the third, P_3=1.2P_2, n_4=3,000 cycles; the fifth level load is 20% larger than the fourth, P_5=1.2P_4, n_5=300 cycles.

(2) Gradually decreasing block loading

The specimen No. 3: the first level load: P_1=15.12 kN, n_1=30 cycles; the second level load is 20% smaller than the first level load P_2=1.2P_1, n_2=300 cycles; the third level load is 20% smaller than the second, P_3=1.2P_2, n_3=600 cycles; the fourth level load is 20% smaller than the third, P_3=1.2P_2, n_4= 600 cycles; the fifth level load is 20% smaller than the fourth, P_5=1.2P_4, n_5=400 cycles.

The specimen No. 4: the first level load: P_1=15.08 kN, n_1=300 cycles; the second level load is 20% smaller than the first level load P_2=1.2P_1, n_2=3,000 cycles; the third level load is 20% smaller than the second, P_3=1.2P_2, n_3=6,000 cycles; the fourth level load is 20% smaller than the third, P_3=1.2P_2, n_4= 6,000 cycles; the fifth level load is 20% smaller than the fourth, P_5=1.2P_4, n_5=4,000 cycles.

The fourth specimen underwent 100 block loading cycles of 30-300-600-600-400 firstly, then 300-3,000-6,000-6,000-4,000, cyclic loading.

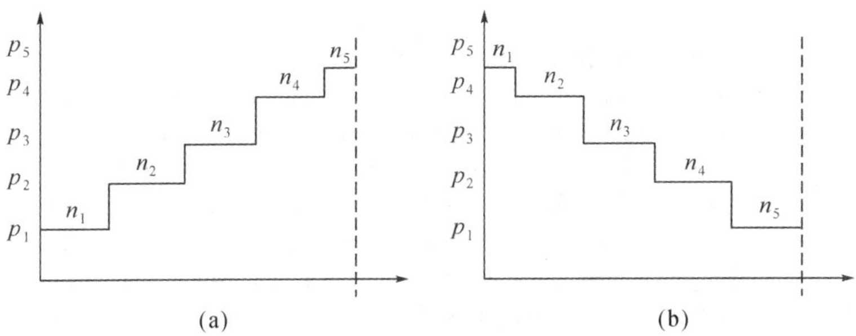

Fig. 7.16 Multilevel loading spectrum

The fatigue crack growth results predicted by the UFLP method and the test crack growth curves under above four multi-level block loadings are shown in Fig.7.17. The line consisting of coarse points is the test fatigue crack growth curve, and slender lines show the predicted results. The test results show that the fatigue lives for the above four block loadings are quite different although the corresponding loading amplitudes are the same but with a different cycle number for each level. The fatigue life of the No. 1 specimen is longer than that of the No. 2 specimen, and the fatigue life of the No. 3 specimen is a bit longer than that of the No. 4 specimen. The reason is that when the cycle number of the load spectrum is low, the chance of overload is large, and the fatigue life of the No. 1 specimen is longer than that of the No. 2 specimen because of the overload retardation effect. At the same time, the UFLP model can simulate the above four loading conditions well.

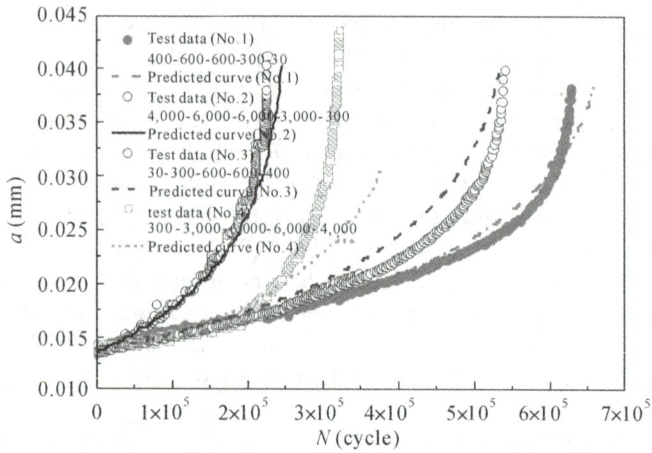

Fig. 7.17 Comparison of predicted results with test data of *a-N* curves under multilevel spectrum loading

All of the above examples show that the overload frequency and amplitude have an important influence on fatigue crack growth. The loading sequence effect can be predicted by regulating the exponent γ in the fatigue crack prediction. The UFLP model can predict the fatigue crack growth under different loading sequences with different overload, underload, overload-underload and underload-overload, *etc.* by regulating the exponent γ. The exponent γ can be determined based on serial fatigue test data of the same material under different loading conditions. It provides an effective way to study fatigue crack growth under random loading. UFLP can be used as a kind of effective method for FCP analysis of marine structures.

7.4 FLP of Cracked Deck of an Oil Tanker

The selected case is the fatigue test of a large stiffened plate with a through-thickness crack simulating the deck cracks of a tanker by Mahmoud and Dexter (2005). Based on the calculated SIFs of the center penetrated crack in the stiffened plate using the finite element method, the UFLP model was used to predict the fatigue crack growth of the stiffened plate considering the effect of the residual stress. The predicted results were compared with the test data. The results show that the UFLP method can predict well the fatigue crack life of an actual structure.

7.4.1 Geometry of the Stiffened Plate

Most of the stiffened plates of a marine structure are welded structures. There is no simple effective method to calculate the SIF of a crack in the stiffened plate, especially for the cracks located both in the plate and stiffener. There are two situations for welded stiffened plates. The first situation is that there are no interfaces between the stiffeners and plate with whole continual welding. So the crack will extend to the stiffener, and continue to grow in the stiffener when the crack tip goes through the stiffener. In this way, the stiffener crack is part of the crack, which can be regarded as the spatial crack. The other situation is a semi-circular notch which is slotted at the intersection of the stiffened plate and semi-transverse welds in order to avoid a cross-weld, and the method is usually used in welded marine structures. The example considers the serious situation of a crack. The crack in a stiffened plate usually emerges at the welds of plate and stiffener. Generally, there are three forms of crack in the plate (Paik, 2008): (a) crack parallel to the loading direction; (b) crack perpendicular to the loading direction and (c) crack with a certain angle to the loading direction, as shown in Fig. 7.18.

In general, the crack perpendicular to the loading direction will be affected mostly by the stiffener, and the stiffened plates with a crack perpendicular to the loading direction will be discussed in the fatigue crack growth prediction of the stiffened plate.

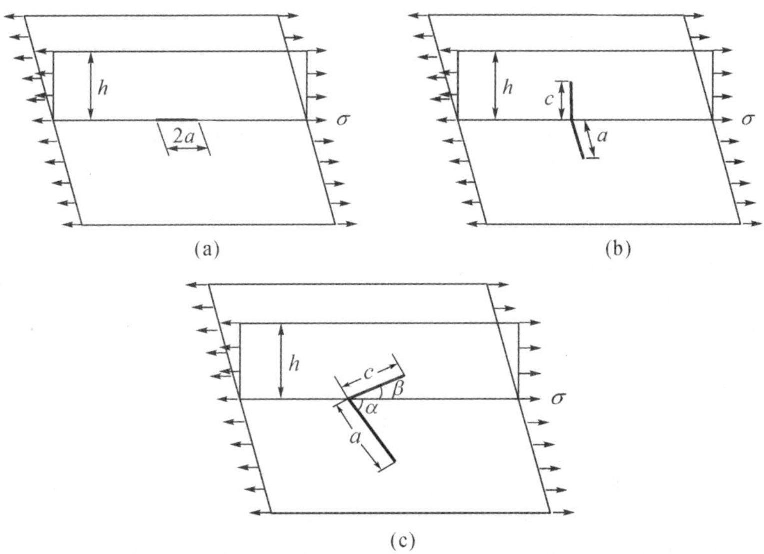

Fig. 7.18 Three types of cracks in stiffened plate. (a) Parallel crack; (b) Vertical crack; (c) Inclined crack (Paik, 2008) (with the permission of Pergamon)

7.4.2 The Crack Growth Pattern in the Stiffened Plate

For the welded stiffened plate with the crack perpendicular to the loading direction, the fatigue crack growth along the plate will be affected by the stiffener and the weld residual stress when the crack tip reaches or surpasses the stiffeners. Situations of fatigue crack growth are as follows.

Case 1: The stiffeners have a small effect on the crack growth. The stiffness and the restrained effect of the stiffeners on the crack are small if the dimension of the stiffener is relatively small. The crack growth will not be affected by the stiffeners significantly. So, in such a situation, the SIF curve will not drop significantly around the stiffener, and will continue to grow gradually.

Case 2: The crack will not grow in the stiffener. This situation appears when the stiffeners are strong enough or there is an interface between the plate and the stiffener. In this situation, it can be thought that the restrained effect on crack growth is mainly due to the stiffener, so the SIF curve of cracks with the crack tip around the stiffeners will reduce dramatically.

Case 3: The crack will grow in the plate and the stiffeners. This is a common situation. The crack will grow in the plate and stiffener and grow in two directions respectively (forwards in the plate, upwards in the stiffener), so the two crack tips can be regarded as two penetrated cracks. Here, the stiffness of the stiffener is reduced when the crack grows in the stiffener and its effect on crack growth varies with the crack length.

7.4.3 Determination of SIFs of the Cracked Stiffened Plate by FEA

7.4.3.1 The stiffened plate model

The geometric parameters of the stiffened plate are shown in Fig. 7.19. The material is high-strength and low-carbon steel ASTM A572 with a yield strength of 345 MPa. The Young's modulus is 2.1×10^5, and Poisson ratio is 0.3. The size of the plate is (legth×width×height) $3{,}431\times1{,}626\times13$ mm^3, and the distance between the stiffeners is 381 mm. There are four stiffeners made of HP bulb steel of 160 mm×9 mm along the width direction, and the initial length of the center penetrated crack is 305 mm. The fatigue tests are conducted at room temperature (25 °C). The fatigue load is applied along the length direction with a stress ratio of 0.2, the maximum load is 68.75 MPa, the minimum load is 13.75 MPa, and the load frequency is 0.7 Hz. The left of the specimen is denoted as the *S* direction and the right is the *N* direction.

7.4.3.2 SIF calculation of the stiffened plate with a penetrated crack

In the finite element model, the mesh density near the crack affects the calculated magnitude of stress or strain directly, and the computation time increases with the increasing mesh density. The finite elements of the model were meshed by the adaptive technology to determine the size of the mesh. In the stiffened plate model, both the plate and stiffener were meshed with SHELL 93 (ANSYS 9.0). The mesh size near the crack is 10 mm, and that of the far field zone is 50 mm. The stiffeners are replaced by the plate with a width of 160 mm and a thickness of 13 mm, and the crack length in the stiffener is *c* which is shown in Fig. 1.18(b).

The singularity of the crack tip stress field was simulated with a singular element around the crack tip. In order to ensure the accuracy of the results, it is necessary to make the mesh of the singular element more intensive when the crack tip goes near the junction of the stiffener and plate or the crack tip in the stiffener goes near the edge of the stiffened plate. Fig. 7.20 shows the mesh of the crack tip of the stiffener and plate and the stress distribution. It is obvious to see that the gradient of the crack tip stress field changes a lot, and singularity appears in the stress field.

For the symmetry of the geometry and the loading, taking the N direction as an example, the crack growing from the center of the plate to the right side should surpass two identical stiffeners. The first stiffener is regarded as A, and the second is B. The initial half crack length is 152.5 mm. The calculated SIFs of cracks of different length are listed in Table 7.9 and shown in Fig. 7.21.

SIFs of the through-thickness cracks in the stiffened plate can be calculated by Eq. (7.2) which is determined by fitting the calculated SIFs listed in Table 7.9. The results calculated by Eq. (7.2) are compared with the data of stiffened and unstiffened plates by FEA shown in Fig. 7.21. From the comparison, it can be seen that stiffeners can reduce SIF effectively, and accordingly reduce the crack growth rate. The fitting Eq. (7.2) can represent well the results by FEA.

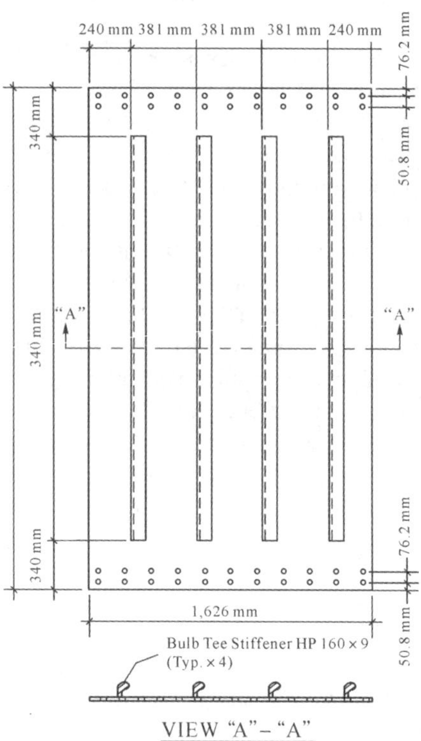

Fig. 7.19 Geometry of the large specimen (Mahmoud and Dexter, 2005) (with the permission of Elsevier)

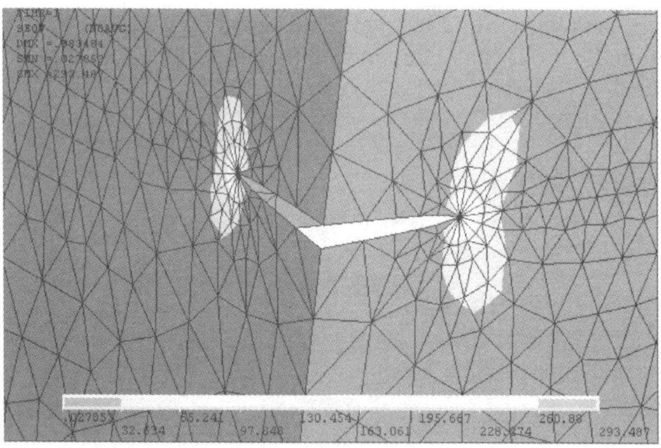

Fig. 7.20 The mesh and stress distribution around the crack tip

Table 7.9 SIFs of cracks with different length in the stiffened plate

a/w	K_I/σ_p	c/d	K_I/σ_S	a/w	K_I/σ_P	c/d	K_I/σ_S
0.188	19.28	—	—	0.443	36.57	—	—
0.197	19.54	—	—	0.467	37.97	—	—
0.209	19.67	—	—	0.492	39.42	—	—
0.221	19.28	—	—	0.554	43.28	—	—
0.228	18.45	—	—	0.615	47.33	—	—
0.231	17.25	—	—	0.640	48.80	—	—
0.234	15.68	—	—	0.677	49.62	—	—
0.236	12.83	0.009	12.79	0.689	48.38	—	—
0.240	15.91	0.028	15.86	0.695	46.53	—	—
0.246	16.76	0.059	16.59	0.699	45.13	—	—
0.271	21.06	0.184	19.51	0.701	40.66	—	—
0.308	25.79	0.372	20.37	0.707	35.15	0.022	35.77
0.344	29.65	0.559	18.58	0.713	40.01	0.053	39.50
0.369	31.80	0.684	16.06	0.738	50.76	0.178	47.60
0.394	33.63	0.809	12.50	0.763	59.96	0.303	51.41
0.418	35.19	0.934	6.79	0.800	73.88	0.491	52.14
0.424	35.54	0.966	5.60	0.836	89.34	0.678	47.13
0.428	35.74	0.984	2.63	0.861	100.94	0.803	39.02
0.429	35.81	0.991	1.58	0.886	115.38	0.928	25.01
0.431	35.88	0.997	0.28	0.892	118.18	0.959	19.56
0.433	36.02	—	—	0.898	122.21	0.991	10.68
0.437	36.23	—	—	0.899	123.04	0.997	7.83

* Note: *P* stands for plate and *S* stands for stiffener

Based on the data in Table 7.9, the fitting equations of SIF are given as follows.

$$K = M\sigma \tag{7.2}$$

$$M = \begin{cases} -582677(a/w)^5 + 351182(a/w)^4 - 78035(a/w)^3 + 7379(a/w)^2 - 177(a/w) + 9 \\ \qquad\qquad\qquad\qquad\qquad\qquad\qquad\qquad (0 < a/w \le 0.234) \\ -38446(a/w)^5 + 83090(a/w)^4 - 68871(a/w)^3 + 27118(a/w)^2 - 4936(a/w) + 345 \\ \qquad\qquad\qquad\qquad\qquad\qquad\qquad\qquad (0.234 < a/w \le 0.703) \\ -29414(a/w)^4 + 101764(a/w)^3 - 130627(a/w)^2 + 74238(a/w) - 15770 \\ \qquad\qquad\qquad\qquad\qquad\qquad\qquad\qquad (0.703 < a/w < 1) \end{cases}$$

where a is the half crack length, w is the half width of the plate.

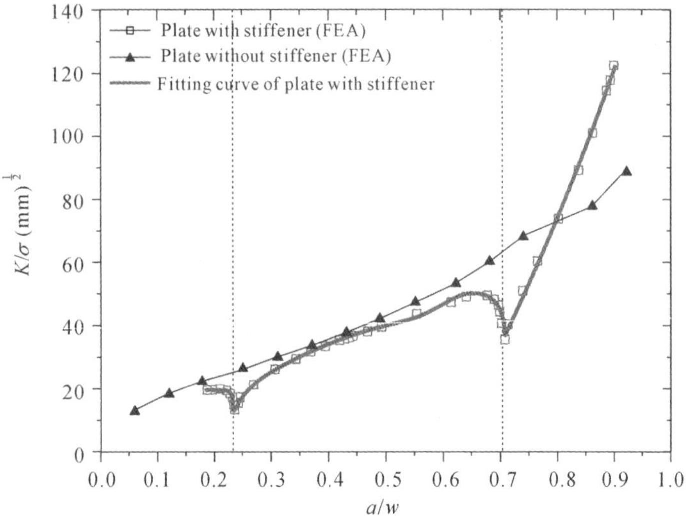

Fig. 7.21 K/σ versus a/w-curves

7.4.4 SIF of the Crack in Stiffened Plate by Weld Residual Stress

The weld residual stress distribution along the transection of the stiffened plate, which is taken from the reference (Mahmoud and Dexter, 2005), is shown in Fig. 7.22. The SIFs of the through-thickness cracks in the stiffened plate caused by weld residual stress can be calculated by Eq. (7.3) (Tan, *et al.*, 2007).

$$K = \int_{-a}^{a} \sigma(x) G(x,a) \mathrm{d}x = \frac{1}{\sqrt{\pi a}} \int_{-a}^{a} \sigma(x) \sqrt{\frac{a \mp x}{a \pm x}} \mathrm{d}x \tag{7.3}$$

where a is the half crack length, $\sigma(x)$ is the weld residual stress distribution along the crack face. The SIF of the through-thickness crack caused by weld residual stress shown in Fig. 7.22 was calculated and shown in Fig. 7.23.

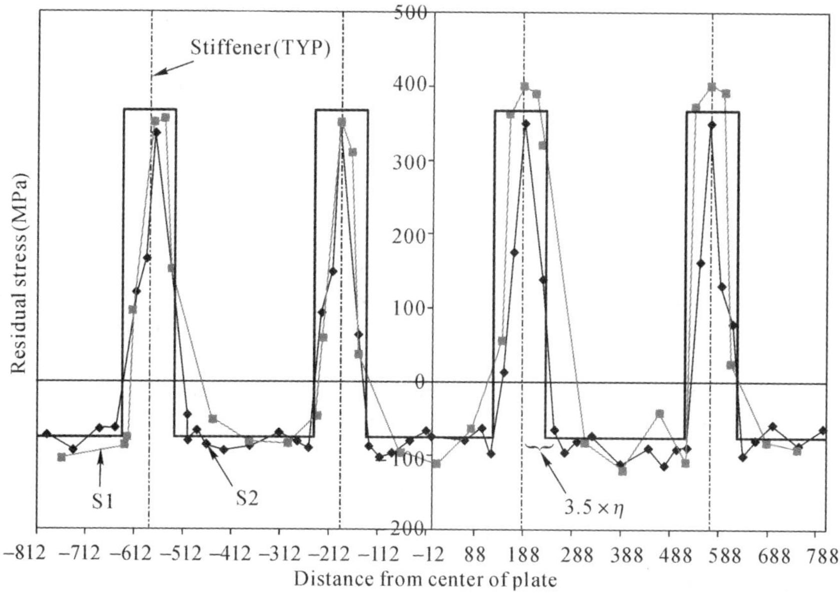

Fig. 7.22 Residual stresses measurements in S1 and S2 (Mahmoud and Dexter, 2005) (with the permission of Elsevier)

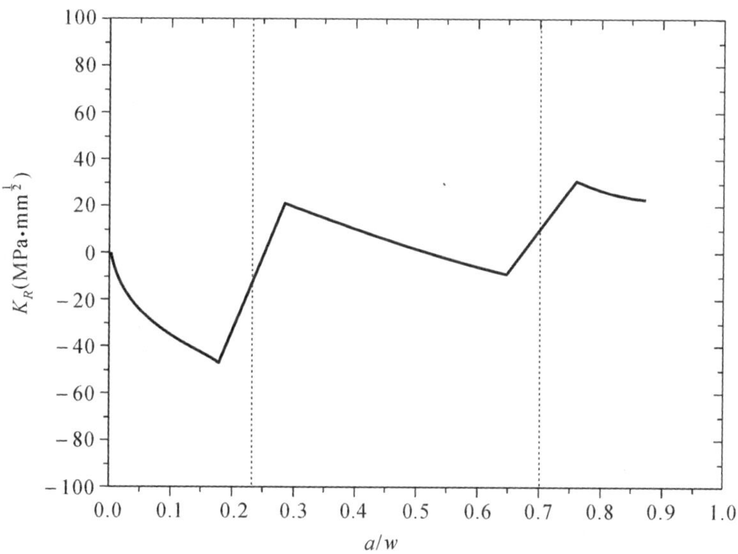

Fig. 7.23 SIFs caused by weld residual stresses

7.4.5 Fatigue Crack Growth Prediction of the Stiffened Plate

The SIFs of cracks in the stiffened plate by applied load can be calculated by

Eq. (7.2), and the SIFs caused by weld residual stress can be calculated by Eq. (7.3). Based on calculated SIFs of the cracks with different crack length, the UFLP model can be used to compute the fatigue crack growth of the stiffened plate. The load ratio is defined by Eq. (7.4)

$$R = \frac{K_{\min} + K_{\mathrm{Res}}}{K_{\max} + K_{\mathrm{Res}}} \tag{7.4}$$

For predicting the crack growth life by UFLP model, the material constants for the crack growth rate model are determined and listed in Table 7.10. A program running in the Matlab environment is used to compute the crack growth based on cycle-by-cycle integration, and the calculated crack lengths versus the loading cycles are compared with the experimental data and that of the unstiffened plate, as shown in Fig. 7.24. The experimental data in Fig. 7.24 are the crack growth data of the right side and left side of specimen S1, respectively, and there is a slight difference between the test data and the predicted results. Obviously, the phenomenon of crack growth in a stiffened plate can be well predicted by UFLP.

Table 7.10 Material constants for UFLP of steel ASTM A572

σ_y (MPa)	A	m	$\Delta K_{\mathrm{th\,0}}$ $(\mathrm{MPa \cdot m^{\frac{1}{2}}})$	K_{IC} $(\mathrm{MPa \cdot m^{\frac{1}{2}}})$	n	r_e (m)
345	7.21E-10	2.1	7.2	120	6	1E-6

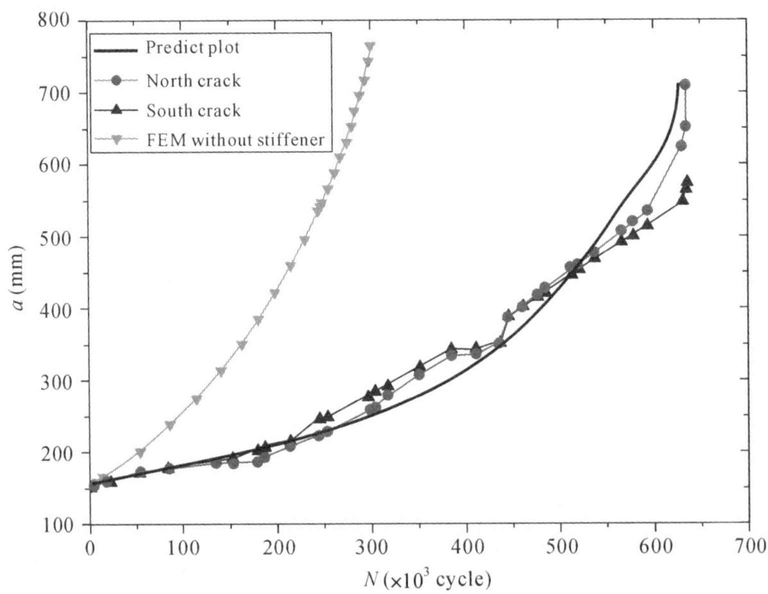

Fig. 7.24 Comparison of crack growth curves in stiffened plate (Mahmoud and Dexter, 2005) (with the permission of Elsevier)

7.5 FLP of Submarine Hull Under Different Fatigue Loading Sequence

The fatigue strength of submarine structures has received increasing attention due to the use of high tensile strength steels, the increase in pressure hull diameter and the diving depth. In giving a precise prediction of the fatigue life, the effect of compressive stress, residual stress and the loading sequence should be considered. A general equation for calculating the SIF of a surface crack at the weld toe incorporates the crack growth rate equations of UFLP that have been used to calculate the fatigue life of a submarine hull under different loading sequences. The loading sequences were rebuilt by the same fatigue loading serial data which are generated from a long term distribution of the fatigue loading described by a two-parameter Weibull function. The fatigue lives of submarine hulls under different loading sequences are predicted by UFLP procedure and these predicted results are compared. The procedure is able to account for the influence of compressive stress, stress concentration, weld residual stress and the loading sequence.

7.5.1 General Equations for Calculating the SIF of a Surface Crack at Weld Toe

The SIF magnification coefficient M_K is used to express the increase in SIF due to weld notch stress concentration. When the effect of weld residual stress is also considered, general SIF calculation equations of a surface crack at the weld toe can be calculated by the following equations (Huang, 2005).

$$K = (M_{\mathrm{K}}^{\mathrm{T}}\sigma + M_{\mathrm{K}}^{\mathrm{B}}H\sigma_b)\frac{\sqrt{\pi a}}{\Phi}F\left(\frac{a}{c},\frac{a}{t},\frac{2c}{w},\varphi\right) + K_{\mathrm{Res}} \tag{7.5}$$

$$\Phi = \sqrt{1.0 + 1.464\left(\frac{a}{c}\right)^{1.65}}$$

$$F\left(\frac{a}{c},\frac{a}{t},\frac{2c}{W},\varphi\right) = \left[M_1 + M_2\left(\frac{a}{t}\right)^2 + M_3\left(\frac{a}{t}\right)^4\right] f_\varphi \cdot g \cdot f_w$$

$$M_1 = 1.13 - 0.09\left(\frac{a}{c}\right)$$

$$M_2 = -0.54 + 0.89/\left(0.2 + \frac{a}{c}\right)$$

$$M_3 = 0.5 - 1/\left(0.65 + \frac{a}{c}\right) + 14\left(1 - \frac{a}{c}\right)^{24}$$

$$f_\varphi = \left[\sin^2\varphi + \left(\frac{a}{c}\right)^2 \cos^2\varphi\right]^{1/4}$$

$$g = 1 + \left[0.1 + 0.35\left(\frac{a}{t}\right)^2\right](1 - \sin\varphi)^2$$

$$f_w = \left[\sec\left(\frac{\pi}{2}\frac{2c}{w}\sqrt{\frac{a}{t}}\right)\right]^{1/2}$$

$$H = H_1 + (H_2 - H_1)\sin^p \phi$$

$$H_1 = 1 - 0.34\left(\frac{a}{t}\right) - 0.11\left(\frac{a}{c}\right)\left(\frac{a}{t}\right)$$

$$H_2 = 1 + G_1\left(\frac{a}{t}\right) + G_2\left(\frac{a}{t}\right)^2$$

$$G_1 = -1.22 - 0.12\left(\frac{a}{c}\right)$$

$$G_2 = 0.55 - 1.05\left(\frac{a}{c}\right)^{0.75} + 0.47\left(\frac{a}{c}\right)^{1.5}$$

$$p = 0.2 + a/c + 0.6\left(\frac{a}{t}\right)$$

where K_I—SIF; σ—tension (compression) stress; σ_b—bending stress; a—depth of the crack; c—half crack length; w—the width of plate; M_K^T—SIF magnification coefficient of welded structure under tension stress; M_K^B—SIF magnification coefficient of welded structure under bending stress; K_{Res}—SIF due to weld residual stress.

7.5.2 SIF of Surface Crack Caused by Weld Residual Stress at Weld Toe

For a surface crack parallel to the weld seam, the weld residual stress perpendicular to the weld seam will influence the SIF of the surface crack. The distribution of the weld residual stress should be described in calculating the SIFs of cracks of different depths by weld residual stress. The magnitude of the residual stress on the surface at the weld toe is usually set as (Huang, *et al.*, 2002).

$$\sigma^R = 0.3\sigma_y \tag{7.6}$$

where σ^R—residual stress (MPa); σ_y—yield stress (MPa).

The distribution of the weld residual stress along the plate thickness and the width direction is not uniform. Simplified distributions of the weld residual stress along the plate thickness and width were shown in Fig. 7.25. The residual stress along the plate thickness can be expressed as

$$\begin{cases} \sigma^R(x) = \sigma^R\left(1 - 4\dfrac{x}{t}\right) & (0 \le x \le t/2) \\ \sigma^R(x) = \sigma^R\left(4\dfrac{x}{t} - 3\right) & (t/2 \le x \le t) \end{cases} \tag{7.7}$$

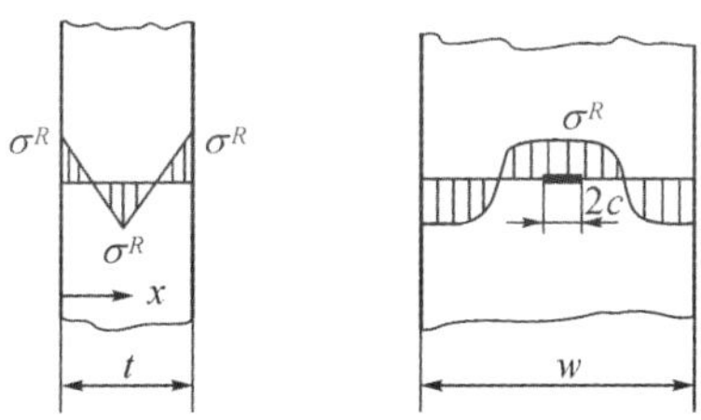

Fig. 7.25 Welded residual stress distribution along the thickness and the breadth

On the surface x=0, and at the deepest point of a surface crack x=a. SIF K_{Res} due to residual stress can be computed by weight function method (Wang and Lambert, 1995) based on the above residual stress distribution. For conservative calculation, the residual stress distribution along plate thickness can be treated as uniform, with equivalent magnitude to the residual stress on the surface, *i.e.* $K_{\text{Res}}(\sigma^R(x)) = K_{\text{Res}}(\sigma^R(0))$.

$$K(\sigma^R) = \int_0^a m(a,x)\cdot\sigma^R(x)\mathrm{d}x \tag{7.8}$$

where $\sigma^R(x)$ is residual stress distribution along the crack surface.

Deepest point:

$$m_D(a,x)=\frac{2}{\sqrt{2\pi(a-x)}}\left[1+M_{D1}\cdot\left(1-\frac{x}{a}\right)^{0.5}+M_{D2}\cdot\left(1-\frac{x}{a}\right)+M_{D3}\cdot\left(1-\frac{x}{a}\right)^{1.5}\right]$$

$$M_{D1}=\frac{2\pi}{\sqrt{2Q}}(2Y_0-3Y_1)-4.8$$

$$M_{D2}=3$$

$$M_{D3}=\frac{6\pi}{\sqrt{2Q}}(2Y_1-Y_0)+1.6$$

$$Q=1+1.464(a/c)^{1.65}$$

$$Y_0=A_0+A_1\left(\frac{a}{B}\right)^2+A_2\left(\frac{a}{B}\right)^4+A_3\left(\frac{a}{B}\right)^6$$

$$A_0=1.0929+0.2581(a/c)-0.7703(a/c)^2+0.4393(a/c)^3$$

$$A_1=0.456-3.045(a/c)+2.007(a/c)^2+1/(0.147+(a/c)^{0.688})$$

$$A_2=0.995-1/(0.027+(a/c))+22(1-(a/c))^{9.953}$$

$$A_3=-1.459+1/(0.014+(a/c))-24.211(1-(a/c))^{8.071}$$

$$Y_1=B_0+B_1\left(\frac{a}{B}\right)^2+B_2\left(\frac{a}{B}\right)^4+B_3\left(\frac{a}{B}\right)^6$$

$$B_0=0.4537+0.1231(a/c)-0.7412(a/c)^2+0.46(a/c)^3$$

$$B_1=-1.652+1.665(a/c)-0.534(a/c)^2+1/\left(0.198+(a/c)^{0.846}\right)$$

$$B_2=3.418-3.126(a/c)-1/(0.041+(a/c))+17.259(1-(a/c))^{9.286}$$

$$B_3=-4.228+3.643(a/c)+1/(0.02+(a/c))-21.924(1-(a/c))^{9.203}$$

Surface point:

$$m_s(a,x)=\frac{2}{\sqrt{\pi x}}\left[1+M_{s1}\cdot\left(\frac{x}{a}\right)^{0.5}+M_{s2}\cdot\left(\frac{x}{a}\right)+M_{s3}\cdot\left(\frac{x}{a}\right)^{1.5}\right]$$

$$M_{s1}=\frac{3\pi}{\sqrt{Q}}(5F_1-3F_0)-8$$

$$M_{s2}=\frac{15\pi}{\sqrt{Q}}(2F_0-3F_1)+15$$

$$M_{s3} = \frac{3\pi}{\sqrt{Q}}(10F_1 - 7F_0) - 8$$

$$F_0 = \left[C_0 + C_1 \left(\frac{a}{t} \right)^2 + C_2 \left(\frac{a}{t} \right)^4 \right] \cdot \sqrt{\frac{a}{c}}$$

$$C_0 = 1.2972 - 0.1548(a/c) - 0.0185(a/c)^2$$

$$C_1 = 1.5083 - 1.3219(a/c) + 0.5128(a/c)^2$$

$$C_2 = -1.101 + 0.879/(0.157 + (a/c))$$

$$F_1 = \left[D_0 + D_1 \left(\frac{a}{t} \right)^2 + D_2 \left(\frac{a}{t} \right)^4 \right] \cdot \sqrt{\frac{a}{c}}$$

$$D_0 = 1.2687 - 1.0642(a/c) + 1.4646(a/c)^2 - 0.725(a/c)^3$$

$$D_1 = 1.1207 - 1.2289(a/c) + 0.5876(a/c)^2$$

$$D_2 = 0.19 - 0.608(a/c) + 0.199/(0.035 + (a/c))$$

7.5.3 *Fatigue Crack Growth Prediction of Surface Crack at Weld Toe of Submarine Structure*

7.5.3.1 Fatigue loading by hydraulic pressure

The fatigue of a submarine structure is mainly caused by the external hydraulic pressure which varies with the diving depth. A two-parameter Weibull distribution is suitable for describing the probability feature of the diving depth distribution of submarines or submersibles and the long term distribution of the diving depth of a submarine or submersible is discussed by Huang and Cui (2003). A long term distribution of the diving depth of submarine with an extreme depth of 450m was selected. The corresponding probability density function and probability distribution function of the depth is as follows,

$$f(h) = 0.00003709 \times h \times \mathrm{e}^{-\left(\frac{h}{232.3}\right)^2} \quad (0 \le h \le 450 \text{ m}) \tag{7.9}$$

$$F(h) = 1 - \mathrm{e}^{-\left(\frac{h}{232.3}\right)^2} \tag{7.10}$$

The weld connection of a cylinder with a conical hull is the fatigue hot spot in the submarine structure and the hot spot stress induced by external pressure can be calculated by Eqs. (7.11) and (7.12).

$$\sigma_1 = \sigma + \sigma_b = \frac{\rho g h R_1}{2t}\left[1 - 1.165\sqrt{\frac{R_1}{t}}\tan(\theta)\right] \tag{7.11}$$

$$\sigma_2 = \frac{\rho g h R_1}{t}\left[1 - 0.321\left(1 + 0.55\sqrt{\frac{R_1}{t}}\tan(\theta)\right)\right] \tag{7.12}$$

where σ_1, σ_2—are axial stress and hoop stress respectively; ρ, h—density and depth of the sea water; R_1, t, θ—the radius, thickness and connection angle of the submarine hull.

7.5.3.2 Fatigue crack growth prediction of submarine under different block loadings

Because the diving depth is random, a series of random data of the diving depth following the Weibull distribution are obtained by Matlab and the data is shown in Fig. 7.26. By substituting the data into Eqs. (7.11) and (7.12), the corresponding stresses can be determined. We assume each dive is independent and starts from the surface of the sea to the corresponding depth and back to the surface. The stress range is such that the minimum stress is zero and the maximum stress corresponds to each depth h. Part of the stress caused by hydraulic pressure plus the residual stress on the surface at the weld toe varying with time is shown in Fig. 7.27(a). The ascending sequence and descending sequence are shown in Fig. 7.27(b) and Fig. 7.27(c) respectively. For comparison of the crack growth under these three loading sequences, two surface cracks with initial crack length a=0.2 mm, half crack length c=5 and 7 mm were selected respectively.

The fatigue crack growth of these two surface cracks at the weld toe under the above three loading sequences, which obey the same Weibull distribution, are calculated by UFLP procedure. The material constants needed in UFLP are $A=2.18\times10^{-10}$, m=2.15, n=6. The calculated crack depth varying with loading cycles is shown in Fig. 7.28. From Fig. 7.28, it can be seen that the effect of the loading sequence is obvious; the crack depth under the descending loading sequence is the shortest, and that of the ascending loading spectrum is the longest after 40,000 cycle loading. But if the crack depth was set as the check parameter at any time, the order of the fatigue crack growth life of these three fatigue loading sequences under the same increase in crack depth is ascending sequence, random loading and descending sequence.

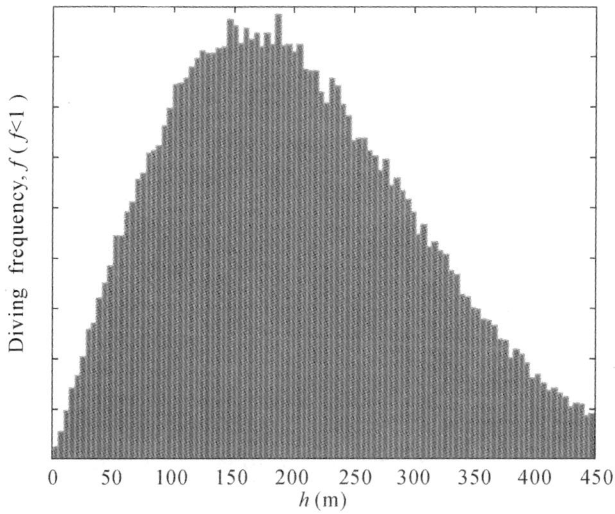

Fig. 7.26 Random diving depth obey the Weibull distribution

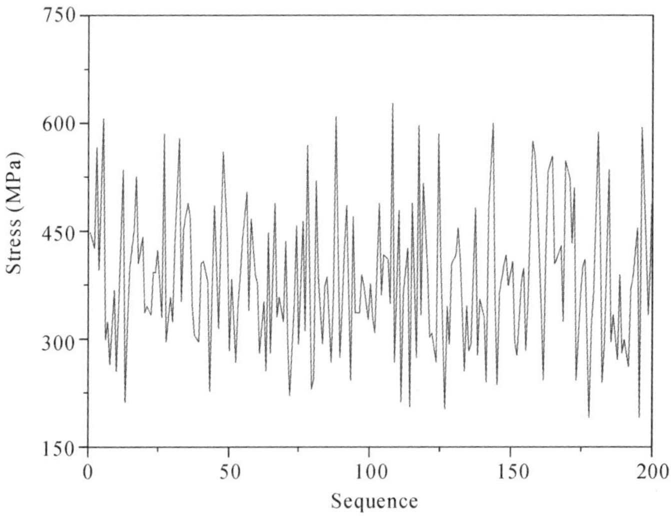

(a)

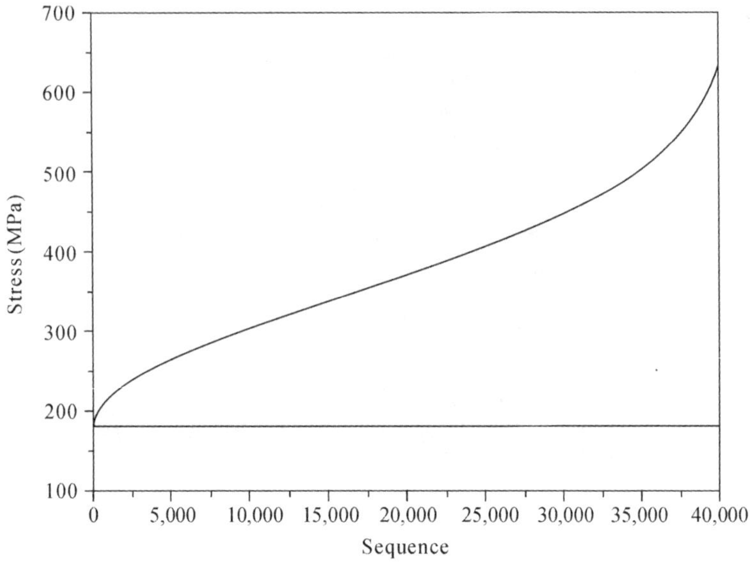

(b)

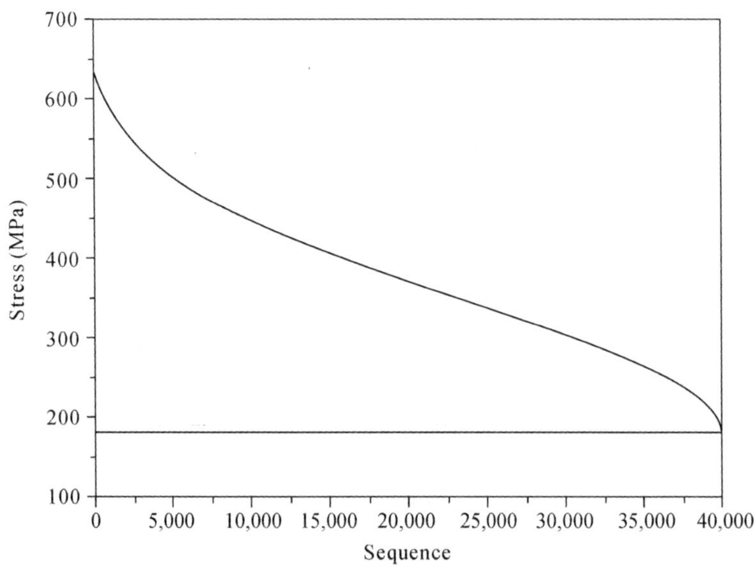

(c)

Fig. 7.27 Fatigue loading spectrum. (a) Random loading sequence; (b) Ascending sequence; (c) Descending sequence

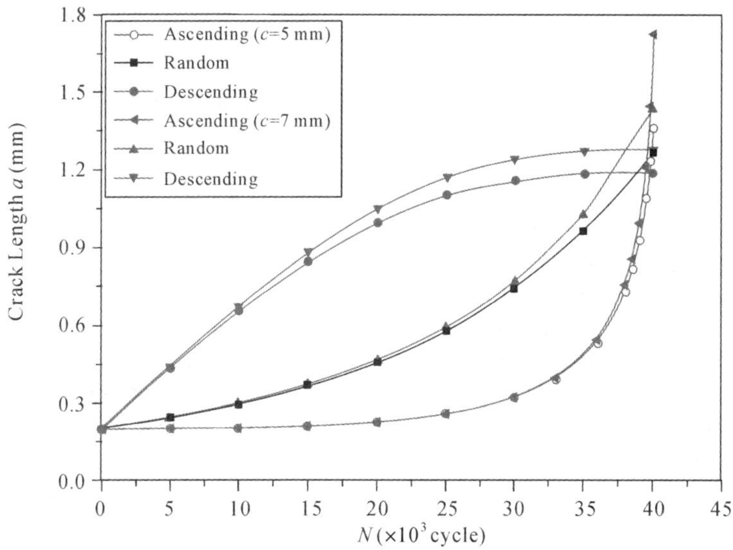

Fig. 7.28 Comparison of the calculated results under different loading sequences

7.6 Summary

In this chapter, the UFLP model is used to predict the fatigue lives of specimens with central through-thickness crack under different fatigue loadings, the fatigue life of the cracked stiffened deck of an oil tanker and the fatigue life of a surface crack at the weld toe in a submarine hull under different loading sequences. From these demonstrations, some conclusions can be drawn.

The fatigue crack growth life of a cracked structure under complex loading situations can be predicted well by the UFLP model.

It can be found that the larger the overload is, the more obvious is the effect of the load sequence. The shaping exponent for loading sequence interaction has a great effect on the predicted results. The UFLP model can predict the fatigue crack growth under different loading sequences with different overload, underload, overload-underload and underload-overload, *etc.* by regulating the exponent γ. The effects of stress concentration and residual stress on fatigue are considered in this model. It provides an effective way to study the fatigue crack growth of marine structures under random loading. The shaping exponent for loading sequence interaction is correlated with the overload sequence and overload range for the same material and it can be determined based on serial fatigue test data of the same material under different loading conditions and needs further study.

References

Christopher, D. G. & Robert R. S. (2006). "Fatigue crack growth and life predictions under variable amplitude loading for a cast and wrought aluminum alloy", International Journal of Fatigue, 28: 53–60.

Cui, W. C., Wang, F. & Huang, X. P. (2010). "Towards a unified fatigue life prediction method for marine structures: an overview", in: Proceedings of the ASME 2010 29th International Conference on Ocean, Offshore and Arctic Engineering, OMAE 2010, June 6–11, Shanghai, China.

Cui, W. C., Wang, F. & Huang, X. P. (2011). "A unified fatigue life prediction method for marine structures", Marine Structures, 24: 1–29.

Huang, X. P. (2007). "Numerical simulation and spectrum experimental study of fatigue behavior of deep-sea structures", Technical Report of Shanghai Jiao Tong University (in Chinese).

Huang, X. P. & Cui, W. C. (2003). "Probability model of fatigue loading for submarines and submersibles", The Ocean Engineering, 21(3): 18–23 (in Chinese).

Huang, X. P., Cui, W. C. & Leng, J. X. (2005). "A model of fatigue crack growth under various load spectra", in: Proc of Sih GC, 7th Int conf of MESO, August 1–4, Montreal, Canada, 303–308.

Huang, X. P., Cui, W. C. & Shi, D. X. (2002). "Calculation of fatigue life of surface cracks at weld toe of submarine cone-cylinder shell", Journal of Ship Mechanics, 6: 62–68 (in Chinese).

Mahmoud, H. N. & Dexter, R. J. (2005). "Propagation rate of large cracks in stiffened plates under tension loading", Marine Structures, 18: 265–288.

Mcmillan, J. C. & Pelloux, R. M. N. (1967). "Fatigue crack propagation under program and random loads, fatigue crack propagation", ASTM STP 415 also Boeing Space Research Laboratory (BSRL) Document D1-82-0558, 1996, 505–532.

Paik, J. K. (2008). "Residual ultimate strength of steel plates with longitudinal cracks under axial compression—experiments", Ocean Engineering, 35: 1775–1783.

Porter, T. R. (1972). "Method of analysis and prediction for variable amplitude fatigue crack growth", Engineering Fracture Mechanics, 4: 717–736.

Ray, A. & Patankar, R. (2001). "Fatigue crack growth under variable-amplitude loading (Part II: Code development and model validation)", Applied Mathematical Modelling, 25: 995–1013.

Schijve, J., Skorupa, M., Skorupa, A., Machniewicz, T. & Gruszczynski, P. (2004). "Fatigue crack growth in the aluminium alloy D16 under constant and variable amplitude loading", International Journal of Fatigue, 26: 1–15.

Skorupa, M. (1998). "Load interaction effects during fatigue crack growth under variable amplitude loading—a literature review (Part I: Empirical trends)", Fatigue and Fracture of Engineering Materials and Structures, 21(8): 987–1006.

Skorupa, M. (1999). "Load interaction effects during fatigue crack growth under variable amplitude loading—a literature review (Part II: Qualitative interpretation)", Fatigue and Fracture of Engineering Materials and Structures, 22(10): 905–926.

Skorupa, M., Machniewicz, T., Schijve, J. & Skorupa, A. (2007). "Application of the strip-yield model from the NASGRO software to predict fatigue crack growth in aluminium alloys under constant and variable amplitude loading", Engineering Fracture Mechanics, 74: 291–313.

Taheri, F., Trask, D. & Pegg, N. (2003). "Experimental and analytical investigation of fatigue characteristics of 350WT steel under constant and variable amplitude loading", Marine Structures, 16: 69–91.

Tan, J. M. L., Fitzpatrick, M. E. & Edwards, L. (2007). "Stress intensity factors for through-thickness cracks in a wide plate, derivation and application to arbitrary weld residual stress fields", Engineering Fracture Mechanics, 74: 2030–2054.

Wang, F. & Cui, W. C. (2009). "Approximate method to determine the model parameters in a new crack growth rate model", Marine Structures, 22(4): 744–757.

Wang, F. & Cui, W. C. (2010). "On the engineering approach to estimating the parameters in an improved crack growth rate model for fatigue life prediction", Ships and Offshore Structures, 5: 227–241.

Wang, X. & Lambert, S. B. (1995). "Stress intensity factors for low aspect ratio semi-elliptical surface cracks in finite-thickness plates subjected to nonuniform stresses", Engineering Fracture Mechanics, 51(4): 517–532.

8

Code Development Based on UFLP for Marine Structures

8.1 Introduction

Fatigue strength assessment has become more important in the structural design of ships and marine structures due to the increase in structure size and the strength of the constructed structural steel. The cracks or crack-like defects are usually pre-existing or initiated in weld details in these welded structures. Fracture mechanics is recommended for use in the assessment of acceptable defects, evaluation of acceptance criteria for fabrication and for planning in-service inspection.

Fatigue failure is an extremely complex physical process which is governed by a great number of parameters related to, for example, fatigue loading behavior, local geometry and material properties of the structural region surrounding the crack growth path. Most of the classification societies have issued recommendations for fatigue strength analysis, but the procedures of the classification societies are mainly "rule based" methods or direct calculation methods which are all based on *S-N* curves and the linear cumulative damage rule (Kukkanen, *et al.*, 2004), and few of the classification societies have issued the fatigue life assessment methods based on fracture mechanics. For fatigue crack assessment methods based on fracture mechanics, there are some rules or guides for fitness-for-service assessment, such as BS7910, API579, SINTAP, IIW, *etc.*

In this chapter, the preliminary code development based on the UFLP method for marine structures, following the similar procedure described in reference BS7910, is presented. Section 8.2 presents the crack growth life prediction procedure of the UFLP method for marine structures. Fatigue loading for marine structures is generally described in Section 8.3. In Section 8.4, the issues related to initial flaw or initial crack size and SIF calculation are discussed. The recommended fatigue crack growth rate material constants of the model for marine structural steels are presented in Section 8.5. Estimation of the material constants of UFLP is given in Section 8.6. A summary of the present chapter is given in Section 8.7. The SIF calculation equations for some typical types of cracks are presented in the appendix.

8.2 Procedure of UFLP

In order to take into account fatigue in marine structural analysis, the wave induced fatigue loading, the effective SIF range, the material constants of the crack growth rate and finally the fatigue strength or fatigue life have to be determined. This means that the operational conditions of the marine structures and the cyclic loading in marine structures service time should be defined, and the structural responses have to be determined in order to obtain stresses in critical structural details. Furthermore, the fatigue strength or fatigue life of the structural detail has to be calculated and assessed. The purpose of such analysis is to calculate that the fatigue cracks which might occur during service life will not exceed the critical crack size corresponding to unstable fracture. General procedure of the UFLP method is similar to that of Part 8 "Assessment for Fatigue" in BS7910, but the crack growth rate equations and the corresponding material constants are different. The fatigue loadings of marine structures are mainly wave induced, which have the random characteristics of wave loading and with about 10^8 cycles in their design life. It implies that most of the fatigue crack growth life of marine structures is consumed around the crack growth threshold and the effect of the loading sequence on fatigue is significant. It has been observed that components in service spend approximately 80% of their lifetime in the region of short crack growth where the crack length is less than 1mm. Since flaws of this size are difficult to detect, it is important to understand high cycle fatigue in order to prevent small cracks growing into failure. Many authors have been investigating the phenomenon in order to estimate the total fatigue life of welded structures. More details are presented in Chapters 2 and 3. The value of the crack growth threshold and initial crack size will be more important for fatigue life assessment of marine structures.

In the FCP life prediction of marine structures, the crack initiation site, the analysis of the response fatigue loading in the marine environment, SIF calculation, determination of the material constants for UFLP and the effective SIF range ΔK_{eff} considering the effect of the loading sequence should be calculated. The basic components of the fatigue crack growth prediction procedure for marine structures, which are outlined in Fig. 8.1, are as follows.

(1) Global structural analysis: Determination of components of nominal stress corresponding to wave load can be calculated by code formula or by direct numerical calculation.

(2) Determination of fatigue loading spectrum: Generation of the stress range serials for the cracked joint can be derived from the external loading spectrum by appropriate transform function or determined by direct numerical simulation in the

time domain.

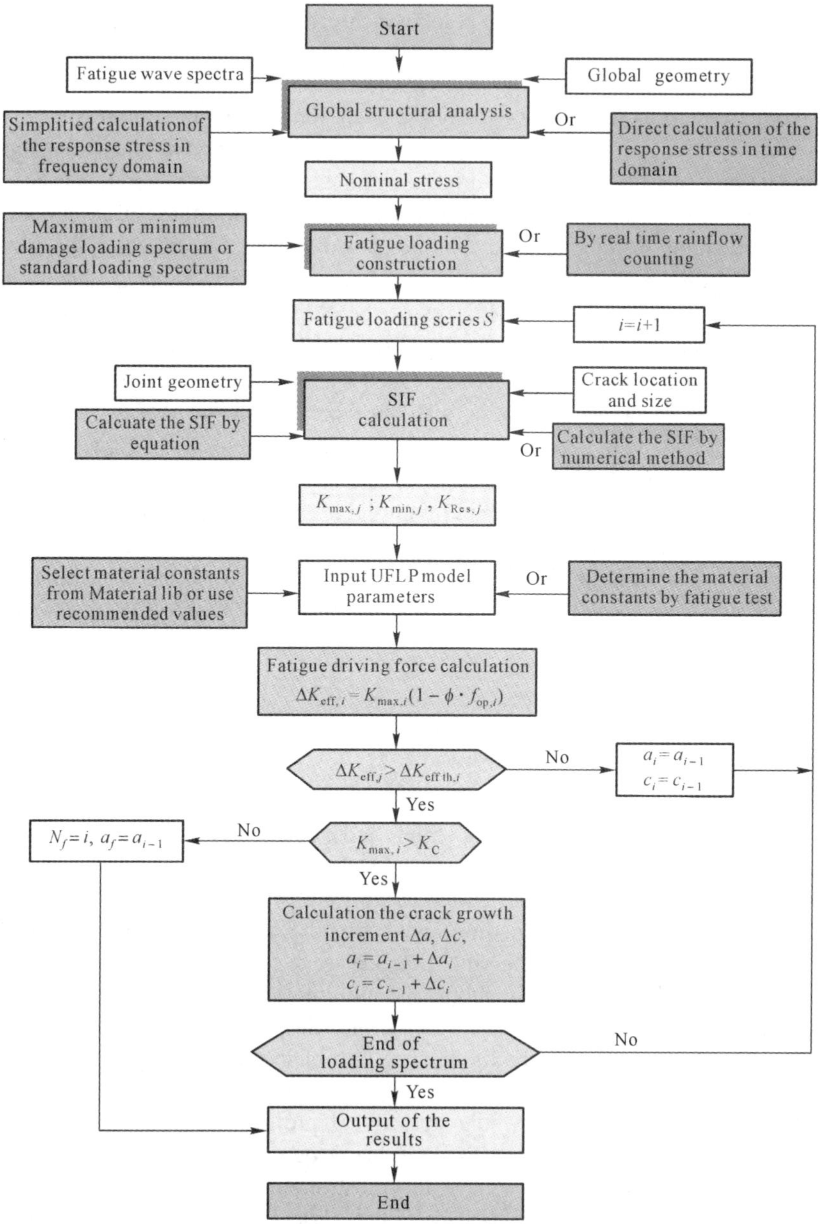

Fig. 8.1 Flow diagram for fatigue life calculation procedure of the UFLP model

(3) Local joint stress analysis and SIF calculation: Determination of the local stress distribution around the crack location and the SIFs can also be done by the

simplified method or by numerical simulation.

(4) Fatigue crack growth analysis: The UFLP approach assumes that a flaw may be idealized as a sharp tipped crack which propagates in accordance with the law relating the crack growth rate, $\frac{da}{dN}$, to the effective range of the SIF, ΔK_{eff}, for the material or component containing a crack flaw. When the initial crack size, the crack located joint geometry and the fatigue loading are known, the proper material constants for UFLP should be determined. The SIF ΔK, corresponding to the applied stress range, can be calculated considering the effect of compressive stresses including residual stresses and then calculated for the effective SIF range. For a partial-thickness surface or embedded flaws, ΔK should be determined at the ends of both the minor and major axes of the elliptical idealization of the flaw. The growth of the crack, Δa, should be estimated for one cycle from the value of ΔK_{eff}. The dimension, a, should be increased by Δa. Similarly, the growth Δc at the ends of a breaking part of a wall surface or embedded flaw should be estimated for one cycle from the value of ΔK_{eff} calculated for those flaw ends. The length of the flaw $2c$ should be increased by $2\Delta c$. The incremented crack dimension or SIF should be compared with the limiting value of failure criteria. If acceptable, the next stress cycle should be considered. If the specified fatigue design life is reached and the limit to growth is not exceeded, the flaw should be regarded as acceptable. The final crack size, remaining fatigue life, *etc.* will be outputted as fatigue assessment results.

8.3 Fatigue Loading

The service life of a component or a structure depends on both the load conditions and the fatigue strength. Hence, in order to obtain a proper fatigue design, it is important to consider the real structural loading history, $P(t)$. Since fatigue failure is a future event, the fatigue loading history for the designed structure can never be known *a priori*. Many engineering methods are based on finding the worst case scenario, where "worst" often should be interpreted as a certain severe load condition. A detailed description of fatigue loading can be found in Chapter 6 while, in the following, only the fundamental problem related to fatigue loading in code development is described.

In a fatigue analysis the whole operational profile of a marine structure is required to obtain all stress cycles that will be encountered during the service time. The most frequent cyclic loads have the most significant effects on the fatigue strength of the hull structure (ISSC, 2009). For the current fatigue strength assessment methods for conventional ship types implemented in the ship classification societies' rules, only linear frequency domain methods for calculation of wave-induced loadsare used to predict fatigue strength, by applying stochastic procedures to estimate lifetime fatigue accumulation. For example, linear spectral methods can be used, requiring transfer functions of stress responses to be determined in several operational and environmental conditions. This is rather tedious if several structural details need to be analysed. Simplified methods have been developed to represent the long term distributions of stresses or loads applying stochastic methods based on, for example Weibull-distribution. Application of linear procedures is basically straightforward but includes several sources of uncertainty, such as calculations of loads and structural responses as well as predictions of the lifetime operational and environmental conditions. Huther, *et al.* (2004) discussed the uncertainties and simplifications in fatigue prediction methods, including loads and stresses, and fatigue strength and failure criteria. They concluded that progress had been made but validation of the calculation methods and results was still needed.

For fatigue prediction assessment of ships and marine structures, it is difficult to obtain an accurate loading history. For the fatigue loading prediction of marine structures there are several kinds of methods, such as code simplified nominal stress calculation by equations, the spectral analysis method, the design wave approach, time domain calculation method (ISSC, 2006). Then, global structural analysis is needed to be carried out. This is the determination of components of nominal stress corresponding to wave load. It can be calculated by code formula or by direct numerical calculation. Details can be found in the ship society rules (ABS, BV, CCS, DNV, LR, NK, *etc.*). For illustrating the fatigue loading calculation of a marine structure and fatigue loading, which were described in DNV note 30.7, fatigue assessment of ship structures is briefly summarized as follows. Details can also be found in corresponding ship society rules.

8.3.1 Simplified Fatigue Loading Analysis

This section outlines a simplified approach to determine the distributions of long-term stress ranges for closed or semi-closed hull cross sections, expressed as Weibull distributions. Simple formats for a combination of global and local stress components are given, and alternative models for simplified calculations of stress response in ship structures are given in ship society rules, such as DNV note 30.7 fatigue assessment of ship structures.

Stress response may be calculated by different levels of accuracy. Calculation of hull girder stresses is the simplest way of getting reasonable approximations to the stress level in longitudinal hull girder elements and connections and can be used for quick evaluation of stress levels in detail.

Dynamic pressure load analysis combined with manual calculation of stiffener bending response is the simplest way of determining the stress response of longitudinal and transverse frames due to external and internal pressure loads. The member end restraints/moments must, however, be evaluated with great care in order to arrive at reliable results. A frame analysis is generally more reliable and should be performed if there is uncertainty about used restraints/moments. Frame analysis should be applied in order to assess stresses in hull elements like transverse frames, floors and girders. In order to obtain a more precise stress estimation of the response in the hull, finite element analysis should be applied. Both 2D and 3D models can be used. The long term distribution of stress ranges at local points may be described by Weibull distribution. The fatigue loading spectrum can be obtained by generating serial data that obey the Weibull distribution or are serial data which consist of all of the data which conform to every short term distribution likely to be encountered in the design life.

8.3.1.1 Nominal stress range

The combined global and local stress range may be taken as

$$\Delta\sigma_0 = f_m f_e \max\begin{cases}\Delta\sigma_g + b\Delta\sigma_l \\ a\Delta\sigma_g + \Delta\sigma_l\end{cases} \tag{8.1}$$

where $\Delta\sigma_0$ is the reference stress range value at the local point exceeded once in n_o cycles; f_m is the reduction factor derived from the combined stress range taking into account the long-term sailing routes of the ship considering the average wave conditions the vessel will be exposed to during its lifetime; f_e is the reduction factor derived from the combined stress range, taking into account the effect of mean stresses; a, b, are load combination factors, taking into account the correlation between the wave induced local and global stress ranges.

$\Delta\sigma_g$ is combined global stress range may in general be taken as

$$\Delta\sigma_g = \sqrt{\Delta\sigma_V{}^2 + \Delta\sigma_{hg}{}^2 + 2\rho_{vh}\Delta\sigma_V\Delta\sigma_{hg}} \tag{8.2}$$

where $\Delta\sigma_V$ stress range is due to wave induced vertical hull girder bending; $\Delta\sigma_{hg}$ stress range is due to wave induced horizontal hull girder bending; ρ_{vh} is the average correlation between vertical and horizontal wave induced bending stress.

$\Delta\sigma_l$ combined local stress range due to external and internal pressure loads may be taken as

$$\Delta\sigma_l = \sqrt{\sigma_e{}^2 + \sigma_i^2 + 2\rho_p\sigma_e\sigma_i} \tag{8.3}$$

where σ_e is total local stress amplitude due to the dynamic sea pressure loads; σ_i is total local stress amplitude due to internal pressure loads; ρ_p is average correlation between sea pressure loads and internal pressure loads.

8.3.1.2 Long term distribution of stress range

The long term distribution of stress ranges at local points may be described by Weibull distribution,

$$Q(\Delta\sigma) = \exp\left[-\left(\frac{\Delta\sigma}{q}\right)^h\right] \tag{8.4}$$

where Q is the probability of exceeding the stress range $\Delta\sigma$; h is Weibull shape parameter; q is Weibull scale parameter, defined as

$$q = \frac{\Delta\sigma_0}{(\ln n_0)^h} \tag{8.5}$$

The stress range distribution may also be expressed as

$$\Delta\sigma = \Delta\sigma_0\left(\frac{\ln n}{\ln n_0}\right)^{1/h} \tag{8.6}$$

where n_0 is total number of cycles associated with the stress range level $\Delta\sigma_0$.

8.3.2 Fatigue Loading by Direct Calculation

8.3.2.1 Stress range calculation

This section gives a brief description of the necessary steps in a direct load analysis for fatigue assessment. The loads computed by direct computation may substitute the simplified load components defined in Section 8.4 and is mainly intended for use in combination with finite element analysis.

A linear modeling of the ship's response is in general sufficient for fatigue assessment purposes. The response is then described by a superposition of the response of all regular wave components that make up an irregular sea, leading to a frequency domain analysis. The resulting stress may be established as a summation of all contributing dynamic loads/load effects.

The linear frequency domain results should normally be applied without any corrections for large wave effects as most of the fatigue damage is related to moderate wave heights.

Transfer function values must be determined for a sufficient number of frequencies (>15) and headings (≥8, or ≥5 when considering symmetry). The length of the model should at least extend over Lpp. The mass model should reflect the steel weight distribution and the distribution of cargo both in vertical, longitudinal and transverse directions. The fatigue damage should be calculated so that the effects of all the main load conditions are accounted for. For tankers, the ballast condition and fully loaded condition are normally sufficient. A vessel speed set to 2/3 of the service speed in full load and ballast condition should be applied in the modeling. For load effects which are unsymmetrical with respect to the ship's centerline (*e.g.* torsion induced stresses of container ships), the fatigue damage must be determined for both halves of the ship. The fatigue damage of any members due to unsymmetrical load effect may then be determined as the mean value of fatigue damage calculated for port and starboard side members.

In the evaluation of the ship's response due to external wave induced loading, the effect of wave diffraction and radiation should be accounted for.

8.3.2.2 The long term distribution of the stress range

The long term distribution of load responses for fatigue analyses may be estimated using the wave climate, represented by the distribution of Hs and Tz, as described in Table 8.1, representing the North Atlantic, or Table 8.2 for worldwide operation. As guidance to the choice between these data sets one should consider the average wave environment that the vessel is expected to encounter during its design life. The world wide sailing routes will therefore normally apply. For shuttle tankers and vessels that will sail frequently on the North Atlantic, or in other harsh environments, the wave data given in Table 8.1 should be applied, if not otherwise specified. The scatter diagrams are equal for all wave directions and specified at class midpoint values (DNV No. 30.7).

Table 8.1 Scatter diagram for the North Atlantic

TZ (s)	Hs (m)														
	3.5	4.5	5.5	6.5	7.5	8.5	9.5	10.5	11.5	12.5	13.5	14.5	15.5	16.5	17.5	Sum
1.0	0	72	1,416	4,594	4,937	2,590	839	195	36	5	1	0	0	0	0	14,685
2.0	0	5	356	3,299	8,001	8,022	4,393	1,571	414	87	16	3	0	0	0	26,167
3.0	0	0	62	1,084	4,428	6,920	5,567	2,791	993	274	63	12	2	0	0	22,196
4.0	0	0	12	318	1,898	4,126	4,440	2,889	1,301	445	124	30	6	1	0	15,590
5.0	0	0	2	89	721	2,039	2,772	2,225	1,212	494	162	45	11	2	1	9,775
6.0	0	0	1	25	254	896	1,482	1,418	907	428	160	50	14	3	1	5,639
7.0	0	0	0	7	85	363	709	791	580	311	131	46	14	4	1	3,042
8.0	0	0	0	2	27	138	312	398	330	197	92	35	12	3	1	1,547
9.0	0	0	0	1	8	50	128	184	171	113	58	24	9	3	1	750
10.0	0	0	0	0	3	17	50	80	82	59	33	15	6	2	1	348
11.0	0	0	0	0	1	6	18	33	37	29	17	8	3	1	0	153
12.0	0	0	0	0	0	2	7	13	15	13	8	4	2	1	0	65
13.0	0	0	0	0	0	1	2	5	6	6	4	2	1	0	0	27
14.0	0	0	0	0	0	0	1	2	2	2	2	1	1	0	0	11
15.0	0	0	0	0	0	0	0	1	1	1	1	0	0	0	0	4
16.0	0	0	0	0	0	0	0	0	0	1	0	0	0	0	0	1
Sum	0	77	1,849	9,419	20,363	25,170	20,720	12,596	6,087	2,465	872	275	81	20	6	100,000

Table 8.2 Scatter diagram for world wide trade

TZ (s)	Hs (m)															
	3.5	4.5	5.5	6.5	7.5	8.5	9.5	10.5	11.5	12.5	13.5	14.5	15.5	16.5	17.5	Sum
1.0	311	2,734	6,402	7,132	5,071	2,711	1,202	470	169	57	19	6	2	1	0	26,287
2.0	20	764	4,453	8,841	9,045	6,020	3,000	1,225	435	140	42	12	3	1	0	34,001
3.0	0	57	902	3,474	5,549	4,973	3,004	1,377	518	169	50	14	4	1	0	20,092
4.0	0	4	150	1,007	2,401	2,881	2,156	1,154	485	171	53	15	4	1	0	10,482
5.0	0	0	25	258	859	1,338	1,230	776	372	146	49	15	4	1	0	5,073
6.0	0	0	4	63	277	540	597	440	240	105	39	13	4	1	0	2,323
7.0	0	0	1	15	84	198	258	219	136	66	27	10	3	1	0	1,018
8.0	0	0	0	4	25	69	103	99	69	37	17	6	2	1	0	432
9.0	0	0	0	1	7	23	39	42	32	19	9	4	1	1	0	178
10.0	0	0	0	0	2	7	14	16	14	9	5	2	1	0	0	70
11.0	0	0	0	0	1	2	5	6	6	4	2	1	1	0	0	28
12.0	0	0	0	0	0	1	2	2	2	2	1	1	0	0	0	11
13.0	0	0	0	0	0	0	1	1	1	1	0	0	0	0	0	4
14.0	0	0	0	0	0	0	0	0	1	0	0	0	0	0	0	1
Sum	331	3,559	11,937	20,795	23,321	18,763	11,611	5,827	2,480	926	313	99	29	9	0	100,000

8.4 SIF Calculation

When the applied stress on the fatigue hot spot, the geometry and the crack size are known, the SIF of the crack at the fatigue hot spot in a marine structure can be determined or calculated. The experimental data and engineering experience show that the effect of the initial crack size on fatigue crack growth life of a component is very significant, especially when the initial crack is small. It is therefore an important parameter to be considered in the fatigue life calculation procedure. The initial crack size to be used in the calculation should be considered in each case, taking account of imperfections or defect sizes for various weldments, geometries, access and reliability of the inspection method. The initial crack size determination by nondestructive detection procedure in BS7910 is outlined here. Details can be found in BS7910.

8.4.1 Planar Flaws and Their Initial Size

Flaws of the type known as "imperfect shape" include some which may be treated as planar flaws. These are caused by 1) cracks; 2) lack of fusion or penetration; 3) undercut, root undercut, concavity and overlap (on some occasions, undercut and root undercut in welds are treated as shape imperfections).

The following should be assessed as planar flaws:

(1) Surface breaking non-planar flaws;

(2) Weld toe undercut deeper than 1 mm;

(3) Embedded flaws that cannot be positively identified as non-planar;

(4) Any flaws, if it is necessary to assess fatigue crack growth from them (*e.g.* to estimate inspection intervals).

8.4.1.1 Initial crack size determined by nondestructive detection

Planar flaws should be characterized by the height and length of their containment rectangles. These dimensions are as follows: $2a$ for through-thickness flaws; a and $2c$ for surface flaws; and $2a$ and $2c$ for embedded flaws, shown in Fig. 8.2.

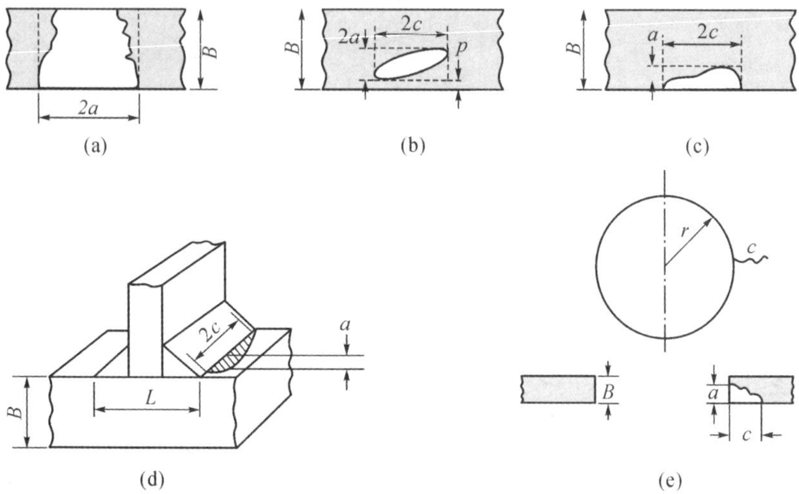

Fig. 8.2 Flaw dimensions. (a) Through-thickness flaw (Required dimensions: $2a$, B); (b) Embedded flaw (Required dimensions: $2a$, p, $2c$, B); (c) Embedded flaw (Required dimensions:a, $2c$, B); (d) Flaw at toe of fillet weld (Required dimensions: $2c$, a, B); (e) Flaw at hole (Required dimensions: a, c, B) (BSI, 2005)

For two or more coplanar flaws, the interactions should be considered. For through-thickness co-linear flaws shown in Fig. 8.3, the equivalent crack length is $a+b$ when a/b is greater than 0.785. Otherwise, the SIF of the interacted flaws can be calculated by Eq. (8.7).

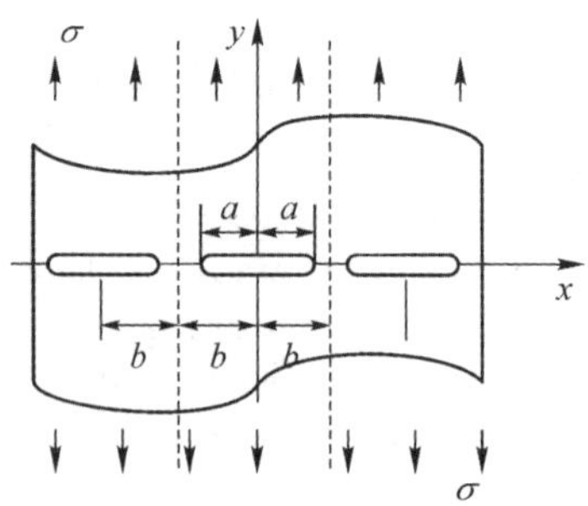

Fig. 8.3 Multi co-planar or co-linear flaws

$$K_I = \sigma\sqrt{\pi a}\sqrt{\frac{2b}{\pi a}\tan\frac{\pi a}{2b}} \tag{8.7}$$

For surface or embedded coplanar flaws shown in Fig. 8.4, the equivalent crack dimensions are also shown in Fig. 8.4.

Schematic flaws	Criteria for interaction	Effective dimensions after interaction
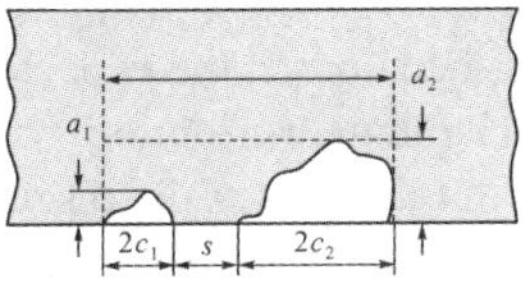 i) Coplanar surface flaws	$s \le 2c_1$ for a_1/c_1 or $a_2/c_2 > 1$ $s = 0$ for a_1/c_1 or $a_2/c_2 < 1$ $(c_1 < c_2)$	$a = \max\{a_1, a_2\}$ $2c = 2c_1 + 2c_2 + s$
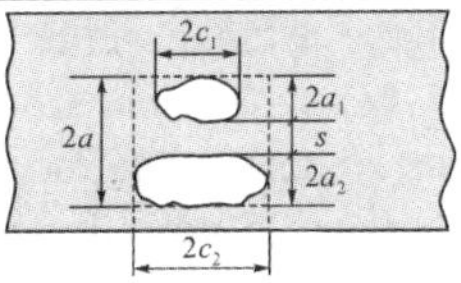 ii) Coplanar embedded flaws (interaction in thickness direction)	$s \le a_1 + a_2$	$2a = 2a_1 + 2a_2 + s$ $2c = \max\{2c_1, 2c_2\}$
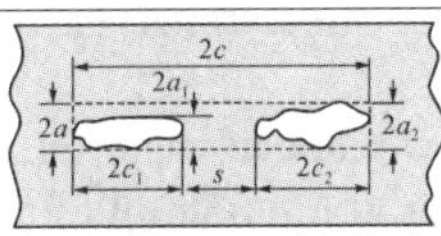 iii) Coplanar embedded flaws (interaction in width direction)	$s \le 2c_1$ for a_1/c_1 or $a_2/c_2 > 1$ $s = 0$ for a_1/c_1 or $a_2/c_2 < 1$ $(c_1 < c_2)$	$2a = 2a_2$ $2c = 2c_1 + 2c_2 + s$
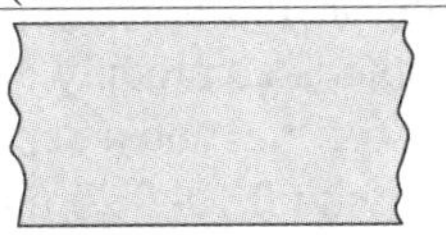 iv) Coplanar surface and embedded flaws (interaction in thickness direction)	$s \le a_1 + a_2$	$a = 2a_1 + a_2 + s$ $2c = \max\{2c_1, 2c_2\}$
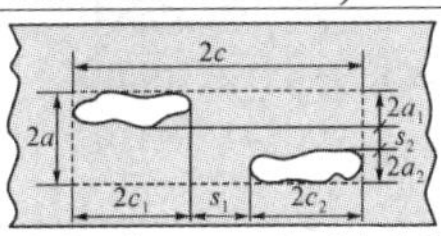 v) Coplanar embedded flaws (interaction in both thickness and width direction)	$s_1 \le 2c_1$ for a_1/c_1 or $a_2/c_2 > 1$ $s_1 = 0$ for a_1/c_1 or $a_2/c_2 < 1$ and $s_2 \le a_1 + a_2$ $(c_1 < c_2)$	$2a = 2a_1 + 2a_2 + s_2$ $2c = 2c_1 + 2c_2 + s_1$
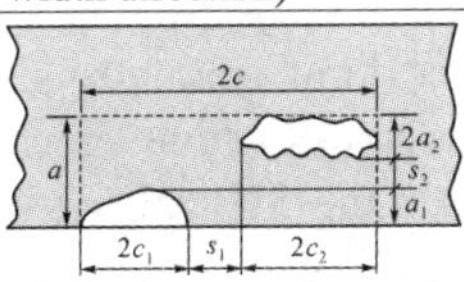 vi) Coplanar surface and embedded flaws (interaction in both thickness and width direction)	$s_1 \le 2c_1$ for a_1/c_1 or $a_2/c_2 > 1$ $s_1 = 0$ for a_1/c_1 or $a_2/c_2 < 1$ and $s_2 \le a_1 + a_2$ $(c_1 < c_2)$	$a = a_1 + 2a_2 + s_2$ $2c = 2c_1 + 2c_2 + s_1$

Fig. 8.4 Coplanar flaw interactions (BSI, 2005)

8.4.1.2 Recommended initial crack size when lack of NDT data

The initial crack size of a surface crack in welded joints induced by the welding process itself is very important herein. Reliability-based crack growth analysis is very sensitive to the mean value of initial crack size (Moan, 2005). According to Bokalrud and Karlsen (1981), the initial crack size obtained from actual welded details in ship hull structures was found to follow an exponential distribution with a mean value of 0.11 mm, and a rate of 16 cracks per m. Then, assuming two independent initial cracks in a specimen as the type described above, the expected largest value would be 0.17 mm, characterized by a Gamma distribution with standard deviation equal to 0.12 mm. Another study of initial crack sizes on real structures was reported in Moan, *et al.* (2000), where an extensive database of cracks detected in tubular joints of a jacket located in the North Sea has been analyzed. The initial crack size derived was estimated to have a mean value of 0.19 mm exponentially distributed for an individual crack, whereas a mean value of 0.38 mm per hot spot was found. Carrying out fatigue tests on small specimens, Otegui, *et al.* (1991) reported an average of 8.8 initiation sites per 25 mm of weld toe. Their findings also indicate that for specimens under relatively high stress ranges up to 13 initiation sites were reported from undercut defects, whereas only 1 site was found in a specimen subjected to 2.5 times lower stress range (Efren and Moan, 2007).

For surface cracks starting from transitions between weld/base material, a crack depth of 0.5 mm (*e.g.* due to undercuts and microcracks at the bottom of the undercuts) may be assumed if other documented information about crack depth is not available. For the semi-elliptical surface cracks initiating at the notch root, taking into account the precision, non-destructive testing methods commonly are used, surface crack initiation is defined as a crack of length of 0.5 mm and a depth of 0.15 mm is appropriate (Beevers, 1982). Initial crack sizes with 0.10–0.25 mm are recommended in ABS (2005), and 0.5 mm is in DNV (2005).

For conservative estimation, a surface crack with initial crack size of 0.5 mm depth and 1.5 mm length or an edge crack with initial crack length of 0.5 mm is recommended for welded marine structures when there is a lack of NDT data.

8.4.2 SIF Range Calculation

The SIF can be calculated when the crack size, the cracked joint geometry size and the applied load are known. For through-thickness cracks, edge cracks, elliptical surface cracks and surface cracks at welded joints, the SIF calculation equations are presented in Appendix A. For elliptical flaws, it can normally be assumed that the same relationship applies for crack growth in both a and c directions. The relevant stress range should be determined. For spectrum loading, unless the order of application of cycles is known, cycles, or blocks of cycles, should be assumed to occur in the most damaging order. This is not always clear. It may be necessary to perform the calculations for various orders to determine the worst case.

8.5 Fatigue Crack Growth Law of UFLP for Marine Structural Materials

The fatigue crack growth rate model is the most important part of fatigue crack growth life prediction. The overall relationship between $\frac{da}{dN}$ and ΔK_{eff} is normally a sigmoidal curve in a $\log\left(\frac{da}{dN}\right)$ versus $\log(\Delta K_{eff})$ plot and can be described well by UFLP. The material constants of the fatigue crack growth rate in UFLP are different from those in Paris' law.

The crack growth law of UFLP is based on the effective SIF range as Eq. (8.8). More detail can be found in Chapter 5.

$$\frac{da}{dN} = \frac{A\left[K_{max} \cdot (1-\Phi f_{op}) - f\Delta K_{th\,0}\right]^m}{1-(K_{max}/K_C)^n} \tag{8.8}$$

The material constants of UFLP, which can consider the effect of R, are not the same as that of Paris' law. Some basic material constants of UFLP are A, m, n, $\Delta K_{th\,0}$, K_c. These constants can be determined by fitting corresponding fatigue test data, or be taken from a database for materials. It is very important to set up a database which consists of these material mechanical property parameters for commonly used commercial materials. It is a time and financial cost project. Fortunately, there is a lot of fatigue test data for some materials available. The material constants of some metals which are used in the demonstration examples in Chapter 7 are listed in Table 8.3.

Table 8.3 Material constants for the UFLP of some metals

Material	σ_y (MPa)	A	M	$\Delta K_{th\,0}$ ($\mathrm{MPa \cdot m^{\frac{1}{2}}}$)	K_{IC} ($\mathrm{MPa \cdot m^{\frac{1}{2}}}$)	n	r_e (m)
Al7075-T6	520	7.2E-10	2	2.2	28	6	1E-6
Al D16	355	7.38E-10	2.068	3.6	51	6	1E-6
Steel 350WT	350	7.38E-10	2.068	13.24	106	6	1E-6
Steel HTS-A	800	2.18E-10	2.267	4.2	140	6	1E-6

In fracture mechanics assessment it is usual to adopt conservative estimates of the various parameters required. Ferritic steels, which are commonly used as construction materials in marine structures, tested in a seawater environment showed R dependence in a similar manner to that found in air.

Freely corroding steels tested in seawater have higher crack growth rates than those tested in air, except at very high values of $\Delta K(100\ \mathrm{MPa \cdot m^{\frac{1}{2}}})$ where environmental effects are minimized and the rate of crack growth is dominated by mechanical factors. The fatigue crack growth rates for steels in seawater with cathodic protection showed a large increase at intermediate values of ΔK compared to an air environment. This is believed to be due to the effect of hydrogen within and immediately ahead of the crack tip plastic zone.

8.5.1 Recommended Fatigue Crack Growth Material Constants for Steels in Marine Environment

In the "Guide to methods for assessing the acceptability of flaws in metallic structures" (BS7910, 2005) , the values of the constants C and m in the Paris' equation given in Table 8.2 (in air), Table 8.3 (in a marine environment, freely corroding), Table 8.4 (in a marine environment with cathodic protection at −850 mV) and Table 8.5 (in a marine environment with cathodic protection at –1100 mV) are recommended for assessing low strength steels. They are applicable to steels (excluding austenitic and duplex stainless steels) with a yield or 0.2% proof strength ⩽600 MPa, when operating in a marine environment at temperatures up to 20 °C.

The two-stage fatigue crack growth rate, both the mean and mean plus two standard deviations (mean+2SD) of $\log\left(\frac{da}{dN}\right)$ versus $\log(\Delta K)$ relationships for $R<0.5$ and $R\geqslant 0.5$ are given. However, for conservatism and to allow for the influence of residual stresses, the mean+2SD laws for $R\geqslant 0.5$ should normally be used to assess welded components. For determining the material constants of UFLP from these available Paris' material constants, the laws for $R\geqslant 0.5$ will be used and compared. They are based on the understanding of the difference between the background of UFLP and Paris' law. The threshold $\Delta K_{\mathrm{th}\,0}$ and fracture toughness K_C are set to $2.0\ \mathrm{MPa \cdot m^{\frac{1}{2}}}$ and $100\ \mathrm{MPa \cdot m^{\frac{1}{2}}}$ respectively in the following comparisons.

8.5.1.1 Fatigue crack growth constants if ferrous metals in air

For low strength steels (ferritic, austenitic or duplex ferritic-austenitic) with yield or 0.2% proof strengths ⩽700 MPa when operating in air or other non-aggressive environments at temperatures up to 100 °C; the recommended values of the constants A and m in the Paris' equation for $R \geqslant 0.5$ regime taken from BS7910 are listed in Table 8.4.

Unless justification is provided, the upper bound (mean+2SD) values for $R \geqslant 0.5$ should be used for all assessments of flaws in welded joints.

The material constants of UFLP can be determined from the above recommended crack growth rate information and listed in Table 8.8. The comparison of the crack growth rate curves by UFLP and by Paris' law are shown in Fig. 8.5.

Table 8.4 Recommended fatigue crack growth laws for steels in air (BS7910)

	Stage 1		Stage 2		Transition point ΔK ($N/mm^{3/2}$)
	A	m	A	m	
Mean curve	4.80×10^{-18}	5.10	5.86×10^{-13}	2.88	196
Mean + 2SD	2.10×10^{-17}	5.10	1.29×10^{-12}	2.88	144

In the table, SD means standard deviation:

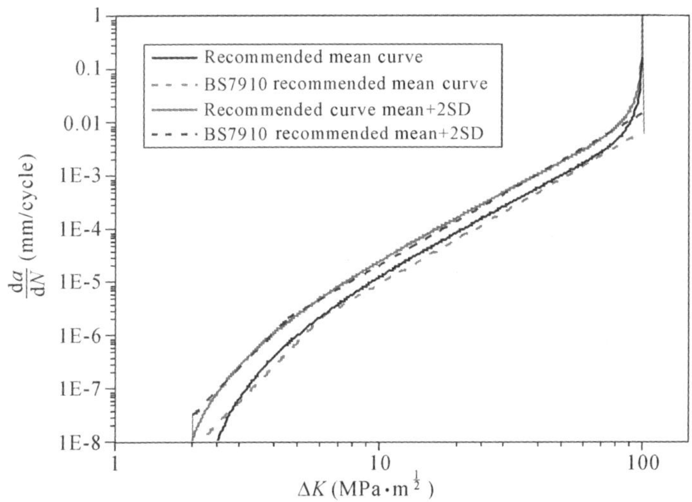

Fig. 8.5 Recommended fatigue crack growth laws for steels in air

8.5.1.2 Fatigue crack growth constants of ferrous metals in marine environment freely corroding

For low strength steels (excluding austenitic and duplex stainless steels) with yield or 0.2% proof strength$\leqslant$600 MPa when operating freely corroding in marine environment at temperatures up to 20 °C; the recommended values of the constants A and m in the Paris' equation for $R \geqslant 0.5$ regime taken from BS7910 are listed in Table 8.5. Unless justification is provided, the upper bound (mean + 2SD) values for $R \geqslant 0.5$ should be used for all assessments of flaws in welded joints.

The material constants of UFLP can be determined from the above recommended crack growth rate information and listed in Table 8.8. The comparison of the crack growth rate curves by UFLP and by Paris' law are shown in Fig. 8.6.

Table 8.5 Recommended fatigue crack growth laws for steels in a marine environment (Steel freely corroding in marine environment) (BS7910)

	Stage 1		Stage 2		Transition point ΔK
	A	m	A	m	($N/mm^{3/2}$)
Mean curve	5.37×10^{-14}	3.42	5.67×10^{-7}	1.11	1098
Mean + 2SD	1.72×10^{-13}	3.42	7.48×10^{-7}	1.11	748

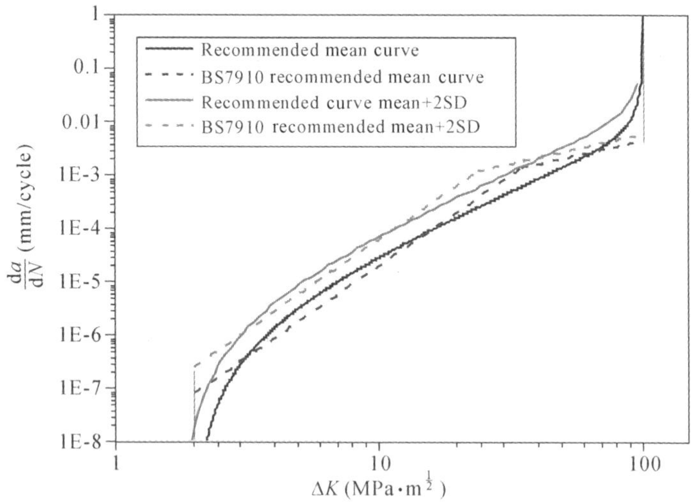

Fig. 8.6 Recommended fatigue crack growth laws for steels in a marine environment

8.5.1.3 Fatigue crack growth constants of ferrous metals in marine environment with cathodic protection at –850 mV

For low strength steels (excluding austenitic and duplex stainless steels) with yield or 0.2% proof strength $\leqslant$600 MPa when operating in marine environment with cathodic protection at –850 mV at temperatures up to 20 °C.The recommended values of the constants A and m in the Paris' equation for $R \geqslant 0.5$ regime taken from BS7910 are listed in Table 8.6.

Unless justification is provided, the upper bound (mean + 2SD) values for $R \geqslant 0.5$ should be used for all assessments of flaws in welded joints.

The material constants of UFLP can be determined from above recommended crack growth rate information and listed in Table 8.8. The comparison of the crack growth rate curves by UFLP and by Paris' law are shown in Fig. 8.7.

Table 8.6 Recommended fatigue crack growth laws for steels in a marine environment (Steel in marine environment with cathodic protection at –850 mV (Ag/AgCl)) (BS7910)

	Stage 1		Stage 2		Transition point ΔK ($N/mm^{3/2}$)
	A	m	A	m	
Mean curve	4.80×10^{-18}	5.10	6.0×10^{-12}	2.67	323
Mean + 2SD	2.10×10^{-17}	5.10	2.02×10^{-11}	2.67	290

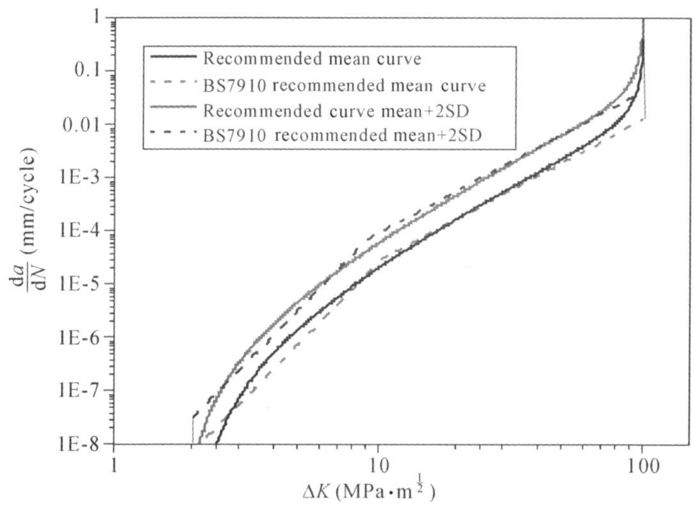

Fig. 8.7 Recommended fatigue crack growth laws for steels in a marine environment with cathodic protection at –850 mV (Ag/AgCl)

8.5.1.4 Fatigue crack growth constants of ferrous metals in marine environment with cathodic protection at –1100 mV

For low strength steels (excluding austenitic and duplex stainless steels) with yield or 0.2% proof strength ⩽600 MPa when operating in marine environment with cathodic protection at –1100 mV at temperatures up to 20 °C.

The recommended values of the constants A and m in the Paris equation for $R \geqslant 0.5$ regime taken from BS7910 are listed in Table 8.7.

Unless justification is provided, the upper bound (mean + 2SD) values for $R \geqslant 0.5$ should be used for all assessments of flaws in welded joints.

The material constants of UFLP can be determined from above recommended crack growth rate information and listed in Table 8.8. The comparison of the crack growth rate curves by UFLP and by Paris' law is shown in Fig. 8.8.

Table 8.7 Recommended fatigue crack growth laws for steels in a marine environment (Steel in marine environment with cathodic protection at –1100 mV (Ag/AgCl)) (BS7910)

	Stage 1		Stage 2		Transition point ΔK ($N/mm^{3/2}$)
	A	m	A	m	
Mean curve	4.80×10^{-18}	5.10	5.25×10^{-8}	1.40	517
Mean + 2SD	2.10×10^{-17}	5.10	1.02×10^{-7}	1.40	415

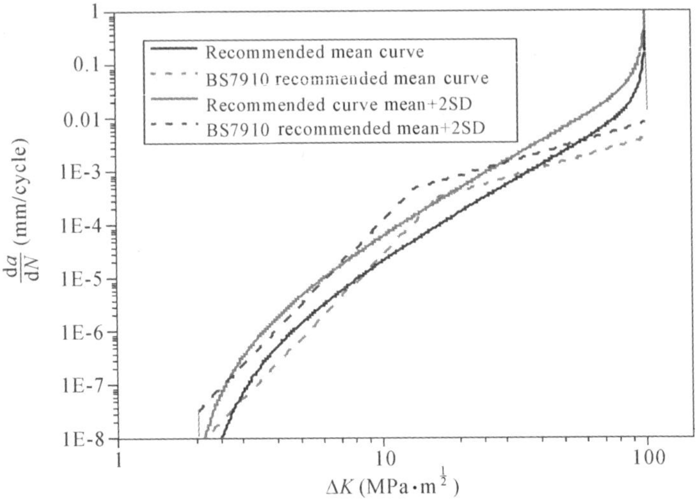

Fig. 8.8 Recommended fatigue crack growth laws for steels in a marine environment with cathodic protection at –1100 mV (Ag/AgCl)

Table 8.8 Recommended material constants of UFLP for steels in air, marine environment (free corrosion, with cathodic protection at –850 mV and –1100 mV (Ag/AgCl)) ($MPa\cdot m^{\frac{1}{2}}$)

	A	m	N	$\Delta K_{th\,0}$	K_C	Environment
Mean curve	6.8×10^{-8}	2.5	6	2.0	100	Air
Mean+2SD	1.2×10^{-7}	2.5	6	1.8	100	Air
Mean curve	2.8×10^{-7}	2.2	6	2.0	100	Marine, corrosion free
Mean+2SD	6.5×10^{-7}	2.2	6	1.8	100	Marine, corrosion free
Mean curve	8×10^{-8}	2.67	6	2.0	100	Marine, –850 mV
Mean+2SD	2.2×10^{-7}	2.67	6	1.8	100	Marine, –850 mV
Mean curve	8.5×10^{-8}	2.67	6	2.0	100	Marine, –1100 mV
Mean+2SD	2.5×10^{-7}	2.67	6	1.8	100	Marine, –1100 mV
Simplified	6.8×10^{-8}	2.67	6	1.8	100	Air
Simplified	2.5×10^{-7}	2.67	6	1.8	100	Marine environment

8.5.2 Simplified Fatigue Crack Growth Law and Threshold (BS7910)

8.5.2.1 Simplified fatigue crack growth law and threshold for steels

For preliminary screening assessments, simple and conservative laws are recommended for steels with yield or 0.2% proof strengths up to 600 MPa, for steels, including austenitics, operating in air or other non-aggressive environments at temperatures up to 100 °C.

The recommended values of the constants m and A in Paris' equation are as follows,

$$m = 3;\ A = 1.648\times10^{-8}(5.21\times10^{-13}) \tag{8.9}$$

$\frac{\mathrm{d}a}{\mathrm{d}N}$ in mm/cycle and ΔK in $\mathrm{MPa}\cdot\mathrm{m}^{\frac{1}{2}}$ (or $\mathrm{N/mm}^{3/2}$ for $A = 5.21\times10^{-13}$).

For elevated temperatures, up to 600 °C, the following values are recommended,

$$m = 3;\ A = 1.648\times10^{-8}(E_{\mathrm{RT}} / E_{\mathrm{ET}})^3 \tag{8.10}$$

where E_{RT} is the elastic modulus at room temperature (say 20 °C) and E_{ET} is the elastic modulus at the elevated temperature.

For steels (excluding austenitic stainless steels) operating in marine environments at temperatures up to 20 °C, with or without cathodic protection, the following values are recommended,

$$m = 3;\ A = 7.273\times10^{-8}(2.3\times10^{-12}) \tag{8.11}$$

Simplified fatigue crack growth constants for UFLP can be determined and the comparison is shown in Fig. 8.9.

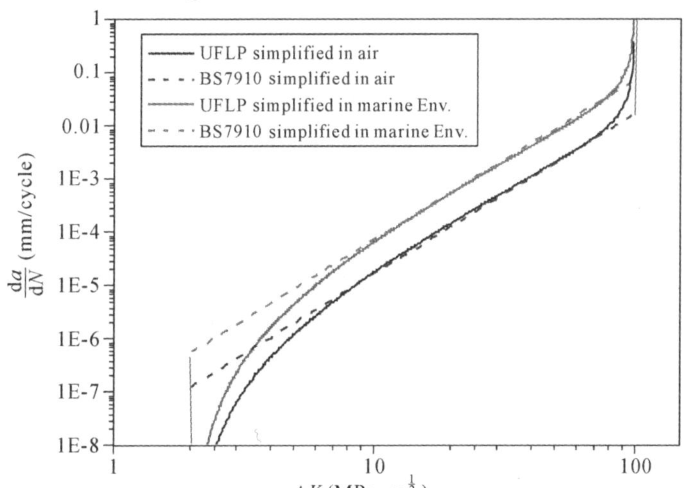

Fig. 8.9 Simplified fatigue crack growth laws for steels in air and in marine environment

The threshold SIF, ΔK_{th} is strongly dependent on environment and load ratio, R. ΔK_{th} is found to increase with a decrease in R. Recommended values for some conditions are given in Table 8.9. It is recommended that the lower bound value obtained at high R values in the relevant environment is adopted for all assessments of flaws in welded joints.

The values of $\Delta K_{th\,0}$ in Table 8.9 for austenitic steels and unprotected steels operating in a marine environment are also applicable for assessing unwelded components. However, for unwelded steel components, account may be taken of R. Based on published data for steels (excluding austenitic) in air and with cathodic protection in marine environments at temperatures up to 20 °C, the following values of $\Delta K_{th\,0}$ (in $N/mm^{3/2}$) are recommended: However, the value used should not exceed 2 $MPa \cdot m^{\frac{1}{2}}$ (63 $N/mm^{3/2}$) for assessments of surface breaking flaws less than 1 mm deep.

Table 8.9 Recommended fatigue crack growth threshold $\Delta K_{th\,0}$ values for assessing welded joint

Material	Environment	$\Delta K_{th\,0}$ ($MPa \cdot m^{\frac{1}{2}}$)
Steels, including austenitic	Air or other non-aggressive environments up to 100 °C	2 (63)
Steels, excluding austenitic	Marine with cathodic protection, up to 20 °C	2 (63)
Steels, including austenitic	Marine, unprotected	0 (0)
Aluminum alloys	Air or other non-aggressive environments up to 20 °C	0.7 (21)

8.5.2.2 Fatigue crack growth law and crack growth threshold in non-ferrous metals

For aluminum alloys, multi-branch crack growth relationships for a range of alloys are given, while, for approximate assessments, the recommended crack growth constants of m=3 and A=1.648×10^{-8}(5.21×10^{-13}) for steels could be used for another material with elastic modulus E by using m=3 and the following value for A given by

$$A = 1.648\times10^{-8}\left(\frac{E_{steel}}{E}\right)^{3} \tag{8.12}$$

Similarly, the threshold SIF can be obtained from Eq. (8.3).

$$\Delta K_{th0} = \Delta K_{th\,0,\,steel}\left(\frac{E}{E_{steel}}\right) \tag{8.13}$$

8.6 Summary

The UFLP procedure for a marine structure is outlined. The fatigue loading of marine structures are mainly wave induced, which has the random characteristic of wave loading. The FLP of a marine structure component requires information of the fatigue loading history or spectrum over the whole design life. The fatigue loading, the initial crack size definition and the SIF range calculation are briefly described. The material constants of UFLP for marine structural materials are also presented. Due to the difficulty of fatigue loading, more research should be carried out, especially related to SLHs discussed in Chapter 6.

Appendix A: SIF Solutions for Some Typical Cracks

A.1 Through-Thickness Flaws in Plates

$$K = \sqrt{\pi r_e \left(\sec \frac{\pi}{2} \frac{\sigma}{\sigma_v} + 1 \right)} \left(1 + Y(a) \sqrt{\frac{a}{2r_e}} \right) \sigma \tag{A.1}$$

A.1.1 Center Through-Thickness Flaw (Fig. A.1a)

$$Y(a) = \sqrt{\sec(\pi a/w)} \tag{A.2}$$

A.1.2 Edge Flaw (Fig. A.1b)

$$Y(a) = 1.12\sqrt{\sec(\pi a/w)} \tag{A.3}$$

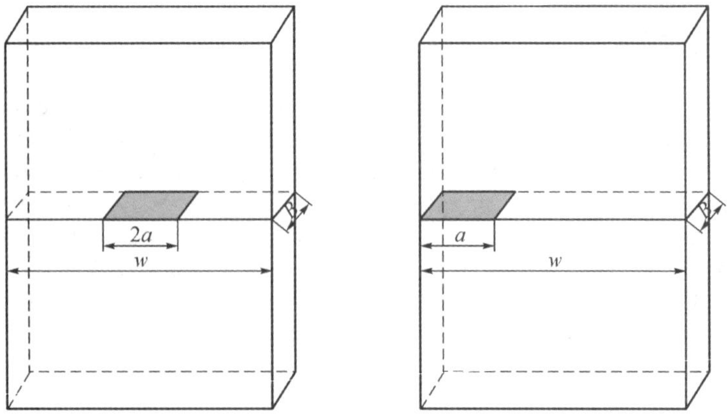

A.1a Center through-thickness flaw   A.1b Edge flaw

Fig. A.1 Through-thickness flaw geometry

A.2 Surface Flaws in Plates

When crack depth a is very small (*i.e.* less than 0.5 mm) , the SIF at the deepest point can be calculated by edge crack Eqs.(A.1), (A.3). The SIF at the surface point can be calculated by the center through-thickness crack Eqs.(A.1), (A.2). The stress should include stress concentration, otherwise calculated by the following equations.

$$K=(\sigma+H\sigma_{\mathrm{b}})\frac{\sqrt{\pi a}}{\Phi}F\left(\frac{a}{c},\frac{a}{B},\frac{2c}{w},\varphi\right) \tag{A.4}$$

where

$$\Phi=\{1+1.464(a/c)^{1.65}\}^{0.5} \quad \text{for} \quad 0\le a/2c\le 0.5$$

$$\Phi=\{1+1.464(c/a)^{1.65}\}^{0.5} \quad \text{for} \quad 0.5\le a/2c\le 1.0 \tag{A.5}$$

$$F\left(\frac{a}{c},\frac{a}{B},\frac{2c}{W},\varphi\right)=\left[M_1+M_2\left(\frac{a}{B}\right)^2+M_3\left(\frac{a}{B}\right)^4\right]f_\varphi\cdot g\cdot f_w \tag{A.6}$$

where

$M_1=1.13-0.09(a/c)$ for $0\le a/2c\le 0.5$

$M_1=(c/a)^{0.5}\{1+0.04(c/a)\}$ for $0.5\le a/2c\le 1.0$

$M_2=[0.89/\{0.2+(a/c)\}]-0.54$ for $0\le a/2c\le 0.5$

$M_2=0.2(c/a)^4$ for $0.5\le a/2c\le 1.0$

$M_3 = 0.5 - 1/\{0.65 + (a/c)\} + 14\{1-(a/c)\}^{24}$ for $0 \le a/2c \le 0.5$

$M_3 = -0.11(c/a)^4$ for $0.5 \le a/2c \le 1.0$

$g = 1 + \{0.1 + 0.35(a/B)^2\}(1-\sin\theta)^2$ for $0 \le a/2c \le 0.5$

$g = 1 + \{0.1 + 0.35(c/a)(a/B)^2\}(1-\sin\theta)^2$ for $0.5 \le a/2c \le 1.0$

$f_\theta = \{(a/c)^2\cos^2\theta + \sin^2\theta\}^{0.25}$ for $0 \le a/2c \le 0.5$

$f_\theta = \{(c/a)^2\sin^2\theta + \cos^2\theta\}^{0.25}$ for $0.5 \le a/2c \le 1.0$

$$H = H_1 + (H_2 - H_1)\sin^q\theta \tag{A.7}$$

where

$q = 0.2 + (a/c) + 0.6(a/B)$ for $0 \le a/2c \le 0.5$

$q = 0.2 + (c/a) + 0.6(a/B)$ for $0.5 \le a/2c \le 1.0$

$H_1 = 1 - 0.34(a/B) - 0.11(a/c)(a/B)$ for $0 \le a/2c \le 0.5$

$H_1 = 1 - \{0.04 + 0.41(c/a)\}(a/B) + \{0.55 - 1.93(c/a)^{0.75} + 1.38(c/a)^{1.5}\}(a/B)^2$ for $0.5 \le a/2c \le 1.0$

$H_2 = 1 + G1(a/B) + G2(a/B)^2$

$G1 = -1.22 - 0.12(a/c)$ for $0 \le a/2c \le 0.5$

$G1 = -2.11 + 0.77(c/a)$ for $0.5 \le a/2c \le 1.0$

$G2 = 0.55 - 1.05(a/c)^{0.75} + 0.47(a/c)^{1.5}$ for $0 \le a/2c \le 0.5$

$G2 = 0.55 - 0.72(c/a)^{0.75} + 0.14(c/a)^{1.5}$ for $0.5 \le a/2c \le 1.0$

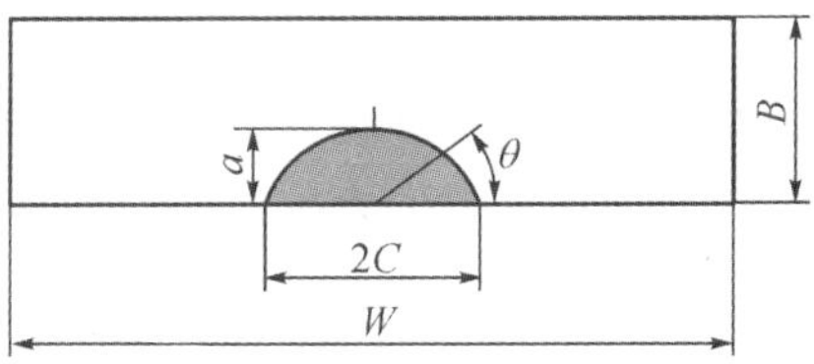

Fig. A.2 Surface flaw geometry

A.3 Surface Flaw at Typical Welded Toe

When crack depth a is very small (*i.e.* less than 0.5 mm) , the SIF at the deepest point can be calculated by edge crack Eqs.(A.1), (A.3). The SIF at the surface point can be calculated by the center through-thickness crack Eqs.(A.1), (A.2).

The stress should include stress concentration, otherwise calculated by the following equations.

$$K=(M_{\mathrm{K}}^{\mathrm{T}}\sigma+M_{\mathrm{K}}^{B}H\sigma_{\mathrm{b}})\frac{\sqrt{\pi a}}{\Phi}F\left(\frac{a}{c},\frac{a}{t},\frac{2c}{w},\varphi\right)+K(\sigma^{R}) \tag{A.8}$$

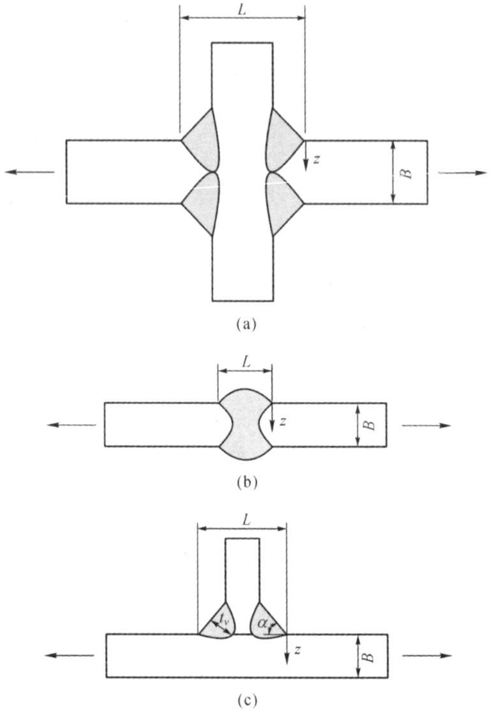

Fig. A.3 Surface flaw and welded joint geometries (Rhee, *et al.*, 1991) (with the permission of ASME)

Deepest point:

$$M_K^T=f_1\left(\frac{a}{B},\frac{a}{c}\right)+f_2\left(\frac{a}{B}\right)+f_3\left(\frac{a}{B},\frac{L}{B}\right).\ \text{If}\ \ M_K^T<1.0,\ \text{then}\ \ M_K^T=1.0 \tag{A.9}$$

$$f_1\left(\frac{a}{B},\frac{a}{c}\right)=0.43358(a/B)\left[g_1+\{g_2(a/B)\}^{g_3}\right]+0.93163\exp\left\{(a/B)^{-0.050966}\right\}+g_4$$

$$g_1=-1.0343(a/c)^2-0.156(a/c)+1.3409;$$

$$g_2=1.3218(a/c)^{-0.61153};$$

$$g_3=-0.87238(a/c)+1.2788;$$

$$g_4=-0.46190(a/c)^3-0.67090(a/c)^2-0.37571(a/c)+4.6511$$

$$f_2\left(\frac{a}{B}\right)=-0.21521\{1-(a/B)\}^{176.4199}+2.8141(a/B)^{-0.10740(a/B)};$$

$$f_3\left(\frac{a}{B},\frac{L}{B}\right)=0.33994(a/B)^{(g_5)}+1.9493(a/B)^{0.23003}+\left\{g_6(a/B)^2+g_7(a/B)+g_8\right\};$$

$$g_5=-0.015647(L/B)^3+0.090889(L/B)^2-0.17180(L/B)-0.24587;$$

$$g_6=-0.20136(L/B)^2+0.93311(L/B)-0.41496;$$

$$g_7=0.20188(L/B)^2-0.97857(L/B)+0.068225;$$

$$g_8=-0.027338(L/B)^2+0.12551(L/B)-11.218.$$

$$M_K^B=f_1\left(\frac{a}{B},\frac{a}{c}\right)+f_2\left(\frac{a}{B}\right)+f_3\left(\frac{a}{B},\frac{L}{B}\right).\text{ If } M_K^B<1.0,\text{ then } M_K^B=1.0 \tag{A.10}$$

for $0.005\le a/B\le 0.5$
where

$$f_1\left(\frac{a}{B},\frac{a}{c}\right)=0.065916(a/B)\left[g_1+\left\{g_2(a/B)\right\}^{g_3}\right]+0.52086\exp\left\{(a/B)^{-0.10364}\right\}+g_4;$$

$$g_1=-0.014992(a/c)^2-0.021401(a/c)-0.23851;$$

$$g_2=0.61775(a/c)^{-1.0278};$$

$$g_3=0.00013242(a/c)-1.4744;$$

$$g_4=-0.28783(a/c)^3+0.58706(a/c)^2-0.37198(a/c)-0.89887;$$

$$f_2\left(\frac{a}{B}\right)=-0.219950\left\{1-(a/B)\right\}^{2.8086}+0.021403(a/B)^{g_5};$$

$$g_5=-17.195(a/B)^2+12.468(a/B)-0.51662;$$

$$f_3\left(\frac{a}{B},\frac{L}{B}\right)=0.23344(a/B)^{(g_6)}-0.14827(a/B)^{-0.20077}$$
$$+\left\{g_7(a/B)^2+g_8(a/B)+g_9\right\};$$

$$g_6=-0.059798(L/B)^3+0.38091(L/B)^2-0.8022020(L/B)+0.31906;$$

$$g_7=-0.35848(L/B)^2+1.3975(L/B)-1.7535;$$

$$g_8=0.31288(L/B)^2-1.3599(L/B)+1.6611;$$

$$g_9=-0.001470(L/B)^2-0.0025074(L/B)-0.0089846.$$

Surface point :

$$M_K^T=f_1\left(\frac{a}{B},\frac{c}{a},\frac{L}{B}\right)\times f_2\left(\frac{a}{B},\frac{a}{c}\right)\times f_3\left(\frac{a}{B},\frac{a}{c},\frac{L}{B}\right) \tag{A.11}$$

where

$$f_1\left(\frac{a}{B},\frac{c}{a},\frac{L}{B}\right)=g_1(a/B)\left\{g_2\left(\frac{c}{a}\right)^2+g_3\left(\frac{c}{a}\right)+g_4\right\}$$
$$+g_5\{1-(a/B)\}\left\{g_6\left(\frac{c}{a}\right)^2+g_7\left(\frac{c}{a}\right)+g_8\right\}$$

$g_1=0.007815(c/a)^2-0.070664(c/a)+1.8508$

$g_2=-0.000054546(L/B)^2+0.00013651(L/B)-0.00047844;$

$g_3=0.000491(L/B)^2+0.001359(L/B)+0.011400;$

$g_4=0.0071654(L/B)^2-0.033399(L/B)-0.25064;$

$g_4=0.0071654(L/B)^2-0.033399(L/B)-0.25064;$

$g_5=0.018640(c/a)^2+0.24311(c/a)-1.7644$

$g_6=-0.0016713(L/B)^2+0.0090620(L/B)-0.016479;$

$g_7=-0.0031615(L/B)^2-0.010944(L/B)+0.13967;$

$g_8=-0.045206(L/B)^3+0.32380(L/B)^2-0.68935(L/B)+1.4954;$

$$f_2\left(\frac{a}{B},\frac{a}{c}\right)=\left\{-0.28639(a/c)^2+0.35411(a/c)+1.6430\right\}(a/B)^{g_9}$$
$$+0.27449\{1-(a/B)\}^{g_{10}};$$

$g_9=-0.25473(a/c)^2+0.40928(a/c)+0.0021892;$

$g_{10}=37.423(a/c)^2-15.741(a/c)+64.903;$

$$f_3\left(\frac{a}{B},\frac{a}{c},\frac{L}{B}\right)=g_{11}(a/B)^{0.75429}+g_{12}\exp\left\{(a/B)^{g_{13}}\right\};$$

$g_{11}=-0.10553(L/B)^3+0.59894(L/B)^2-1.0942(L/B)-1.2650;$

$g_{12}=0.043891(L/B)^3-0.24898(L/B)^2+0.44732(L/B)+0.60136;$

$g_{13}=-0.011411(a/c)^2+0.004369(a/c)+0.51732;$

$$M_K^B=f_1\left(\frac{a}{B},\frac{c}{a},\frac{L}{B}\right)\times f_2\left(\frac{a}{B},\frac{a}{c}\right)\times f_3\left(\frac{a}{B},\frac{a}{c},\frac{L}{B}\right) \tag{A.12}$$

where

$$f_1\left(\frac{a}{B},\frac{c}{a},\frac{L}{B}\right)=g_1(a/B)\left\{g_2\left(\frac{c}{a}\right)^2+g_3\left(\frac{c}{a}\right)+g_4\right\}$$
$$+g_5\{1-(a/B)\}\left\{g_6\left(\frac{c}{a}\right)^2+g_7\left(\frac{c}{a}\right)+g_8\right\}+g_9;$$

$g_1 = 0.0023232(c/\alpha)^2 - 0.00037156(c/\alpha) + 45985;$

$g_2 = -0.000044010(L/B)^2 + 0.00014425(L/B) + 0.00086706;$

$g_3 = 0.00039951(L/B)^2 - 0.0013715(L/B) + 0.014251;$

$g_4 = 0.0046169(L/B)^2 - 0.017917(L/B) - 0.16335;$

$g_5 = -0.018524(c/a)^2 + 0.27810(c/a) - 5.4253;$

$g_6 = -0.00037981(L/B)^2 + 0.0025078(L/B) + 0.00014693;$

$g_7 = -0.0038508(L/B)^2 + 0.0023212(L/B) - 0.026862;$

$g_8 = -0.011911(L/B)^3 + 0.082625(L/B)^2 - 0.16086(L/B) + 1.2302;$

$g_9 = 0.27798(a/B)^3 - 1.2144(a/B)^2 - 2.4680(a/B) + 0.099981.$

$$f_2\left(\frac{a}{B},\frac{a}{c}\right) = \left\{-0.35006(a/c)^2 + 0.40768(a/c) + 1.7053\right\}(a/B)^{g_{10}}$$
$$+ 0.24988\{1-(a/B)\}^{g_{11}}$$

$g_{10} = -0.25922(a/c)^2 + 0.39566(a/c) + 0.011759;$

$g_{11} = 6.5964(a/c)^2 + 55.787(a/c) + 37.053;$

$$f_3\left(\frac{a}{B},\frac{a}{c},\frac{L}{B}\right) = g_{12}(a/B)^{0.94761} + g_{13}\exp\left\{(a/B)^{g_{14}}\right\};$$

$g_{12} = -0.14895(L/B)^3 + 0.815(L/B)^2 - 1.4795(L/B) - 0.89808;$

$g_{13} = 0.055459(L/B)^3 - 0.30180(L/B)^2 + 0.54154(L/B) + 0.53433;$

$g_{14} = -0.01343(a/c)^2 + 0.0066702(a/\mathrm{c}) - 0.75939.$

A.4 Calculation of Stress Intensity Factor Caused by Residual Stress (Wang, 1995)

$$K(\sigma^R) = \int_0^a m(a,x)\cdot\sigma^R(x)\mathrm{d}x \tag{A.13}$$

where $\sigma^R(x)$ is residual stress distribution along the crack surface.

Deepest point :

$$m_D(a,x) = \frac{2}{\sqrt{2\pi(a-x)}}\left[1 + M_{D1}\cdot\left(1-\frac{x}{a}\right)^{0.5} + M_{D2}\cdot\left(1-\frac{x}{a}\right) + M_{D3}\cdot\left(1-\frac{x}{a}\right)^{1.5}\right]$$

$$M_{D1} = \frac{2\pi}{\sqrt{2Q}}(2Y_0 - 3Y_1) - 4.8$$

$$M_{D2} = 3$$

$$M_{D3} = \frac{6\pi}{\sqrt{2Q}}(2Y_1 - Y_0) + 1.6$$

$$Q = 1 + 1.464(a/c)^{1.65}$$

$$Y_0 = A_0 + A_1\left(\frac{a}{B}\right)^2 + A_2\left(\frac{a}{B}\right)^4 + A_3\left(\frac{a}{B}\right)^6$$

$$A_0 = 1.0929 + 0.2581(a/c) - 0.7703(a/c)^2 + 0.4393(a/c)^3$$

$$A_1 = 0.456 - 3.045(a/c) + 2.007(a/c)^2 + 1/\left(0.147 + (a/c)^{0.688}\right)$$

$$A_2 = 0.995 - 1/(0.027 + (a/c)) + 22(1 - (a/c))^{9.953}$$

$$A_3 = -1.459 + 1/(0.014 + (a/c)) - 24.211(1 - (a/c))^{8.071}$$

$$Y_1 = B_0 + B_1\left(\frac{a}{B}\right)^2 + B_2\left(\frac{a}{B}\right)^4 + B_3\left(\frac{a}{B}\right)^6$$

$$B_0 = 0.4537 + 0.1231(a/c) - 0.7412(a/c)^2 + 0.46(a/c)^3$$

$$B_1 = -1.652 + 1.665(a/c) - 0.534(a/c)^2 + 1/\left(0.198 + (a/c)^{0.846}\right)$$

$$B_1 = -4.228 + 3.643(a/c) + 1/(0.02 + (a/c)) - 21.924(1 - (a/c))^{9.203}$$

Surface point:

$$m_s(a,x) = \frac{2}{\sqrt{\pi x}}\left[1 + M_{s1}\cdot\left(\frac{x}{a}\right)^{0.5} + M_{s2}\cdot\left(\frac{x}{a}\right) + M_{s3}\cdot\left(\frac{x}{a}\right)^{1.5}\right]$$

$$M_{s1} = \frac{3\pi}{\sqrt{Q}}(5F_1 - 3F_0) - 8$$

$$M_{s2} = \frac{15\pi}{\sqrt{Q}}(2F_0 - 3F_1) + 15$$

$$M_{s3} = \frac{3\pi}{\sqrt{Q}}(10F_1 - 7F_0) - 8$$

$$F_0 = \left[C_0 + C_1\left(\frac{a}{t}\right)^2 + C_2\left(\frac{a}{t}\right)^4\right]\cdot\sqrt{\frac{a}{c}}$$

$$C_0 = 1.2972 - 0.1548(a/c) - 0.0185(a/c)^2$$

$$C_1 = 1.5083 - 1.3219(a/c) + 0.5128(a/c)^2$$

$$C_2 = -1.101 + 0.879/(0.157 + (a/c))$$

$$F_1 = \left[D_0 + D_1\left(\frac{a}{t}\right)^2 + D_2\left(\frac{a}{t}\right)^4\right]\cdot\sqrt{\frac{a}{c}}$$

$$D_0 = 1.2687 - 1.0642(a/c) + 1.4646(a/c)^2 - 0.725(a/c)^3$$

$D_1 = 1.1207 - 1.2289(a/c) + 0.5876(a/c)^2$

$D_2 = 0.19 - 0.608(a/c) + 0.199/\left(0.035 + (a/c)\right)$

A.5 Calculation of SIF of surface crack at welded tubular joints (Rhee, et al., 1991, adapted in SINTAP)

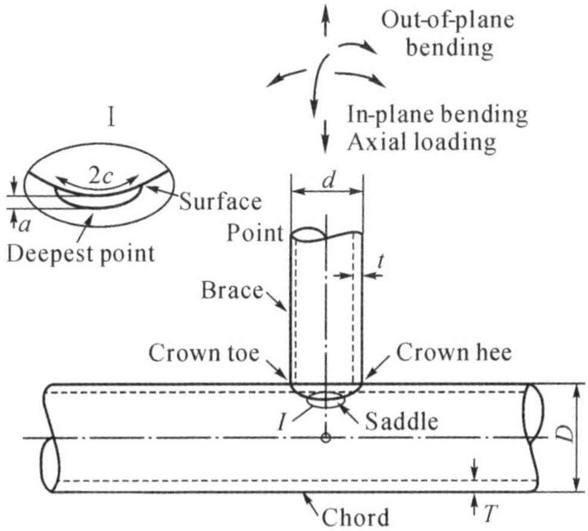

Fig. A.4 T-joint geometries

Notation:

a—crack depth

$2c$—surface crack length

d—brace diameter

D—chord diameter

L—chord length

t—brace thickness

T—chord thickness

$\alpha = 2L/D$

$\beta = d/D$

$\alpha = D/2T$

$\alpha = t/T$

$\alpha =$ brace nominal stress

Limits to SIF solution:

$\alpha = 12$

$0.4 \le \beta \le 0.8$

$10 \le \gamma \le 20$

$0.1 \le \tau \le 1.0$

$$0.05 \le a/T \le 0.80$$
$$0.05 \le 3c/d \le 1.2$$

Surface crack at the saddle point of T-joints under axial loading (deepest point) K_{eaa}

$$K_{eaa} = \sigma_n F_g F_i F_s \sqrt{\pi a} \tag{A.14}$$

$$F_g = 0.2749\beta^{(-0.6225-1.2685\ln\beta)}\gamma^{(1.3191-0.1661\ln\tau)}\tau^{(1.6621+0.3704\ln\beta)}$$
$$F_i = \beta^{(0.3561A-0.0956C)}\gamma^{(0.0983A+0.2298C+0.0817C^2)}\tau^{-0.0762A}$$
$$F_s = (a/T)^p(3c/d)^r$$
$$p = -0.8669 - 0.2198A - 0.0162A^2 - 0.475C^2 - 0.1667C^3 - 0.0193C^4$$
$$r = 0.0777 + 1.0531A + 0.582A^2 + 0.081A^3 - 0.07C - 0.0604C^2 - 0.006C^3$$
$$A = \ln(a/T)$$
$$C = \ln(3c/d)$$

Surface crack at the saddle point of T-joints under axial loading (surface point) K_{eac}

$$K_{eac} = \sigma_n F_g F_i F_s \sqrt{\pi a} \tag{A.15}$$

$$F_g = 204.08\beta^{(-0.5858-0.7492\ln\beta)}\gamma^{(-2.6713-0.2884\ln\beta+0.5646\ln\gamma)}\tau^{(1.1491-0.2936\ln\gamma-0.5043\ln\tau)}$$
$$F_s = (a/T)^p(3c/d)^r$$
$$F_s = (a/T)^p(3c/d)^r$$
$$p = 1.0787 + 0.6397A + 0.1569A^2 + 0.0186A^3$$
$$\quad -(0.077 + 0.0478A + 0.0099A^2)C^2$$
$$r = 0.8617 + 0.4888A + 0.1816A^2 + 0.0123A^3 - 0.3252C - 0.221C^2 - 0.0275C^3$$
$$A = \ln(a/T)$$
$$C = \ln(3c/d)$$
$$\sigma_n = \frac{4P}{\pi[d^2 - (d-2t)^2]}$$

Surface crack at the saddle point of T-joints under in-plane bending (deepest point) K_{eia}

$$K_{eia} = \sigma_n F_g F_i F_s \sqrt{\pi a} \tag{A.16}$$

$$F_g = 0.0887\beta^{(1.3433-0.4798\ln\beta)}\gamma^{(5.2247-0.5555\ln\beta-0.831\ln\gamma)}\tau^{(0.6928-0.4302\ln\beta)}$$
$$F_i = 0.0887\beta^{(-0.0758A-0.2391C)}\gamma^{(0.1411A+0.4341C+0.1543C^2)}\tau^{-0.1771A}$$
$$F_s = (a/T)^p(3c/d)^r$$
$$p = 1.8586+2.2859A+0.9035A^2+0.1215A^3-1.0918C-0.4785C^2$$
$$r = -1.0298-0.304A^2+0.4834C+0.703C^2+0.113C^3-0.1207A^2C$$
$$A = \ln(a/T)$$
$$C = \ln(3c/d)$$

Surface crack at the saddle point of T-joints under in-plane bending (surface point) K_{eic}

$$K_{eic} = \sigma_n F_g F_i F_s\sqrt{\pi a} \quad \text{(A.17)}$$

$$F_g = 0.1395\beta^{(-0.6498-1.1883\ln\beta)}\gamma^{(1.0779-0.3414\ln\beta)}\tau^{(0.8168-0.2149\ln\beta)}$$
$$F_i = \beta^{(0.0422A-0.2452C)}\gamma^{(1.4558A+0.4173A-0.9276C-0.3297C^2)}\tau^{(-0.0905A-0.0338C)}$$
$$F_s = (a/T)^p(3c/d)^r$$
$$p = -2.4921-0.0063A+0.2056A^2+0.9804C+0.3916C^2$$
$$+0.062C^3-0.011C^4$$
$$r = 2.8298+0.5682A^2+0.0704A^3+0.6562C-0.0453C^2-0.0022C^3$$
$$+(0.1621+0.0384C)A^2C$$
$$A = \ln(a/T)$$
$$C = \ln(3c/d)$$

For In-Plane Bending (IPB) $\sigma_n = \dfrac{32dM_i}{\pi[d^4-(d-2t)^4]}$

Surface crack at the saddle point of T-joints under out-of-plane bending (deepest point) K_{eoa}

$$K_{eoa} = \sigma_n F_g F_i F_s\sqrt{\pi a} \quad \text{(A.18)}$$

$$F_g = 0.1718\beta^{(0.9626-0.5003\ln\beta)}\gamma^{1.5274}\tau^{(0.6488+0.3353\ln\beta-0.2962\ln\tau)}$$
$$F_i = \beta^{(0.3066A-0.0598C)}\left(\frac{a}{T}\right)^{(0.1315\ln\gamma-0.0775\ln\tau)}$$
$$F_s = (a/T)^p(3c/d)^r$$
$$p = -1.313-0.4253A-0.0584A^2+0.9843C-0.3278C^2-0.0308C^3$$

$$r = 0.7184 + 0.5401A^2 + 0.0889A^3 - 0.4186C - 0.0496C^2$$
$$+ 0.113C^3 - 0.0421A^2C$$
$$A = \ln(a/T)$$
$$C = \ln(3c/d)$$

Surface crack at the saddle point of T-joints under out-of-plane bending (surface point) K_{eoc}

$$K_{\text{eoc}} = \sigma_n F_g F_i F_s \sqrt{\pi a} \tag{A.19}$$

$$F_{\text{g}} = 4.7016\beta^{(0.7362-0.9523\ln\beta)}\gamma^{(0.2227-0.7169\ln\beta)}\tau^{(0.6663-0.104\ln\gamma-0.3802\ln\tau)}$$
$$F_{\text{i}} = \beta^{(0.1388A-0.2143C)}\gamma^{(0.0573A-0.5026C-0.1175C^2)}\tau^{-0.1548A}$$
$$F_{\text{s}} = (a/T)^p(3c/d)^r$$
$$p = 1.5044 + 0.835A + 0.1258A^2 + 0.6624C - 0.0202C^2$$
$$r = 0.954 + 0.3328A^2 + 0.0453A^3 - 0.699C - 0.3648C^2 + 0.113C^3 - 0.0473C^3$$
$$A = \ln(a/T)$$
$$C = \ln(3c/d)$$

For Out-of-Plane Bending (OPB) $\sigma_n = \dfrac{32dM_o}{\pi[d^4 - (d-2t)^4]}$

References

ABS (American Bureau of Shipping) (2005). Fatigue Assessment of Offshore Structures. Houston, New Jersey.

API (API Publishing Services) (2007). "Fitness-for-service", The American (2nd ed.). Society of Mechanical Engineers, Jun. 2007, Washington, USA.

Beevers, C. J. (1982). Advances in Crack Length Measurement. West Midlands, UK: Engineering Materials Advisory Services Ltd.

Bokalrud, T. & Karlsen, A. (1981). "Probabilistic fracture mechanics evaluation of fatigue failure from weld defects in butt weld joints", in: Proceedings of Conference on Fitness for Purpose Validation of Welded Constructions, London, UK, Paper 28.

BSI (British Standards Institution) (2005). Guidance on Methods for Assessing the Acceptability of Flaws in Structures. BS7910. London.

DNV (Det Norske Veritas) (2005a). Fatigue Assessment of Ship Structures. Oslo, Norway.

DNV (Det Norske Veritas) (2005b). Fatigue Design of Offshore Steel Structures. Oslo, Norway.

Efren, A. U. & Moan, T. (2007). "Fatigue reliability-based assessment of welded joints applying consistent fracture mechanics formulations", International Journal of Fatigue, 29: 444–456.

Farahmand, B. & Nikbin, K. (2008). "Predicting fracture and fatigue crack growth properties using tensile properties", Engineering Fracture Mechanics, 75: 2144–2155.

Hobbacher, A. (2003). "IIW joint working group xiii–xv, recommendations for fatigue design of welded joints and components", International Institute of Welding, Feb., 2003, Paris, France.

Huther, M., Parmentier, G. & Maherault, S. (2004). "Uncertainties in ship structures cumulative fatigue calculations", in: Proc. 9th Int. Symp. on Practical Design of Ships and other Floating Structures PRADS, 2: 849–854.

ISSC Committee VI.2 (2006). "Loads", in: Proceedings of 16th International Ship and Offshore Structures Congress, Aug. 20–25, 2006, Southampton, UK.

ISSC Committee VI.2 (2009). "Loads", in: Proceedings of 17th International Ship and Offshore Structures Congress, Aug. 16–21, 2009, Seoul, Korea.

ISSC Committee III.2 (2009). "Fatigue and fracture", in: Proc. of the 17th Int. Ship and Offshore Structures Congress.

Kukkanen, T. & Mikkola, T. P. J. (2004). "Fatigue assessment by spectral approach for the ISSC comparative study of the hatch cover bearing pad", Marine Structures, 17: 75–90.

Ling, C. & Zheng, X. L. (1990). "Prediction of fatigue crack propagation rates of aluminum alloys from tensile properties", Journal of Northwestern Polytechnical University, 8(1): 115–120 (in Chinese).

Minakawa, K. & McEvily, A. J. (1981). "On crack closure in the near-threshold region", Scripta Metallurgica, 15(6): 633–636.

Moan, T., Vårdal, O. T., Hellevig, N. C. & Skjoldli, K. (2000). "Initial crack depth and POD values inferred from in-service observations of cracks in North Sea jackets", Journal of Offshore Mechanics and Arctic Engineering, 122: 157–162.

Moan T. (2007). "Reliability-based management of inspection, maintenance and repair of offshore structures", Journal of Structural Engineering, 1(1): 33–62.

Otegui, J. L., Burns, D. J., Kerr, H. W. & Mohaupt, U. H. (1991). "Growth and coalescence of fatigue cracks at weld toes in steel", International Journal of Pressure Vessels and Piping, 48: 129–165.

Rhee, H. C., Han, S. & Gibson, G. S. (1991). "Reliability of solution method and empirical formulas of stress intensity factors for weld toe cracks of tubular joints", in: Proceedings of the 10th Offshore Mechanics and Arctic Engineering Conference, American Society of Mechanical Engineers, New York, Vol. III Part B, 441–452.

Roberts, R. & Newton, C. (1981). "Interpretive report on small scale test correlations with KIC data", Welding Research Council Bulletin No. 265: 1–18.

DNV (2006). Fatigue Methodology for Offshore Ships. Oslo, Norway: Det Norske Veritas.

SINTAP (1999). "Structure integrity assessment procedure for European Industry project BE 95-1426", Final Procedure, British Steel Report, Rotherhan.

Vosikovski, O. (1979). "The effect of stress ratio on fatigue crack growth rates in steels", Engineering Fracture Mechanics, 11(3): 595–602.

Wang, X. & Lambert, S. B. (1995). "Stress intensity factors for low aspect ratio semi-elliptical surface cracks in finite-thickness plates subjected to non-uniform stresses", Engineering Fracture Mechanics, 51(4): 517–532.

Index